D1698268

Naval Architecture for
Marine Engineers

Naval Architecture for Marine Engineers

W. MUCKLE
M.Sc., Ph.D.(DUNELM), D.Sc.(N'CLE),
C.ENG., F.R.I.N.A., F.I.MAR.E.

*Formerly Professor of Naval Architecture,
University of Newcastle upon Tyne*

NEWNES-BUTTERWORTHS
LONDON   BOSTON
Sydney   Wellington   Durban   Toronto

The Butterworth Group

United Kingdom **Butterworth & Co (Publishers) Ltd**
London: 88 Kingsway, WC2B 6AB

Australia **Butterworths Pty Ltd**
Sydney: 586 Pacific Highway, Chatswood, NSW 2067
Also at Melbourne, Brisbane, Adelaide and Perth

Canada **Butterworth & Co (Canada) Ltd**
Toronto: 2265 Midland Avenue, Scarborough,
Ontario M1P 4S1

New Zealand **Butterworths of New Zealand Ltd**
Wellington: T & W Young Building,
77-85 Customhouse Quay, 1, CPO Box 472

South Africa **Butterworth & Co (South Africa) (Pty) Ltd**
Durban: 152-154 Gale Street

USA **Butterworth (Publishers) Inc**
Boston: 19 Cummings Park, Woburn, Mass. 01801

First published 1975
Reprinted 1978

© Butterworth & Co (Publishers) Ltd, 1975

ISBN 0 408 00169 0

Printed in Great Britain by
Fletcher & Son Ltd, Norwich

Preface

The marine engineer who is concerned with the design or operation of machinery in ships requires some knowledge of certain aspects of naval architecture and it will be found that naval architecture is included in most courses leading to a degree or some other qualification in marine engineering. The branches of the subject concerned are mainly resistance, propulsion and vibration and are dealt with in great detail in specialist books. The present book gives an overall account of these problems and sufficient theory is developed to enable the student to grasp fundamental principles.

Whilst resistance, propulsion and vibration are the branches of naval architecture which are the main interest of the marine engineer it is desirable that he should have some knowledge of other branches. Thus, chapters have been included dealing with ship calculations, stability and trim, ship motions, and structural strength and brief reference has also been made to ship design.

The book is intended primarily for marine engineers but it is hoped that deck officers may find it useful and it should be of value to students in naval architecture who are studying for degrees or diplomas and who are approaching the subject for the first time. The practising naval architect should also find the book useful.

Lists of references have been included at the ends of some of the chapters and these should be of use for further study.

W. Muckle

Contents

The function of the ship; ship types

The ship is one of the oldest methods of transport and the modern ship incorporates many developments which have taken place over hundreds if not thousands of years. The two major developments which distinguish the present-day ship from that built 150 or more years ago are the use of steel in place of wood as the material of construction and the employment of mechanical means of propulsion instead of sails. Many new problems have of course arisen with the passage of time in the design of ships; there are, however, many which are common to ships of all ages. Great developments have taken place in the practical construction of ships and there has been a parallel increase in understanding of the factors which go to make a successful design. The subject of naval architecture is concerned with all these developments in design and construction and although not an exact science it is nevertheless a science. In the past few decades knowledge has expanded very rapidly and naval architecture has divided itself naturally into a number of branches requiring specialist study. These different branches will be considered in more detail in later chapters, but it is fitting to consider here in broad terms the problems which arise in the design of ships and to describe the requirements which they must fulfil. Many are common to ships of all types, whether warships, cargo or passenger ships, or ships designed for some special purpose. The features of different types of merchant ships will also be discussed briefly in this chapter.

The function of the ship

Disregarding ships built for special purposes, merchant ships may be said to be part of a transport system. Their function is therefore to transport commodities or people from one place to another. It will be realised that the ship is only one part of such a system. The overall problem involves a study of the facilities to be provided at the terminal ports for loading and discharging cargo, as well as the means for collecting cargo at the loading port and for the distribution of the cargo at the port where it is discharged.

The ship should be capable of carrying out its part in the transport system as quickly and efficiently as possible, and this is the problem which the ship designer has to solve. The requirements which the ship has to fulfil may be stated as follows.

The dimensions of the ship should be such that it provides sufficient buoyancy to support the load for which it is designed. It will be seen later how dimensions are fixed to satisfy this requirement. Another important consideration is that the ship shall be stable in all normal conditions of loading. This simply means that if the ship is displaced by some external force from its equilibrium position when floating in still water, it will return to that position when the force is removed. The most important problem in this field is that of transverse stability when the ship rotates about a longitudinal axis. It will be shown that transverse stability is governed largely by the ratio of breadth to draught. The ship is also capable of rotating about a transverse axis so that there could be a problem of longitudinal stability. The surface ship is, however, so much more stable in this direction that it is virtually impossible to make it unstable. This is not true of a submarine in the immersed condition, where the stability is the same in both directions. This is one of the reasons why the longitudinal distribution of loads is so important in vessels of this type.

When the ship is at sea forces are generated on it due to the waves through which it is passing and the gravitational forces arising from the loads which it carries. The ship also has six degrees of freedom, which all involve accelerations so that dynamic forces are created. All these forces cause the structure to deform and the ship bends in a longitudinal vertical plane like a beam. It is necessary, therefore, that the ship should have some minimum structural strength to resist this type of bending. In addition to longitudinal bending there are transverse and local deformations of the structure which arise from the forces imposed upon it. The longitudinal strength of the ship is of primary importance and will be dealt with in Chapter 7. This aspect of strength governs to a large extent the ratio of the depth of the ship to the length.

It was stated earlier that a ship should have sufficient buoyancy to support the loads which it is intended to carry. Because the ship is passing through waves when at sea it will more often than not be pitching and heaving and to enable it to 'rise' to the sea it is necessary to have some reserve buoyancy above the waterline. The ship must therefore be designed to have a certain amount of freeboard, i.e. the top deck should be some distance above the waterline. Freeboard is important from several points of view. It increases the range of stability when the ship is rolling, it helps to prevent water coming on board and it provides reserve buoyancy which can enable it to float in the event of damage.

In the foregoing the requirements of the ship to enable it to survive at sea have been discussed. But it has, of course, to move from one place to another, so that it is also important that this should be done with the least expenditure of power. It necessitates that the form of the ship should be suitably designed and Chapter 8 considers this further. It may occur that the proportions of the ship which would result in the least power would not be those which would be acceptable from consideration of stability, carrying capacity, etc. A compromise has therefore often to be made between conflicting requirements and this will be found to be one of the problems always confronting the designer.

The layout of the ship

The layout of the ship is shown on a plan known as the 'general arrangement'. It shows the sizes and positions of the various spaces which are required for the ship to fulfil its duties. Their disposition will depend very much on the type of ship, but there are certain spaces which must be provided in all ships.

Assuming that the ship is intended to carry cargo, then spaces called 'cargo holds' must be provided. They should have sufficient volume to contain the lightest density cargo which it is expected to carry. Often cargo ships have two, three, four or more holds, separated from one another and from other compartments by watertight bulkheads. Most ships have double bottoms, the inner bottom being some 1.0–1.2 m above the outer bottom. This inner bottom forms the lower extremity of the holds, the top of the holds being the lowest deck in the ship. Some ships have more than one continuous deck extending all fore and aft, and the spaces between these decks (the 'tween decks) are also available for the carriage of cargo. These spaces may be 2.4–3.0 m in height. Access to both holds and 'tween decks has usually been made through large openings in the decks called hatches. These access openings have to be closed when the ship is at sea by means of hatch covers. For many years hatch covers were of wood supported on steel hatch beams which could be removed easily. The covers were in sufficiently small pieces to be easily handled and when in place were covered by tarpaulins and battened down. In recent years, however, because of the vulnerability of wooden covers when the ship is at sea, steel covers have been developed. They are usually in sections and because of their weight they require some mechanical means for opening and closing them.

The rapid loading and unloading of cargo from holds is a factor which has considerable bearing on the economic efficiency of the ship. This function has often been performed by derricks attached to the masts or special derrick posts and operated by winches. These derricks are usually capable of lifting relatively light loads of about 5 t. Some ships have, however, been fitted with heavy lift derricks where larger loads have to be catered for.

The time spent in port unloading and loading cargo can be appreciable and has led to the development of other means of loading and discharging cargoes. Thus, side loading of ships has been developed, where loading is through doors in the side shell rather than through hatches. This method is especially suitable for 'tween deck spaces and permits the use of such devices as fork lift trucks. The use of cranes, either on shore or on the ship, is another means for loading and discharging cargo. It will be seen later that the use of containers has speeded up the turn round of ships in port.

Generally in the dry cargo ship the cargo holds were placed fore and aft of the machinery space which was situated at or near the middle of length. Nowadays, however, many such ships have machinery aft with a continuous line of holds from the machinery space to the forward end of the ship.

Adequate space for the safe and efficient working of the machinery is necessary. The trend in modern ships is towards smaller machinery spaces. This is well illustrated by a comparison of the machinery spaces of existing large liners with those of similar ships in the pre-war era. In addition to the space allotted to the machinery it is also necessary to provide space for the carriage of fuel (bunkers). Fuel can be carried in double bottom tanks or in cross bunkers. In the former case this otherwise useless space from the point of view of the carriage of cargo is employed. Cross bunkers reduce the amount of space which can be devoted to cargo. Sometimes it is possible to make use of side tanks within the machinery space itself for carrying fuel, thus saving space outside the machinery compartment.

It is often necessary to adjust the end draughts of a ship so that it will trim correctly. This is achieved by the use of water ballast. Ballast tanks are provided in the double bottom and there are deep tanks at the two ends of the ship, the fore peak and the after peak tanks. The double bottom is divided up into a number of small tanks, all of which are provided with pumping arrangements which enable them to be filled and emptied as desired.

Spaces must also be provided in the ship for machinery which is essential to its operation. Such machinery includes electrical generators which may be housed in the machinery compartment or may have separate spaces allotted to them. The steering gear is also housed in a separate compartment which in the modern ship is situated at the extreme after end directly above the rudder.

The position of accommodation for officers and crew has varied from time to time. In older ships the crew were often housed forward in a forecastle but later regulations prohibited this, so it would be found that the crew were situated either amidships or aft. The engineer and deck officers would be housed in an amidship deckhouse near the navigating bridge. With the modern tendency to have machinery aft it is now usual to have the officers and crew accommodated in a superstructure aft.

Passengers, when carried, will usually be accommodated in a large midship superstructure the size of which would depend mainly on the number of passengers to be catered for and the standard of accommodation required. The large midship superstructure is a feature which distinguishes the passenger ship from the cargo ship. Many cargo ships carried up to twelve passengers, this being the limit beyond which it was necessary to have a passenger certificate involving compliance with very stringent regulations.

Ship types

The types of ships which have been developed over the years have been dictated very largely by the types of cargo. One of the types which may be described as a general cargo carrier was what was known as the 'shelter deck ship'. Briefly, this was a two-deck ship with machinery amidships and with four or five holds, two or three of which were situated forward of the engine room and the other two aft. *Figure 1.1* shows the profile of such a ship. Accommodation for officers was in a central deckhouse structure amidships and the ship usually had a forecastle and a small deckhouse at the after end. Because of a legal decision in the nineteenth century (see Chapter 3) the space between the uppermost continuous deck (called the 'shelter deck') and the second deck was regarded as an open space if there was a small hatch aft in the shelter deck without permanent means of closing and if there were tonnage openings in the bulkheads in the 'tween deck. This exempted the 'tween deck space from inclusion in the tonnage measurement (see Chapter 3).

It was usual in this type for the watertight bulkheads to be stopped at the second deck, although classification societies required the

6

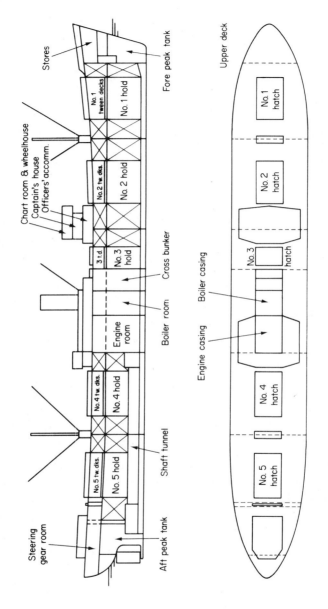

Figure 1.1 (From *Trans. North East Coast Instn. of Engrs. & Shpbdrs.,* **59**, p. 83 (1942–1943))

Stores

Chart room & wheelhouse
Captain's house
Officers' accomm.

Steering
gear room

Fore peak tank

No. 1 tween decks

No. 1 hold

No. 2 tw. dks.

No. 2 hold

3 t.d.

No. 3
hold

Engine
room

No. 4 tw. dks.

No. 4 hold

No. 5 tw. dks.

No. 5 hold

Cross bunker

Boiler room

Shaft tunnel

Aft peak tank

Upper deck

No. 1
hatch

No. 2
hatch

No. 3
hatch

Boiler casing

Engine casing

Boiler casing

No. 4
hatch

No. 5
hatch

foremost bulkhead (the collision or fore peak bulkhead) to be carried up to the weather deck.

Many of these ships were built and were often known as 'tramp' ships, being capable of going to almost any place in the world and carrying a variety of cargoes. Typical cargoes from the UK were coal outwards with grain on the return journey.

Ships of the shelter deck type had limited draughts and the scantlings of the structure were reduced accordingly. Another type which was somewhat similar in appearance was one in which the deepest draught which could be assigned on the dimensions could be obtained. This was known as a 'full scantling ship', all the bulkheads being carried up to the uppermost deck. It was sometimes called a 'three island ship', because it usually had a poop, a bridge and forecastle. All these were erections which extended out to the ship's side, the sides of the erections being a continuation of the side shell of the ship. The ship usually had one or two decks extending all fore and aft. It had a much deeper draught than the shelter deck ship and was therefore suitable for carrying heavier cargoes where deadweight carrying capacity was more important than space.

Carriage of dry cargoes in bulk

The types of ships discussed in the previous section were in a sense bulk carriers, in that they frequently carried coal, grain or some other cargo in bulk. In the post war years, however, a special type has developed for the bulk carriage of a variety of cargoes and it has now become known as a 'bulk carrier'. It should, however, be mentioned that ships designed for the carriage of cargoes in bulk were not unknown before the last war. Ships trading on the Great Lakes in North America were of this type. They had many of the features of the modern bulk carrier and were employed in the grain and ore trades.

A modern bulk carrier is shown in *Figure 1.2*. The characteristics of these ships are that they are single deckers with machinery aft, and generally all the accommodation is aft also. The section through the holds is shown in *Figure 1.3*. The construction is such that the ship can be loaded without having to trim the cargo. This is achieved by the sloping plating which extends from the hatch sides to the ship's side. Similarly sloping plating extends from the ship's side to the tank top as shown. This allows easy unloading of cargo, since the cargo will fall down under the hatchways as unloading proceeds, and permits the use of grabs or other means for discharging the cargo.

The spaces provided at the bilge and the deck can be used as ballast

8

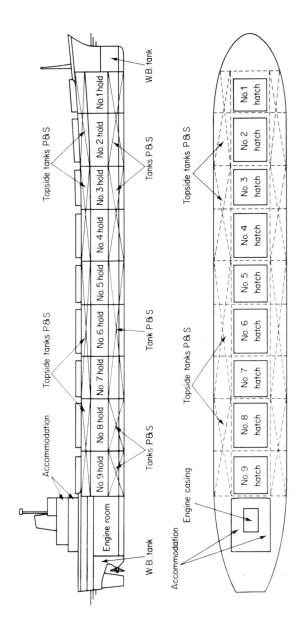

Figure 1.2 (From *Shipping World and Shipbuilder*, p. 306 (Mar. 1973))

tanks. The topside ballast tanks are useful in particular when the ship is travelling in the ballast condition. Ships then usually have excessive stability and the stowage of ballast at a high centre of gravity helps to make the ship easier in a seaway.

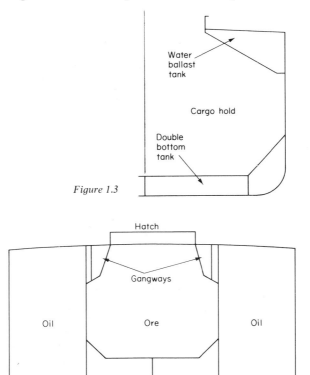

Figure 1.3

Figure 1.4

There are many variations of the hold arrangement in vessels of this type. Some are arranged to carry oil as well as other cargoes such as ore. The section through a ship of this type is shown in *Figure 1.4*. The ore hold in this case is placed relatively high up in the ship. Ore being a dense cargo would have a very low centre of gravity if placed in the hold in a normal ship, and this would lead to excess stability in the fully loaded condition. By putting the cargo in a high position in the ship this excess stability is reduced. Oil can be carried in the ship shown in wing tanks and in tanks underneath the ore holds.

Another variation of the bulk carrier is one which can carry ore, other bulk cargoes and oil—the so-called OBOs.

The carriage of liquid cargoes

The most important liquid cargo which is carried in bulk today is oil. The demand for oil has expanded rapidly during the present century and particularly since 1939. While it is now believed that sources of oil are becoming very limited, large vessels for the carriage of oil continue to be built. Of a total world tonnage of 268 000 000 in 1972, 105 000 000 were oil tankers and 63 000 000 were bulk/ore carriers, including bulk oil. This shows that nearly two thirds of the world tonnage is at present engaged in the carriage of bulk cargoes.

In the early days oil was carried in barrels in normal ships, but with the development of the trade the need for a special type of ship became evident. The oil tanker has many of the features of the bulk carrier but pre-dates that type of vessel considerably. The tankers of the earlier decades of this century were ships with machinery aft, the oil cargo tanks being in a block forward of the machinery space. In the centre of this block a pump room was usually situated. The cargo tanks were separated from the rest of the ship by cofferdams about 1 m long at each end. The ships had poops, a short bridge amidships and a forecastle. Engineers and crew were accommodated aft and the master and officers were situated in the bridge. Some of the earlier ships were of the trunk type as shown in section in *Figure 1.5(a)*. They had centre line bulkheads to prevent loss of stability due to free surface effects (see Chapter 5). At a later stage the two-deck type developed as shown in *Figure 1.5(b)*, separate small tanks being provided for the carriage of oil in the space between the second and upper decks. Finally, the single-deck tanker developed as shown in *Figure 1.5(c)*. In this type more than one longitudinal bulkhead was fitted and up to 1939 two were quite common, and this is the usual practice today. A modern tanker is shown in *Figure 1.6*. The cofferdams fitted in the old riveted tankers have disappeared: the modern ship being all welded there is no need for them. In these large modern ships all the accommodation is aft, as also is the navigating position.

The rapid growth in the sizes of oil tankers dates from about the 1950s. In the pre-war years and for some time after, tankers were of the order of 20 000 ton deadweight as a maximum. Today, however, the 250 000 ton tanker is quite common and the world's present largest tanker (1973) is of 483 664 ton deadweight. It is expected that

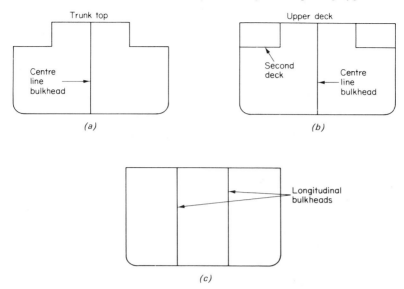

Figure 1.5   (a) *Trunk tanker* (b) *Two-deck oil tanker* (c) *Twin-bulkhead oil tanker*

this will be exceeded in the mid seventies by ships of 540 000 ton deadweight. For normal dry cargo ships this explosion in size would have posed problems as far as port facilities are concerned. The nature of the oil cargo, however, enables these mammoth tankers to unload at deep water quays.

Liquefied gas carriers

In recent years the carriage of liquefied gases has become an important trade and a great deal of study has gone into the development of a special type of ship to carry liquid natural gases. The ships are appropriately called LNGs. Natural gas consists very largely of methane, and in order to carry it in liquid form the temperature has to be reduced to $-165\,°C$. The reason for liquefying these gases is because of the great reduction in volume which is possible, the ratio of the volume of the gas to the liquid being considerable.

Many problems have arisen in the carriage of this type of cargo. In the first place special materials are required for the tanks containing the liquid. Ordinary mild steel is totally unsuitable for this purpose because at the low temperature concerned the material is susceptible to brittle fracture (see Chapter 7). The problem has been

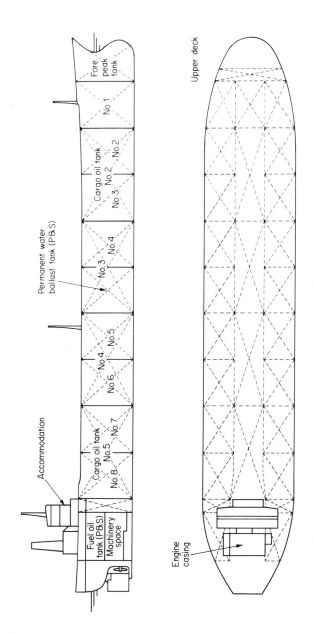

Figure 1.6 (From *Shipping World and Shipbuilder*, p. 596 (June 1973))

solved by the use of steels containing a high percentage of nickel or by the use of aluminium alloys which do not manifest the pheno-menon of brittle fracture at low temperatures.

Unlike the oil tanker the tanks in an LNG ship are not integral parts of the hull, separate tanks being employed for the containment of the cargo. There are basically two different methods for the carriage of the liquid gas. Self-supporting tanks are used, which are designed to take all service loads. The alternative method is to use membrane tanks which are designed to be liquid-tight only, normal pressures being transmitted through insulation to the hull structure.

Because of the low temperature involved the tanks have to be insulated. It is not usual, however, to fit refrigeration machinery, some 'boil off' of the liquid gas in transit being accepted. When self-supporting tanks are employed the tanks themselves are of various shapes, cylindrical, spherical or prismatic tanks having been employed. Liquefied gas carriers have machinery aft and the ships are single-deckers. Because of the low relative density of the liquid (0.474) space is more important than available deadweight. In some vessels the tanks project above the deck to obtain additional cubic capacity.

It has been predicted that something like 200 ships of this type will be required in the next 15 years to handle the increasing trade in liquid gas. There are, however, only about 14 such ships operating at the present time, these having been built between 1964 and 1972.

Container ships

Time in port is a factor which has a considerable influence on the economic efficiency of ships and as has already been noted, attempts have been made to speed up the loading and discharging of cargoes. There are limits to the extent to which this can be achieved when cargo is stowed in conventional holds. The idea of stowing cargo in large 'boxes' or containers of standard sizes has developed con-siderably in recent years. Containers are not new, their use having been considered a long time ago. What is new is the design of special types of ships to deal with them. The advantage of containers is that they can be pre-packed wherever the goods concerned are to be transported from, and then all the containers assembled at the load-ing port. They can then be very readily loaded on to the ship and similarly very rapidly unloaded at the discharge port. Rapid turn rounds are therefore possible.

A typical container ship is shown in *Figure 1.7*. The ship is virtually a single-decker with machinery towards the after end, although it is

14

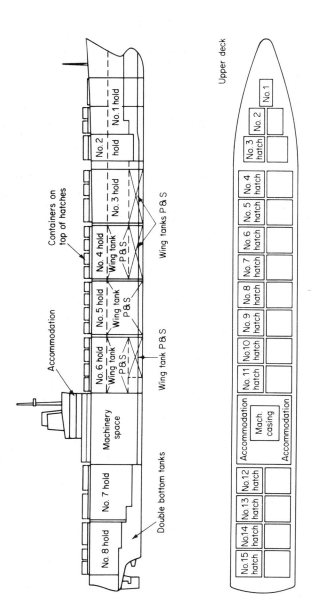

Figure 1.7 (From *Trans. Royal Instn. of Naval Architects*, **114**, p. 243 (1972))

customary to have some space available for cargo abaft the machinery space. Container ships are open deck ships, because of the fact that there are large hatch openings in the deck to permit the stowage of containers in the holds. The hatchways are enclosed by flush hatch covers and additional containers are stowed on the open deck. The fact that a large portion of the deck is cut away creates structural problems, and it is often necessary to use high strength steel to obtain the required structural strength. Since there is space between the hatch coamings and the outside of the ship which cannot be used for the stowage of containers it is usual to use this space for the provision of water ballast. The longitudinal bulkheads forming the inboard sides of the ballast tanks also help in providing additional longitudinal strength. In the holds the containers work in guides which keep them in position and they are stowed in several tiers. The deck containers are also stowed in tiers and are anchored in position by wire ropes.

The container ship is a good example of the need to design suitable port facilities to enable the greatest benefit to be obtained from the rapid unloading of the cargo. It is necessary, therefore, to have considerable space for the storage of containers before dispatch to their destination and also space for collecting containers ready for loading on to the ship. Special cranes are also needed at the container ports with sufficient outreach to be able to load containers into the parts of the holds furthest away from the quayside.

Container ships are much faster than normal cargo ships, speeds of up to 26 knots and even greater being not uncommon. They are ships of the liner type in that they work on fixed schedules and between fixed ports. Because of their high speeds and complicated arrangements they are of necessity very costly, but apparently the large sums of money which have been spent on them are justified by the speed with which cargo can be dispatched.

Refrigerated cargo ships

The refrigerated cargo ship differs from other dry cargo ships in that it carries perishable goods and in consequence has insulated holds and 'tween deck spaces with a refrigerating system which maintains these spaces at a low temperature. Ships of this type are of necessity faster than other cargo vessels because of the nature of the cargo. Generally a large depth of hold is not important in a ship, for example, carrying chilled or frozen meat. All that is required is sufficient height to hang carcasses, so that it will be found that ships of this type have more decks than normal cargo ships. The cargo is essentially one where cubic capacity is more important than dead-weight. It will usually be found then that such ships will not make use of the maximum draught which would be permissible under Load Line Regulations. The refrigerated cargo ship is essentially a cargo liner having fixed schedules and sailing between fixed terminal ports. She may or may not carry passengers as well as cargo, but the passenger element is a fast disappearing one in view of other more rapid means of passenger transport.

Passenger ships

The term 'passenger ship' covers a wide variety of different types, from the fast cross channel ship to the large fast ocean-going liner. During the present century and certainly in the last 20 years the situation regarding passenger transport, especially over long distances, has changed markedly. The development of aircraft, which has reduced the time of journeys from, say, the UK to America, the Far East and Australia from days or weeks to hours, has had a great effect on the passenger liner and few if any such ships are being built or contemplated. Those that are still in business do not always offer an all-the-year round service. They spend a good deal of their time cruising, and many ships have been specially designed for this latter purpose.

There would still appear to be room for the short voyage passenger ship, like those, for example, which sail between the UK and the Continent. The cross channel ship is one type in particular which would still seem to have a useful function to perform. Vessels of this type trading between the channel ports in the UK and the Continent have relatively high speeds for their sizes, somewhere in the region of 21 or 22 knots, and as they are at sea for a short time only ($1\frac{1}{2}$–2 h) sleeping accommodation is not required and the minimum of normal passenger facilities are necessary. They accommodate large numbers of people for their size and in addition nowadays have accommodation for the transport of passenger motor cars. They are ships which must be able to manoeuvre in and out of ports without the use of tugs. They are therefore generally twin screw and are fitted with bow thrusters. They would almost certainly now be fitted with ship stabilising equipment for the comfort of passengers; they carry little or no cargo. Ships of this type making longer journeys, for example UK to Holland and Norway, and short journeys in various places in the world, have many of the same characteristics but would also include sleeping accommodation for passengers.

A wide variety of what may be called intermediate passenger

cargo liners existed on the various trade routes of the world. Such ships carried considerable quantities of cargo and a certain number of passengers. They still exist, but as has already been pointed out, the developments in air travel have lessened the demand for this type of passenger ship transport.

At the other end of the scale is the large ocean-going passenger ship, the numbers of which have now shrunk as already stated. These ships probably reached their peak in the 1930s in the North Atlantic trade. They had speeds of up to 30 knots and provided passenger accommodation of varying grades for large numbers of people. Their future must be regarded as very doubtful and one can anticipate that in course of time they will vanish altogether, having been super-seded by air transport.

Ships of special type

Several special types of ships have been developed in recent years. Some of these are hovercraft, hydrofoil boats and multi-hull ships.

Hovercraft

The hovercraft introduces a new principle as far as ships are con-cerned. A normal ship is supported by the hydrostatic pressure acting on the immersed surface. In the hovercraft the support is generated by an air cushion created by blowing air out of the bottom of the boat. The air gradually leaks out of a curtain or skirt surrounding the boat, and pressure is built up which supports the craft. Propul-sion is usually, but not necessarily, by means of airscrews. The effect of the air cushion is to reduce the resistance to motion through the water so that higher speeds are possible for a given expenditure of power. The system has been used successfully in cross channel services between the UK and the Continent. The behaviour of ships of this sort in rough water is a matter which would need to be given careful consideration if the principle was to be extended to large ocean-going craft. It is, however, one which has great potential.

Hydrofoil boats

Hydrofoil boats make use of the same principle involved in the flight of aircraft. Hydrofoils are attached to the bottom of the ship by means of struts, and when the ship moves through water a lift force is generated just as occurs when an aeroplane wing moves through the air. By correct adjustment of the angle of incidence of the hydrofoils the lift can be made to be just equal to the weight of the ship, which is lifted clear of the water. The resistance is thereby reduced, being then only equal to the drag of the hydrofoils. The result is that high speeds are possible without using unduly large powers. Like the hovercraft, the hydrofoil boat has been used for service in sheltered waters but it is likely that considerable problems would exist in really rough water. Both types of craft have stability problems which are peculiarly their own. The extension of the use of hydrofoil craft in open sea would probably encounter the same problems as hovercraft.

Multi-hull ships

Interest has been shown in recent years in ships in which there are two and sometimes three hulls placed parallel to one another and some distance apart, connected by a bridge structure. The type of craft so produced has very great transverse stability and also great deck area is obtained. This can be used for carrying cargo or for the accommodation of passengers. The hulls in a ship of this sort would be very much narrower than the hull of a normal ship and this could reduce the wave-making resistance. By suitably positioning in a fore and aft direction a third hull between the two outside hulls an interference effect is possible which could further reduce the resistance, so that higher speeds might be possible than in the conventional ship. The potential of this type of craft has not yet been fully explored, but it would appear that it could have useful applications in container ships, ferries and trawlers.

Conclusion

In the foregoing sections some of the problems encountered in the design of ships have been discussed broadly and some of the types of ships which have been developed over the years have been described. The list is by no means complete but sufficient has been said to illustrate the considerations which affect design and type. Succeeding chapters will discuss in much more detail the various aspects of naval architecture with which the ship designer has to be familiar.

Definitions, principal dimensions, form coefficients

Before studying in detail the various technical branches of naval architecture it is important to define various terms which will be made use of in later chapters. The purpose of this chapter is to explain these terms and to familiarise the reader with them. In the first place the dimensions by which the size of a ship is measured will be considered; they are referred to as 'principal dimensions'. The ship, like any solid body, requires three dimensions to define its size, and these are a length, a breadth and a depth. Each of these will be considered in turn.

Principal dimensions

Length

There are various ways of defining the length of a ship, but first the length between perpendiculars will be considered. The length between perpendiculars is the distance measured parallel to the base at the level of the summer load waterline from the after perpendicular to the forward perpendicular. The after perpendicular is taken as the after side of the rudder post where there is such a post, and the forward perpendicular is the vertical line drawn through the intersection of the stem with the summer load waterline. In ships where there is no rudder post the after perpendicular is taken as the line passing through the centre line of the rudder pintals. The perpendiculars and the length between perpendiculars are shown in *Figure 2.1*.

The length between perpendiculars (l.b.p.) is used for calculation purposes as will be seen later, but it will be obvious from *Figure 2.1* that this does not represent the greatest length of the ship. For many purposes, such as the docking of a ship, it is necessary to know what the greatest length of the ship is. This length is known as the

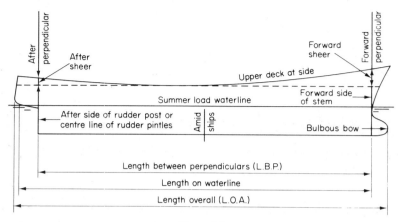

Figure 2.1

'length overall' and is defined simply as the distance from the extreme point at the after end to a similar point at the forward end. This can be clearly seen by referring again to *Figure 2.1*. In most ships the length overall will exceed by a considerable amount the length between perpendiculars. The excess will include the over-hang of the stern and also that of the stem where the stem is raked forward. In modern ships having large bulbous bows the length overall (l.o.a.) may have to be measured to the extreme point of the bulb.

A third length which is often used, particularly when dealing with ship resistance, is the length on the waterline (l.w.l.). This is the distance measured on the waterline at which the ship is floating from the intersection of the stern with the waterline to the inter-section of the stem with the waterline. This length is not a fixed quantity for a particular ship, as it will depend upon the waterline at which the ship is floating and upon the trim of the ship. This length is also shown in *Figure 2.1*.

Breadth

The mid point of the length between perpendiculars is called 'amid-ships' and the ship is usually broadest at this point. The breadth is measured at this position and the breadth most commonly used is called the 'breadth moulded'. It may be defined simply as the distance from the inside of plating on one side to a similar point on the other side measured at the broadest part of the ship.

As is the case in the length between perpendiculars, the breadth moulded does not represent the greatest breadth of the ship, so that to define this greatest breadth the breadth extreme is required (see *Figure 2.2*). In many ships the breadth extreme is the breadth moulded plus the thickness of the shell plating on each side of the ship. In the days of riveted ships, where the strakes of shell plating were overlapped the breadth extreme was equal to the breadth moulded plus four thicknesses of shell plating, but in the case of modern welded ships the extra breadth consists of two thicknesses of shell plating only.

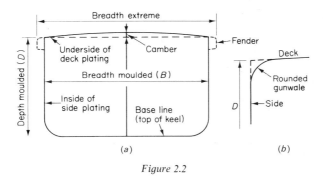

Figure 2.2

The breadth extreme may be much greater than this in some ships, since it is the distance from the extreme overhang on one side of the ship to a similar point on the other side. This distance would include the overhang of decks, a feature which is sometimes found in passenger ships in order to provide additional deck area. It would be measured over fenders, which are sometimes fitted to ships such as cross channel vessels which have to operate in and out of port under their own power and have fenders provided to protect the sides of the ships when coming alongside quays.

Depth

The third principal dimension is depth, which varies along the length of the ship but is usually measured at amidships. This depth is known as the 'depth moulded' and is measured from the underside of the plating of the deck at side amidships to the base line. It is shown in *Figure 2.2(a)*. It is sometimes quoted as a 'depth moulded to upper deck' or 'depth moulded to second deck', etc. Where no deck is specified it can be taken the depth is measured to the uppermost continuous deck. In some modern ships there is a rounded gunwale as shown in *Figure 2.2(b)*. In such cases the depth moulded is measured from the intersection of the deck line continued with the breadth moulded line.

Other features

The three principal dimensions give a general idea of the size of a ship but there are several other features which have to be considered and which could be different in two ships having the same length, breadth and depth. The more important of these will now be defined.

Sheer

Sheer is the height of the deck at side above a line drawn parallel to the base and tangent to the deck line at amidships. The sheer can vary along the length of the ship and is usually greatest at the ends. In modern ships the deck line at side often has a variety of shapes: it may be flat with zero sheer over some distance on either side of amidships and then rise as a straight line towards the ends; on the other hand there may be no sheer at all on the deck, which will then be parallel to the base over the entire length. In older ships the deck at side line was parabolic in profile and the sheer was quoted as its value on the forward and after perpendiculars as shown in *Figure 2.1*. So called 'standard' sheer was given by the formulae:

$$\text{Sheer forward (in)} = 0.2\,L_{ft} + 20$$
$$\text{Sheer aft} \qquad \text{(in)} = 0.1\,L_{ft} + 10$$

These two formulae in terms of metric units would give:

$$\text{Sheer forward (cm)} = 1.666\,L_{m} + 50.8$$
$$\text{Sheer aft} \qquad \text{(cm)} = 0.833\,L_{m} + 25.4$$

It will be seen that the sheer forward is twice as much as the sheer aft in these standard formulae. It was often the case, however, that considerable variation was made from these standard values. Sometimes the sheer forward was increased while the sheer after was reduced. Occasionally the lowest point of the upper deck was some distance aft of amidships and sometimes departures were made from the parabolic sheer profile. The value of sheer and particularly the sheer forward was to increase the height of the deck above water (the 'height of platform' as it was called) and this helped to prevent water being shipped when the vessel was moving through rough sea. The reason for the abolition of sheer in some modern ships is that their depths are so great that additional height of the deck above water at the fore end is unnecessary from a sea keeping point of view.

Deletion of sheer also tends to make the ship easier to construct, but on the other hand it could be said that the appearance of the ship suffers in consequence.

Camber

Camber or round of beam is defined as the rise of the deck of the ship in going from the side to the centre as shown in *Figure 2.2(a)*. The camber curve used to be parabolic but here again often nowadays straight line camber curves are used or there may be no camber at all on decks. Camber is useful on the weather deck of a ship from a drainage point of view, but this may not be very important since the ship is very rarely upright and at rest. Often, if the weather deck of a ship is cambered, the lower decks particularly in passenger ships may have no camber at all, as this makes for horizontal decks in accommodation which is an advantage.

Camber is usually stated as its value on the moulded breadth of the ship and standard camber was taken as one-fiftieth of the breadth. The camber on the deck diminishes towards the ends of the ship as the deck breadths become smaller.

Bilge radius

An outline of the midship section of a ship is shown in *Figure 2.3(a)*. In many 'full' cargo ships the section is virtually a rectangle with the lower corners rounded off. This part of the section is referred to as the 'bilge' and the shape is often circular at this position. The radius of the circular arc forming the bilge is called the 'bilge radius'. Some designers prefer to make the section some curve other than a circle in way of the bilge. The curve would have a radius of curvature which increases as it approaches the straight parts of the section with which it has to link up.

Rise of floor

The bottom of a ship at amidships is usually flat but is not necessarily horizontal. If the line of the flat bottom is continued outwards it will intersect the breadth moulded line as shown in *Figure 2.3(a)*. The height of this intersection above base is called the 'rise of floor'.

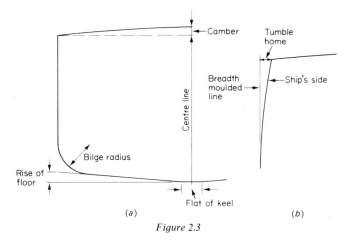

Figure 2.3

The rise of floor is very much dependent on the ship form. In ships of full form such as cargo ships the rise of floor may only be a few centimetres or may be eliminated altogether. In fine form ships much bigger rise of floor would be adopted in association with a larger bilge radius.

Flat of keel

A feature which was common in the days of riveted ships was what was known as 'flat of keel' or 'flat of bottom'. Where there is no rise of floor, of course, the bottom is flat from the centre line to the point where the curve of the bilge starts. If there was a rise of floor it was customary for the line of the bottom to intersect the base line some distance from the centre line so that on either side of the centre line there was a small portion of the bottom which was horizontal, as shown in *Figure 2.3(a)*. This was known as the 'flat of bottom' and its value lay in the fact that a right-angle connection could be made between the flat plate keel and the vertical centre girder and this connection could be accomplished without having to bevel the connecting angle bars.

Tumble home

Another feature of the midship section of a ship which was at one time quite common but has now almost completely disappeared is what was called 'tumble home'. This is the amount which the side of the ship falls in from the breadth moulded line, as shown in *Figure 2.3(b)*. Tumble home was a usual feature in sailing ships and often appeared in steel merchant ships before World War II. Ships of the present day rarely employ this feature since its elimination makes for ease of production and it is of doubtful value.

Rake of stem

In ships which have straight stems formed by a stem bar or a plate the inclination of the stem to the vertical is called the 'rake'. It may be defined either by the angle to the vertical or the distance between the intersection of the stem produced with the base line and the forward perpendicular. When ships have curved stems in profile, and especially where they also have bulbous bows, stem rake cannot be simply defined and it would be necessary to define the stem profile by a number of ordinates at different waterlines.

In the case of a simple straight stem the stem line is usually joined up with the base line by a circular arc, but sometimes a curve of some other form is used, in which case several ordinates are required to define its shape.

Draught and trim

The draught at which a ship floats is simply the distance from the bottom of the ship to the waterline. If the waterline is parallel to the keel the ship is said to be floating on an even keel, but if the waterline is not parallel then the ship is said to be trimmed. If the draught at the after end is greater than that at the fore end the ship is trimmed by the stern and if the converse is the case it is trimmed by the bow or by the head. The draught can be measured in two ways, either as a moulded draught which is the distance from the base line to the waterline, or as an extreme draught which is the distance from the bottom of the ship to the waterline. In the modern welded merchant ship these two draughts differ only by one thickness of plating, but in certain types of ships where, say, a bar keel is fitted the extreme draught would be measured to the underside of the keel and may exceed the moulded draught by 15–23 cm (6–9 in). It is important to know the draught of a ship, or how much water the ship is 'drawing', and so that the draught may be readily obtained draught marks are cut in the stem and the stern. These are figures giving the distance from the bottom of the ship.

In imperial units the figures are 6 in high with a space of 6 in between the top of one figure and the bottom of the next one. When the water level is up to the bottom of a particular figure the draught in feet has the value of that figure. If metric units are used then the figures would probably be 10 cm high with a 10 cm spacing.

In many large vessels the structure bends in the longitudinal vertical plane even in still water, with the result that the base line or the keel does not remain a straight line. The mean draught at which the vessel is floating is not then simply obtained by taking half the sum of the forward and after draughts. To ascertain how much the vessel is hogging or sagging a set of draught marks is placed amidships so that if d_a, $d_\otimes$ and d_f are the draughts at the after end amidships and the forward end respectively then

$$\text{Hog or sag} = \frac{d_a + d_f}{2} - d_\otimes$$

When use is made of amidship draughts it is necessary to measure the draught on both sides of the ship and take the mean of the two readings in case the ship should be heeled to one side or the other.

The difference between the forward and after draughts of a ship is called the 'trim', so that trim $T = d_a - d_f$, and as previously stated the ship will be said to be trimming by the stern or the bow according as the draught aft or the draught forward is in excess. For a given total load on the ship the draught will have its least value when the ship is on an even keel. This is an important point when a ship is navigating in restricted depth of water or when entering a dry dock. Usually a ship should be designed to float on an even keel in the fully loaded condition, and if this is not attainable a small trim by the stern is aimed at. Trim by the bow is not considered desirable and should be avoided as it reduces the 'height of platform' forward and increases the liability to take water on board in rough seas.

Freeboard

Freeboard may be defined as the distance which the ship projects above the surface of the water or the distance measured downwards from the deck to the waterline. The freeboard to the weather deck, for example, will vary along the length of the ship because of the sheer of the deck and will also be affected by the trim, if any. Usually the freeboard will be a minimum at amidships and will increase towards the ends.

Freeboard has an important influence on the seaworthiness of a ship. The greater the freeboard the greater is the above water volume, and this volume provides reserve buoyancy, assisting the ship to rise when it goes through waves. The above water volume can also help the ship to remain afloat in the event of damage. It will be seen later that freeboard has an important influence on the range of stability. Minimum freeboards are laid down for ships under International Law in the form of Load Line Regulations. These will be discussed in Chapter 3.

Ship form

The outside surface of a ship is the surface of a solid with curvature in two directions. The curves which express this surface are not in general given by mathematical expressions, although attempts have been made from time to time to express the surface mathematically. It is necessary to have some drawing which will depict in as detailed a manner as possible the outside surface of the ship. The plan which defines the ship form is known as a 'lines plan'. The lines plan consists of three drawings which show three sets of sections through the form obtained by the intersection of three sets of mutually orthogonal planes with the outside surface.

Consider first a set of planes perpendicular to the centre line of the ship. Imagine that these planes intersect the ship form at a number of different positions in the length. The sections obtained in this way are called 'body sections' and are drawn in what is called the 'body plan' as shown in *Figure 2.4*. When drawing the body plan half-sections only are shown because of the symmetry of the ship. The sections aft of amidships (the after body sections) are drawn on one side of the centre line and the sections forward of amidships (the fore body sections) are drawn on the other side of the centre line. It is normal to divide the length between perpendiculars into a number of divisions of equal length (often ten) and to draw a section at each of these divisions. Additional sections are sometimes drawn near the ends where the changes in the form become more rapid. In merchant ship practice the sections are numbered from the after perpendicular to the forward perpendicular—thus a.p. is 0 and f.p. is 10 if there are ten divisions. The two divisions of length at the ends of the ship would usually be subdivided so that there would be sections numbered $\frac{1}{2}$, $1\frac{1}{2}$, $8\frac{1}{2}$ and $9\frac{1}{2}$. Sometimes as many as 20 divisions of length are used, with possibly the two divisions at each end subdivided, but usually ten divisions are enough to portray the form with sufficient accuracy.

Suppose now that a series of planes parallel to the base and at different distances above it are considered. The sections obtained

28

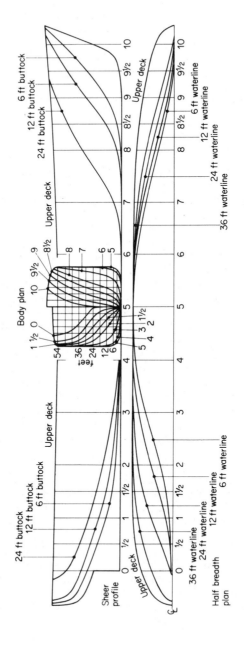

Figure 2.4 Lines plan

by the intersections of these planes with the surface of the ship are called 'waterlines' or sometimes 'level lines'. The lines are shown in *Figure 2.4*. The waterlines like the body sections are drawn for one side of the ship only. They are usually spaced about 1 m (3–4 ft) apart, but a closer spacing is adopted near the bottom of the ship where the form is changing rapidly. Also included on the half-breadth plan is the outline of the uppermost deck of the ship.

A third set of sections can be obtained by considering the inter-section of a series of vertical planes parallel to the centre line of the ship with the outside surface. The resulting sections are shown in a view called the 'sheer profile' (see *Figure 2.4*) and are called 'buttocks' in the after body and 'bow lines' in the fore body or often simply 'buttocks'. The buttocks like the waterlines will be spaced 1 m (3–4 ft) apart. On the sheer profile the outline of the ship on the centre line is shown and this can be regarded as a buttock at zero distance from the centre line.

The three sets of sections discussed above are obviously not independent of one another, in the sense that an alteration in one will affect the other two. Thus, if the shape of a body section is altered this will affect the shape of both the waterlines and the buttocks. It is essential when designing the form of the ship that the three sets of curves should be 'fair' and their interdependence becomes important in this fairing process. What constitutes a fair curve is open to question but formerly the fairing process was done very largely by eye. Nowadays the lines plan is often faired by some mathematical means which will almost certainly involve the use of the computer. However the fairing process is carried out the design of the lines of a ship will normally start by the development of an approximate body plan. The designer when he has such a body plan will then lift offsets for the waterlines and will run the waterlines in the half-breadth plan. This means drawing the best possible curves through the offsets which have been lifted from the sections, and this is done by means of wooden or plastics battens. If it is not possible to run the waterlines through all the points lifted from the body plan then new offsets are lifted from the waterlines and new body sections drawn. The process is then repeated until good agreement is obtained between waterlines and body sections. It is then possible to run the buttocks, and to ensure that these are fair curves it may be necessary to adjust the shape of body sections and waterlines.

The process of fairing is usually done in the drawing office on a scale of 1/4 in to 1 ft or on a 1/50 scale drawing. It is clear that a much more accurate fairing of the form is necessary for production purposes in particular, and this used to be done in the mould loft of the shipyard full size. The procedure was for the drawing office to send to the mould loft offsets from the lines as faired in the office and they were laid out full size on the loft floor. A contracted scale was adopted for the length dimension but waterline and section breadths and buttock heights were marked out full size. The same process of fairing was then adopted as used in the office, the fairing being done by using wood battens of about 25 mm square section pinned to the loft floor by steel pins. To save space the waterlines and buttocks in the forward and after bodies were overlapped in the length direction. This type of full scale fairing enabled sections, waterlines and buttocks to be produced which represented the desired form with considerable accuracy. From the full scale fairing, offsets were lifted which were returned to the drawing office and made the basis of all subsequent calculations for the ship, as will be seen later.

A more recent development has been the introduction of 1/10 scale lofting, which can be done in the drawing office, and the tendency has been to dispense with full scale loft work. Several methods have also been developed for the mathematical fairing of ship form and linking this up with production processes. Discussion of these topics, however, is outside the scope of this work.

The lines drawn on the lines plan representing the ship form are what are called 'moulded lines', which may be taken to represent the inside of the plating of the structure. The outside surface of the ship extends beyond the moulded lines by one thickness of shell plating in an all welded ship. When riveting was the usual method of construction, the shell plating was put on in a series of 'in' and 'out' strakes. In this case the outside surface of the ship extended two thicknesses of plating beyond the moulded lines in way of an outside strake and one thickness beyond the moulded lines in way of an inside strake. Actually the outside surface would be rather more than one thickness or two thicknesses of plating, as the case may be, beyond the moulded line in places where there is considerable curvature of the structure, as for example at the ends of the ship or below the level of the bilge.

In multiple screw merchant ships it is customary to enclose the wing shafts in what is called a 'shaft bossing'. This consists of plating, stiffened by frames, and extending from the point where the shafts emerge from the ship and ending in a casting called a 'shaft bracket'. The bossing is usually faired separately and added on to the main hull form. The bossing is treated as an appendage.

In many ships the shape of the cross section does not change for an appreciable distance on either side of amidships. This portion is called the 'parallel middle body' and may be of considerable extent in full slow ships but may not exist at all in fine fast ships.

Forward of the parallel middle the form gradually reduces in section towards the bow and in like manner the form reduces in section abaft the after end of the parallel middle. These parts of the form are called respectively the 'entrance' and the 'run' and the points where they join up with the parallel middle are referred to as the 'forward' and 'after shoulders'.

Requirements of ship form

The hull form of a ship must be designed to fulfil certain requirements, and the first to be considered is the provision of sufficient buoyancy to carry the various loads such as the weight of the ship itself, plus cargo, fuel, etc. In other words the ship form must provide a certain displacement up to the load waterline. Calling this displacement Δ it follows that

$$\Delta = \rho g V$$

were ρ is the density of the water in which the ship is floating, g is the acceleration due to gravity, and V is the underwater volume. It may be said, therefore, that the designer must so design the form that some underwater volume V is obtained.

Another important requirement of the underwater form is that the centroid of the volume must be in a particular position in the fore and aft direction. The importance of this will be seen in Chapter 5.

Form coefficients

If the ship form consisted simply of a rectangular block of length equal to the length between perpendiculars, breadth equal to the breadth moulded, and depth equal to the draught, then the underwater volume would be given simply by

$$V = L \times B \times d$$

It will, however, be clear that the actual volume is less than the volume of this block, or in other words the ship form can be imagined to have been cut out of this block. What is called the 'block coefficient' is the ratio of the actual volume of the underwater form to the volume LBd. In other words

$$\text{Block coefficient } C_\text{B} = \frac{V}{L \times B \times d} \qquad (2.1)$$

When the ship designer has decided what volume is required he then has four factors to consider: the length, breadth and draught of the ship, and also the block coefficient. There is an infinite number of combinations of these factors which will give the required result and the problem is to decide what are the best values of the four parameters. In the meantime, however, the block coefficient only will be considered. Generally it is governed by resistance considerations, which will be dealt with in Chapter 8. At this stage it may be said that fast ships require low values of block coefficient while in slow ships high values of the block coefficient are permissible. In slow speed ships, say of the bulk carrier type, a high value of block coefficient means a large displacement on given principal dimensions, which means that there is a large amount of displacement available for the carriage of cargo. In fast ships it is essential to keep down the value of the block coefficient, so they normally have lower block coefficients than slow ships. The influence of block coefficient on the shape of the hull form is that in ships with high values of this coefficient considerable parallel middle is likely to be found and the slopes of the waterlines at the ends are steep, whereas with low block coefficients parallel middle is often quite short or may not exist at all and the slopes of the waterlines at the ends will be small also.

Another coefficient which is useful is what is known as the 'prismatic coefficient'. The ship form could be imagined to have been cut from a prism of length equal to the length of the ship and of constant cross section of area equal to the immersed midship area. Thus

$$\text{Prismatic coefficient } C_P = \frac{V}{\text{Midship area} \times L} \qquad (2.2)$$

This particular coefficient has its use in dealing with ship resistance.

A coefficient which is used to express the fullness of the midship section is the midship area coefficient. If the midship section is imagined to be cut out of a rectangle of dimensions breadth × draught then

$$\text{Midship area coefficient } C_m = \frac{\text{Midship area}}{B \times d} \qquad (2.3)$$

The three coefficients so far discussed are related to one another since

$$C_B = \frac{V}{L \times B \times d} = \frac{V}{\text{Midship area} \times L} \times \frac{\text{Midship area}}{B \times d}$$

$$\therefore \quad C_B = C_P \times C_m \qquad (2.4)$$

Generally speaking as the block coefficient becomes finer the midship area coefficient becomes finer, as does the prismatic coefficient.

The waterplane area of a ship, i.e. the area enclosed by any particular waterline, can also be expressed in terms of a coefficient and the area of the circumsecting rectangle. Hence waterplane area coefficient

$$C_W = \frac{\text{Waterplane area}}{L \times B} \qquad (2.5)$$

One other coefficient is sometimes used in defining the ship form. This is the ratio of the underwater volume to the volume of a prism of cross-sectional area equal to the waterplane area, and length equal to the draught so that

$$\text{Vertical prismatic coefficient } C_{PV} = \frac{V}{\text{Waterplane area} \times d} \qquad (2.6)$$

It will be seen that

$$\frac{V}{L \times B \times d} = \frac{V}{\text{Waterplane area} \times d} \times \frac{\text{Waterplane area}}{L \times B}$$

or
$$C_B = C_{PV} \times C_W \qquad (2.7)$$

The values of these coefficients can give useful information about the ship form. It has already been stated that the value of the block coefficient gives an idea of whether the form is full or fine and will indicate whether the waterlines will have large angles of inclination to the centre line at the ends. A large value of the vertical prismatic coefficient will indicate body sections of U-form whilst a small value of this coefficient will be associated with V-shaped sections.

For any particular ship the form coefficients vary with draught, becoming smaller for lower draughts, and hence they have their greatest values at the load draught. Since the coefficients are non-dimensional they are very useful in comparing one ship with another and for geometrically similar ships they will have the same values at corresponding draughts.

The volume of displacement and the areas used in calculating the coefficients are usually taken to the moulded lines of the ship so that moulded dimensions must be used, i.e. length between perpendiculars, breadth moulded and draught moulded. In special cases it might be found desirable to use volumes and areas to the extreme dimensions, that is the volume of displacement may include the displacement of the shell and cruiser stern, and in modern vessels bulbous bow displacement beyond the forward perpendicular.

In such cases extreme dimensions should then be used in calculating the coefficients, i.e. the length on the waterline plus the projection of the bulbous bow forward of the forward perpendicular, the breadth overall and the draught to the bottom of the keel. In like manner where areas are concerned the waterplane area and the midship area may be taken overshell.

3

Classification societies and governmental authorities

Two important groups of organisations which exert considerable influence on the design, construction and safety of ships are classification societies and governmental authorities. The former have a quite long history and have established standards of construction by the production of rules which have done a great deal to ensure the safety of ships. A shipowner is not compelled to build his ship to the rules of such a society but it will be found that the vast majority of ships are so built. Classification is defined as 'a division by groups in order of merit' and this was precisely what was attempted in the early days of ship classification. It was done for the benefit of shipowners, cargo owners and underwriters in order to ascertain if a particular ship represented a reasonable risk. The origin of classification is associated with the name Lloyd's Register of Shipping, which is the oldest society.

Governmental authorities are concerned with the safety of ships and the well being of all who sail in them. In the UK the authority concerned is the Department of Trade (formerly the Board of Trade) and the rules which they produce are compulsory as far as the shipowner is concerned. Should a ship not meet the standards laid down by such an Authority it would not be allowed to sail.

The work of classification societies and governmental authorities overlap to a certain extent and such is the standing of the former that governments often delegate authority to them. For instance, classification societies are concerned to a very considerable extent with the strength of the ship's structures, so that a government authority would accept the strength of a ship as being adequate if it was built to the rules of such a society.

The origin of classification

It was customary in the seventeenth and eighteenth centuries for merchants, shippers and underwriters to meet in coffee houses in London to discuss business. Ship lists were circulated in these establishments which contained information concerning ships and in particular were useful in providing underwriters with information concerning the degree of risk involved in insuring them and their cargoes. Amongst these coffee houses was one owned by an enterprising man called Edward Lloyd who lived in the later seventeenth and early eighteenth centuries. His coffee house was originally in Tower Street but he later moved to Lombard Street. Lloyd provided a list or bulletin about ships as far back as 1702 and after being withdrawn for a time it was issued again in 1734 and has continued to be published until the present day as Lloyd's List.

As time went on the provision of information about ships became more formalised and eventually a Register was published, the first dating from the 1760s. Originally the business of classifying ships and insuring them went on under the same roof but eventually the two activities became entirely separate. Both activities took the name of the coffee house proprietor, the classification side taking the name Lloyd's Register of British and Foreign Shipping. The society is known today simply as Lloyd's Register of Shipping. Lloyd's Register was established in 1760 and has thus existed for more than 200 years.

Lloyd's Register of Shipping

The original system of classification adopted by Lloyd's was to use the notation A E I O U, referring to the quality of the hull, and to use the letters G, M or B (good, middling or bad) to describe the condition of the equipment (anchors, cables, etc.). In the process of time, however, the idea of setting up one uniform standard of construction developed and this became 100 A1, where 100 A referred to the hull when built to the highest standards laid down in the rules of the Society, and 1 referred to the equipment. In addition it will be found that there is a cross placed before 100 A1 so that the classification becomes ✠ 100 A1. The cross indicated that a ship had been built under the supervision of the Society's surveyors. Lloyd's came into existence before the development of mechanical means of propulsion. It is now common practice for machinery to be surveyed as well as the hull, so that the notation LMC (Lloyd's machinery certificate) will be found in the register.

Since Lloyd's Rules cover a wide variety of ship types the type is indicated after the classification symbol. Thus classes such as 100 A1 oil tanker, 100 A1 liquefied gas carrier, 100 A1 ore carrier, etc., will be found in the Register.

Going back to the early days of classification it was the practice to class ships in terms of age and place of building, ships built in the North of England being given a lower classification than those in the South. Shipowners became dissatisfied with this practice and in the early nineteenth century they issued their own register (the Green Book), which was really in competition with Lloyd's Register. Eventually, however, it became evident that the two organisations could not survive separately and amalgamation took place in 1834, from which date Lloyd's Register, as it is known today, really began. The Society is free from control by government and is run by a committee composed of members of the Industry, which operates on behalf of the Industry.

The Register is published annually and gives particulars of ships of 100 tons gross and upwards, whether classed by Lloyd's or not, and is thus an extremely useful index of world shipping. Statistics are published quarterly regarding shipbuilding activities throughout the world.

Activities of Lloyd's Register

Lloyd's Register was originally concerned in the survey of ships' hulls and their equipment. With developments in ships, however, it has become necessary for the Society to deal with other matters as well. It has already been pointed out that the survey of machinery is now included. Other problems with which the Society deals include special types of ships such as oil tankers, liquefied gas carriers, dredgers, hopper barges, etc., and pumping and piping, fire protection, detection and extinction, boilers and other pressure vessels, electrical equipment, refrigerated cargo installations and materials for construction.

With the passage of time and the consequent development in ship technology the rule book produced by the Society, which was originally very simple, has become much more complex and even in recent years considerable developments have taken place. Before 1939, for example, the scantlings of the structural members of the ship could all be determined from the principal dimensions of the ship, L, B and D. These scantlings were governed by two numbers $L \times (B+D)$ and $L \times D$. This procedure proved inadequate for the developments in the post-1945 era and without going into details it can be said that since that time many revisions have taken place and a much more fundamental look has been taken at the problem of structural strength.

The rules originally developed were largely empirical and the scantlings of the structure laid down were those which practice had shown to be adequate. This approach to the determination of scantlings could be said to exist still. Lloyd's Register collects data on ship casualties and analysis of these data suggests areas where modifications should be made. This empirical process is of course backed up by research work carried out by the Society. There is also co-operation with other research organisations.

In order to ensure that a ship which has been built in accordance with the Rules still complies with the highest standards, surveys are to be carried out from time to time during its life. All steel ships are first of all to be surveyed at intervals of approximately one year. These annual surveys deal with a number of relatively minor items which require a yearly check and also require the freeboard marks on the side of the ship to be verified. More comprehensive surveys called 'special surveys' are to be carried out at four-yearly intervals throughout the ship's life. These surveys include the requirements of the annual survey and become progressively more stringent with the age of the ship. Amongst other things which are required to be checked are the scantlings of the structure because of deterioration due to corrosion. Where plating, for example, has been reduced in thickness due to this cause, replacement is required. It is not possible to detail here all the requirements of these special surveys: they are listed in Lloyd's Rules and Regulations for the Construction and Classification of Steel Ships.

A brief outline has been given here of the origin and development of Lloyd's Register of Shipping. More detailed information can be found in a paper by Archer,[1] and the reader is also recommended to study the Rules themselves and also the Register for amplification of many of the matters discussed here.

Lloyd's Register is the oldest classification society and has been considered here, but the development of the other societies which now exist throughout the world has followed a somewhat similar pattern. Some of these societies are Bureau Veritas (France), Det Norske Veritas (Norway), American Bureau of Shipping (USA), **Germanischer Lloyd (Germany)**, **Registro Italiano Navale (Italy)**, and Nippon Kaiji Kyokai (Japan). Societies such as these nowadays consult together in matters of common interest in classification and the development of efficient and improved structural standards. Consultation is carried on through the International Association of Classification Societies (I.A.C.S.). Whilst the primary function of classification societies is still that of classifying ships, they now do very much more than this. They give advice to shipowners and builders on special structural arrangements and are always prepared to vet any new proposals. In recent years they have been instrumental in developing new methods of designing and analysing structures and in so doing are fulfilling an important function in improving ship structures while at the same time contributing towards the design of safe ships.

Activities have been extended to other types of marine vehicles such as oil rigs, and some societies do a considerable amount of work with land based structures.

Governmental authorities

Legislation regarding the safety of ships is the responsibility of the government of the country concerned with registering the ship. In the UK this originally came under the Board of Trade and at the present time is the concern of the Department of Trade. This Department is empowered to draw up rules by virtue of a number of Merchant Shipping Acts extending back more than a hundred years. The Department of Trade employ surveyors who examine ships to verify that they are built in accordance with the regulations. Some of the matters with which the Department of Trade is concerned are:

> Load lines
> Tonnage
> Master and crew spaces
> Watertight subdivision of passenger ships
> Life-saving appliances
> Carriage of grain cargoes
> Dangerous cargoes.

Some of these topics are now the subject of international regulations, e.g. load lines, tonnage, and regulations relative to passenger ships. Some of these will be considered in a little more detail.

Load lines

The question of limiting the depth to which a ship can load is one on which there has been much discussion over the last hundred years. In the nineteenth century Lloyd's Register had some simple rules for limiting draught but there was no enforcement of rules as far as the Government was concerned. Lloyd's gave the rule that a ship should have 3 in freeboard for each foot of depth of hold. It was not until the last quarter of the nineteenth century that the question of introducing legislation through Parliament was seriously considered. Load line limitation is popularly associated with the name of Samuel Plimsoll, who was the Member of Parliament responsible for introducing a bill to limit the draught to which a ship could load. The familiar mark which is now to be seen on ships is often called the Plimsoll line, although its official name is the 'load line mark'.

It is not proposed to consider the history of the development of load line limitation here, but it is clear that there should be some minimum volume of the ship above water, for three reasons, which will become more evident from later chapters. It is sufficient to state here that, first, a minimum freeboard is required so as to provide reserve buoyancy when a ship moves through waves, so that it can rise as the sea passes. This prevents to a large extent water coming on board and thus makes for a dry ship. Secondly, as will be seen later, the more of the ship there is above water the greater will be the range of stability. The third point is that the ship requires reserve buoyancy so that in the event of damage it can remain afloat, at least for a sufficient length of time to allow those on board to get off.

Although the question of a minimum freeboard is really a dynamic one, the rules as set out at present governing the computation of the freeboard are essentially based on static considerations. It is probable that in the future development of regulations concerning freeboards the dynamic behaviour of the ship at sea will be considered.

Freeboard is measured downwards from a deck which is called the 'freeboard deck'. This is defined as the uppermost complete deck exposed to the weather and sea which has permanent means of closing and below which the sides of the ship are fitted with permanent means of watertight closure. Alternatively, a deck lower than this may be considered to be the freeboard deck, subject to its being a permanent deck which is continuous all fore and aft and athwartships.

Basic freeboards are given in the present Load Line Regulations in two tables, one for ships of Type A and one for ships of Type B. These minimum freeboards depend upon the length of the ship. A Type A ship is one which is designed to carry liquid cargoes only in bulk and has only small gasketed openings to the cargo tanks. A further requirement for a ship of this type is that if over 150 m long and designed to have empty compartments it shall be capable of floating with any one of such compartments flooded. A ship of Type

B is a ship other than one of Type A. The difference between the tabular freeboards for the two types can be seen from *Table 3.1*.

Table 3.1 FREEBOARDS OF TYPE A AND TYPE B SHIPS

Length, m	Freeboard, mm	
	Type A	*Type B*
25	208	208
50	443	443
100	1135	1271
150	1968	2315
200	2612	3269
250	3012	4018
300	3262	4630
350	3406	5160

Inspection of the complete tables in the rules will show that above a length of 61 m the Type A freeboard is less than Type B. This is because of the greater integrity of the hull of a Type A ship, which usually has a very high standard of subdivision. Type B ships may qualify for a reduced freeboard if certain conditions are fulfilled with regard to floodability.

The freeboards for these two types of ships are to be corrected for features of the ship form and for the existence of superstructure on the freeboard deck. The features of the form involved are (a) the block coefficient, (b) the length/depth ratio, and (c) the sheer of the freeboard deck.

The block coefficient correction requires that the freeboard shall be multiplied by $(C_B + 0.68)/1.36$ for block coefficients in excess of 0.68. The reasons for this can be readily understood. As block coefficient increases the volume of the underwater form increases at a greater rate than the volume of the above water form. Consequently, to maintain the same ratio of reserve buoyancy to the buoyancy of the underwater form the freeboard should be increased.

The correction for depth consists in increasing the freeboard by an amount $(D - L/15)R$ mm, where D is the depth to the freeboard deck and $R = L/0.48$ for ships less than 120 m long and equals 250 in ships whose lengths are above 120 m. This applies to ships in which D is greater than $L/15$. If the depth is less than $L/15$ the freeboard can be reduced by the same amount, provided there is an enclosed superstructure covering at least $0.6L$, or an efficient trunk extending the full length, or a combination of superstructure and trunks extending the full length of the ship.

The correction for sheer consists in working out the mean sheer of the deck and comparing it with the standard mean sheer laid down in the rules. The correction for sheer is then the deficiency or excess of sheer, multiplied by $0.75 - S/2L$ where S is the length of the superstructures. The basis for the sheer correction is that additional sheer provides extra reserve buoyancy, so that some reduction in freeboard is permissible. On the other hand if the sheer is deficient an increase in freeboard should be made.

The presence of superstructures on the freeboard deck also increases the reserve buoyancy so that a reduction in freeboard should be allowed. This is based on an effective length E, which depends upon the length of the superstructure S and its breadth in relation to the breadth of the ship. If the effective length is $1.0L$ then the allowances on the freeboard are as follows:

$$350 \text{ mm for a length } L = \quad 24 \text{ m}$$
$$860 \text{ mm for a length } L = \quad 85 \text{ m}$$
$$1070 \text{ mm for a length } L = 122 \text{ m}$$

For effective lengths less than $1.0L$ a percentage of these amounts is taken, this percentage being dependent upon the fraction which E is of the length L. It also depends upon the type of ship, i.e. whether A or B. When all these corrections have been made to the basic freeboard the figure calculated is the 'Summer Freeboard'. This is measured downwards from a line on the side of the ship (which denotes the top of the freeboard deck) and a second line is painted on the side with its top edge passing through the centre of a circle, as shown in *Figure 3.1*.

With seasonal variations in weather it is recognised that there should be differences in the freeboard, so that it will be found that

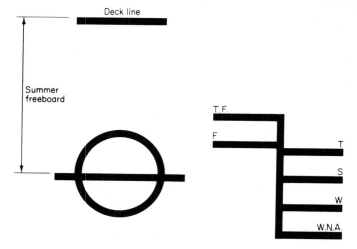

Figure 3.1   Load line mark

there are other marks on the ship's side showing Winter, Winter North Atlantic and Tropical Freeboards. The Winter Freeboard is obtained by adding to the Summer Freeboard 1/48 of the Summer Draught of the ship and the Winter North Atlantic Freeboard is obtained by adding a further 50 mm to this. The Tropical Freeboard is obtained by deducting 1/48 of the Summer Draught from the Summer Freeboard. Finally, a fresh water mark is shown on the ship's side which is obtained by deducting $\Delta/4T$ mm from the Summer Freeboard, where Δ is the displacement in salt water in tonnes and T is the metric tonnes per centimetre immersion, both taken at the Summer Load Waterline. This mark is used when a ship loads in fresh water and should therefore be allowed to load deeper because of the difference in density between salt and fresh water.

A map of the world is provided in the Load Line Regulations, which divides the ocean into zones showing at what times of the year these various zones can be considered to be Summer, Winter, Winter North Atlantic or Tropical. It is important for the ship's officers to ensure that the ship does not exceed the maximum draught allowable for a particular zone when the route passes from one zone to another.

General comments on Load Line Regulations

The basic principles concerning the assignment of load lines have been dealt with above, but no attempt has been made to consider the details of existing Regulations. The reader should consult The Merchant Shipping (Load Line) Rules 1968 for full information on this subject. These Regulations are the interpretation by the Department of Trade in the UK of the International Load Line Convention held in 1966, and represent international law on this subject. All nations which have signed this Convention are therefore assigning load lines to ships on the same basis. The 1966 Conference reviewed the findings of an earlier Conference held in 1930 and considered modifications to the rules then laid down. A comparison of the rules of the two Conventions is to be found in a paper by Smith.[2] Certain ships under the 1966 Rules (particularly tankers) were allowed increased draughts and as far as existing ships were concerned this necessitated that the structural strength of the ships concerned should be investigated to ascertain their adequacy for increased draughts. Wheareas in the 1930 Rules simple formulae for the longitudinal and transverse strengths were included, they were omitted in the new Rules. When a ship is built to the rules of a classification society the government authority of the country concerned accepts this as an adequate standard of structural strength. If a ship is not classed then the government concerned would have to satisfy itself that the standard of strength was adequate for the load line assigned to it.

Tonnage measurement

The calculation of tonnage for a ship is another matter with which the government of the country registering the ship is concerned. Tonnage is a means of measuring the earning capacity of the ship and is used in assessing harbour and port dues and services rendered to the ship.

The history of tonnage measurement goes back a long way, methods having been used for many centuries to assess the earning capacity of ships. The word 'ton' originally came from 'tun', which was a wine cask, and in some cases the capacities of ships were measured by the number of wine casks which they could carry. The 'modern' system of tonnage measurement in the UK really dates from that put forward in the Merchant Shipping Act of 1854 and is associated with the name of Moorsom.[3] Whether or not existing regulations upon which tonnage calculations are based assessed earning capacity fairly and correctly is a matter of long-standing controversy, which it is likely will continue in the future. Setting this aside, the principles of existing regulations of the Department of Trade will be discussed and the findings of the 1969 International Conference on Tonnage outlined.

Tonnage is measured in units of 100 ft^3, so that it has no relation to weight and is really a measurement of volume. What is called the 'gross tonnage' consists of the underdeck tonnage (i.e. the tonnage below the tonnage deck), the tonnage of spaces between the tonnage deck and the upper deck, the tonnage of closed-in spaces above the upper deck, and the tonnage of hatchways. The tonnage deck is the second deck, except in single-deck ships, in which case it is the upper deck. The underdeck tonnage is measured from the inner bottom to the tonnage deck and inside of frames. This means that double-bottom tanks are excluded, which is reasonable since they are not capable of carrying cargo.

In order to obtain the net or register tonnage certain deductions are made from the gross, because of spaces which are necessary for the propulsion and operation of the ship. The spaces so deducted include an allowance for the propelling machinery space, accommodation of the master and crew, the wheelhouse and chart room, chain lockers and space appropriated to the operation of the steering gear, anchor and capstan gear, spaces for storage of safety equipment, workshops, certain storerooms, donkey engine and boiler space, the space occupied by the main pumps if outside the machinery space, and water ballast tanks other than double-bottom tanks.

The allowance for the propelling machinery space was intended to include space occupied by bunkers and is calculated as follows. If the tonnage of the machinery space is 13% or over but less than 20% of the gross tonnage, the allowance is 32% of the gross tonnage. If it is less than 13% of the gross tonnage the allowance is the actual tonnage of the engine room multiplied by 32/13. These figures apply to ships propelled by screws.

In the case of ships propelled by paddles the allowance is 37% of the gross when the machinery space tonnage is 20% or over but less than 30% of the gross. When it is less than 20% the allowance is the actual engine room tonnage multiplied by 37/20. In other cases the allowance for propelling machinery space is $1\frac{3}{4}$ times the actual tonnage of the machinery space in screw vessels, or $1\frac{1}{2}$ times in the case of paddle ships.

Spaces above the upper deck which contain any part of the machinery or which light and ventilate the machinery space may be included in the machinery space, but must also be included in gross tonnage. Such spaces will only be included on the request of the owner.

Formerly it was possible to have certain spaces below the upper deck exempted from tonnage measurement, for example the space between the second and upper decks. The decision to allow this was the result of a court case in which it was decided in a certain ship that this space was not strictly speaking closed in, the cargo in the space being only partially protected from the weather. From this developed the idea of the shelter deck ship, to which reference was made in Chapter 1. The shelter deck ship was one in which the bulkheads stopped at the second deck and the upper or shelter deck above this from a structural strength point of view was the strength deck. The space between the second and upper decks was regarded as being open if a small tonnage hatch, or 'opening' as it was called, was fitted in the shelter deck aft. This opening did not have permanent means of closing and had a very shallow hatch coaming. The bulkheads, as already stated, were stopped at the second deck, or if they continued up to the shelter deck they were non-watertight and had openings in them. Ships of this type had very low freeboards measured to the second deck and in consequence, in the event of damage, had little reserve buoyancy and in most cases would be incapable of floating with a compartment open to the sea. Should an owner decide to carry the watertight bulkheads up to the upper deck tonnage exemption was not permitted for the 'tween deck space, so that this was an instance of tonnage regulations encouraging a type of ship which was not as safe as it might be. Ships of this type were built in large numbers and formed the bulk of the tramp ships of the pre-1939 period. During World War II their vulnerability to enemy attack was recognised and as a temporary measure watertight 'tween deck bulkheads were introduced to improve their subdivision.

In the post war years steps were taken to amend existing regulations to eliminate the 'open' shelter deck ship. This was done on an international basis and in essence consists in permitting a ship to

Optional tonnage
mark for fresh
or tropical waters

Tonnage
mark

Figure 3.2   Tonnage mark

have the 'tween deck space between the second and upper decks exempted from tonnage even if the watertight bulkheads are carried up to the upper deck. The tonnage opening in the upper deck was also eliminated. In order to obtain this exemption, however, the draught of the ship is limited and this limit is shown by a tonnage mark on each side of the ship at amidships. The mark is shown in *Figure 3.2*. The position of the mark in general corresponds to the draught which would be obtained if the second deck of the ship were treated as the freeboard deck. If, however, the ship were designed for the maximum draught which could be obtained, treating the upper deck as the freeboard deck, and was loaded to this draught no exemption would be obtained for the 'tween deck space.

Recent developments in tonnage measurement

Further developments in regard to tonnage measurement have taken place in recent years. In 1969 an International Conference on the Tonnage Measurement of Ships was convened in London by the Inter-governmental Maritime Consultative Organisation (I.M.C.O.). The results of this Conference have been given in a paper by Wilson.[4] The attempt was made to simplify existing tonnage regulations and to reduce the calculation of gross and net tonnages to formulae. The formulae can be stated as follows:

$$\text{Gross tonnage } (GT) = K_1 V \tag{3.1}$$

$$\text{Net tonnage } (NT) = K_2 V_c \left(\frac{4d}{3D}\right)^2 + K_3 \left(N_1 + \frac{N_2}{10}\right) \tag{3.2}$$

where

V = total volume of all enclosed spaces of the ship in cubic metres

$K_1 = 0.2 + 0.02 \log_{10} V$

V_c = total volume of cargo spaces in cubic metres

$K_2 = 0.2 + 0.02 \log_{10} V_c$

$K_3 = 1.25 \dfrac{GT + 10\,000}{10\,000}$

D = moulded depth amidships in metres

d = moulded draught amidships in metres

N_1 = number of passengers in cabins with not more than eight berths

N_2 = number of other passengers

$N_1 + N_2$ = total number of passengers the ship is permitted to carry as indicated on the ship's passenger certificate

When $N_1 + N_2$ is less than 13, N_1 and N_2 shall be taken as zero

GT = gross tonnage of the ship.

In the above the factor $(4d/3D)^2$ is not to be taken as greater than unity and the term $K_2 V_c (4d/3D)^2$ is not to be taken as less than $0.25\ GT$.

The volumes referred to in these formulae are to be calculated to the inside of plating and include the volumes of appendages. Volumes of spaces open to the sea are excluded.

In the proposed new system of calculating tonnage the tonnage mark referred to in the previous section was abolished, although the inclusion of draught in the formulae suggests the variation of the tonnage with the draught for which the ship is designed.

The regulations shown above for determining the tonnage of ships are embodied in a new International Convention on Tonnage. At the time of writing (1973) this Convention has not replaced existing legislation but when a sufficient number of signatures to the Convention have ratified it, national regulations will be prepared by the governments (in the UK through the Department of Trade).

The proposed new regulations appear to simplify greatly what has been for a long period a very complex subject. However, it could be argued that in the process the basic philosophy behind the assessment of tonnage has been lost, or at least is no longer evident.

Other tonnages

The tonnage of a ship calculated according to the existing regulations described in this chapter is accepted for ships today on international voyages, the tonnage of a ship being shown on its Tonnage Certificate. There are, however, special tonnages which are calculated slightly differently and shown on separate certificates. They are for ships trading through the Suez Canal, and the Panama Canal. The former is at the present time of little interest, since the Suez Canal has now been closed for some years, but formerly the charges for passage through the Canal were based on this Suez Canal Tonnage. Similarly the charges for use of the Panama Canal were based on the Panama Tonnage.

Passenger ships

Ships intended for the carriage of passengers are required to comply with very stringent safety regulations. For this purpose a passenger ship is defined as one which carries more than 12 passengers, and such a ship would be issued with a Passenger Certificate on compliance with the regulations. Present day passenger ship regulations are the outcome and interpretation of the findings of various international conferences on this subject which have taken place during the present century. Although the various maritime countries of the world had passenger regulations before 1912, it was the loss of the *Titanic* in that year which focused attention internationally on the safety of passenger ships. The *Titanic* sank with great loss of life on her maiden voyage when she struck an iceberg and was damaged in several compartments. In the UK a Bulkhead Committee was set up by the Board of Trade to investigate the strength and disposition of bulkheads in passenger ships, since the ability of a ship to float after damage was a subject which loomed large in the minds of those concerned with legislation after this disaster. The technical problems involved in flooding after damage will be discussed in Chapter 5.

It was considered that the safety of passenger ships was a subject for study on an international basis, so that after the *Titanic* disaster an international conference was held in 1914. The incidence of World War I interrupted further discussion, although the results of some of the findings were published during the war.[5,6] It was not, however, until 1929 that another Conference was held and it was 1932 before the International Convention for the Safety of Life at Sea was signed by the major maritime nations. As has already been stated in connection with load lines and tonnage, the Convention then had to be ratified by the signatory nations and the findings incorporated in the law of the individual countries. The 1932 Convention was examined at later Conferences in 1948 and 1960 and certain modifications made in the light of experience. Another Conference will probably be held in 1976, when it is expected that major changes will be made in the whole approach to the assessment of safety.

The Safety Convention was not only concerned with the watertight subdivision of passenger ships and the associated problem of safety in the damaged condition, but also laid down regulations governing other aspects of safety, such as fire detection and extinguishing and fire protection, machinery and electrical installation, life saving appliances such as boats and the means for launching them, radiotelegraphy and radiotelephony, safety of navigation, carriage of grain and dangerous cargoes and regulations relative to nuclear ships.

Inter-governmental Maritime Consultative Organisation

The International Conference on the Safety of Life at Sea which took place in 1914 probably represents the first international approach to maritime problems of a technical nature. It has been seen that the subjects of load lines and tonnage have been studied on an international basis and regulations drawn up which are applicable to all countries signing the various conventions. In 1959 a permament body was set up under the aegis of the United Nations to deal with all such matters in the future. This is called the Intergovernmental Maritime Consultative Organisation (I.M.C.O.). The Organisation has its headquarters in London. It has Committees drawn from the various maritime countries which meet periodically to discuss matters of mutual interest. From time to time I.M.C.O. arranges international Conferences such as the International Load Line Conference of 1966, the Tonnage Conference of 1969, and the International Conference on the Safety of Life at Sea, 1960. Conventions such as the Safety Convention may be amended by unanimous agreement between contracting governments and on the request of a government a proposed amendment will be communicated to the other governments. Alternatively, an amendment may be proposed to the Organisation by a contracting government and if adopted by a two-thirds majority of the Assembly of the Organisation upon the recommendation of the Maritime Safety Committee of the Organisation it will be communicated to the contracting government for their acceptance. A conference of governments to consider amendments to Conventions proposed by a contracting government can be convened at any time on the request of one-third of the contracting governments.

The formation of I.M.C.O. has made it easier to obtain amendments of existing Conventions and it is certain that this Organisation will play a big part in the future development of international legislation with regard to shipping.

REFERENCES

1. Archer, S., 'The Classification Society: some thought on its functions and contribution to marine engineering', *Trans. North East Coast Instn. of Engrs. & Shpbdrs*, 1972–1973
2. Murray Smith, D. R., 'The 1966 International Conference on Load Lines', *Trans. Royal Instn. of Naval Architects*, 1969
3. Moorsom, G., 'On the new tonnage law as established in the Merchant Shipping Act of 1854', *Trans. Instn. of Naval Architects*, 1860
4. Wilson, E., 'The 1969 International Conference on Tonnage Measurement of Ships', *Trans. Royal Instn. of Naval Architects*, 1970
5. Welch, J. J., 'The watertight subdivision of ships', *Trans. Instn. of Naval Architects*, 1915
6. Denny, Sir Archibald, 'Subdivision of merchant ships: reports of the Bulkhead Committee 1912–15', *ibid.*, 1916.

4

Ship calculations

The form of a ship is generally expressed by a number of curves of non-mathematical form, and in carrying out the various calculations which will be discussed later in this chapter it is necessary to be able to determine the areas enclosed by such curves and the moments and second moments (or moments of inertia) of these areas. It is also necessary to calculate volumes and moments of volume. Although in recent years attempts have been made to produce mathematical hull forms, generally this has not reached the stage where the areas, moments and volumes can be obtained by direct mathematical integrations. The naval architect has recourse, therefore, to approximate rules for the calculation of these quantities and even when the actual numerical work is carried out by computer the program usually makes use of one of the approximate rules. It is not intended to deal exhaustively with this subject, but some indication of the basis of these rules will be given. Most rules depend upon the substitution of the actual curve to be integrated by some curve of simple mathematical form. The accuracy of the result then depends upon the accuracy with which the real curve is represented by the mathematical curve.

Trapezoidal rule

Figure 4.1 shows a curve ABC. The area underneath this curve could be approximated to by replacing the part AB by a straight line joining A to B, and similarly BC could be replaced by a straight line joining B to C. Then if the ordinates y_1, y_2 and y_3 are spaced a distance h apart:

$$\text{Area} = \frac{y_1 + y_2}{2} \times h + \frac{y_2 + y_3}{2} \times h = h\left\{\frac{y_1}{2} + y_2 + \frac{y_3}{2}\right\} \quad (4.1)$$

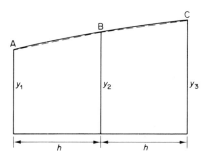

Figure 4.1

on the assumption that the areas of the trapezoid represent sufficiently accurately the area under the actual curves. For the curve shown the area determined from equation 4.1 will under-estimate the actual area. If the curve was concave upwards instead of downwards then the area obtained from the trapezoidal rule would overestimate the true area. To increase the accuracy the length of the curve to be integrated should be divided up into smaller portions. If, for example, ordinates were taken at $h/2$ and $3h/2$, then calling these $y_{1\frac{1}{2}}$ and $y_{2\frac{1}{2}}$ the area A would be given by:

$$A = \frac{h}{2} \times \left\{ \frac{y_1 + y_{1\frac{1}{2}}}{2} + \frac{y_{1\frac{1}{2}} + y_2}{2} + \frac{y_2 + y_{2\frac{1}{2}}}{2} + \frac{y_{2\frac{1}{2}} + y_3}{2} \right\}$$

$$= \frac{h}{2} \times \left\{ \frac{y_1}{2} + y_{1\frac{1}{2}} + y_2 + y_{2\frac{1}{2}} + \frac{y_3}{2} \right\} \tag{4.2}$$

To obtain some idea of the accuracy which might be achieved by the use of the trapezoidal rule consider the parabolic curve

$$y = 10 + 4x - x^2 \text{ between } x = 0 \text{ and } x = 2$$

If three ordinates are taken, then

$$y_0 = 10$$
$$y_1 = 10 + 4 \times 1 - 1^2 = 13$$
$$y_2 = 10 + 4 \times 2 - 2^2 = 14$$

Using the trapezoidal rule:

$$A = 1 \times \left\{ \frac{10}{2} + 13 + \frac{14}{2} \right\} = 25 \text{ units}$$

If the area is obtained by integration then

$$A = \int_0^2 y \, dx = \int_0^2 (10 + 4x - x^2) \, dx$$

$$= \left(10x + 2x^2 - \frac{x^3}{3} \right)_0^2 = 25.33 \text{ units}$$

The error is $\frac{4}{3}\%$

If now ordinates are taken at $x = \frac{1}{2}$ and $1\frac{1}{2}$ the five ordinates would be:

$$
\begin{aligned}
y_0 &= 10 \\
y_{\frac{1}{2}} &= 10 + 4 \times \tfrac{1}{2} - (\tfrac{1}{2})^2 = 11.75 \\
y_1 &= 13 \\
y_{1\frac{1}{2}} &= 10 + 4 \times \tfrac{3}{2} - (\tfrac{3}{2})^2 = 13.75 \\
y_2 &= 14
\end{aligned}
$$

$$
\text{Area} = \frac{1}{2} \times \left\{ \frac{10}{2} + 11.75 + 13 + 13.75 + \frac{14}{2} \right\} = 25.25 \text{ units}
$$

The error has now been reduced to 1%, so that it would seem that many more ordinates would be required to obtain great accuracy by the use of this rule.

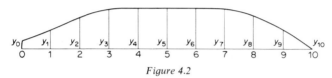

Figure 4.2

The trapezoidal rule can be applied to long curves such as the waterlines of a ship. If, for example, the area of a ship's waterplane were required, the length would be divided up into a number of equal divisions as in *Figure 4.2*. Calling this number of divisions n, the area would be given by:

$$
A = h \left\{ \frac{y_0}{2} + y_1 + y_2 + y_3 \ldots + y_{n-1} + \frac{y_n}{2} \right\} \tag{4.3}
$$

In many cases it may be sufficiently accurate to use ten divisions of length, because often the curve between two divisions of length is almost indistinguishable from a straight line. In fact in many modern ships of full form the waterlines over the midship portion are straight lines. If greater accuracy was required in regions of considerable curvature, i.e. at the ends of the ship, ordinates could be taken at half-divisions.

Before leaving the discussion of the trapezoidal rule, its application to the determination of the moment of an area will be discussed.

Referring once again to *Figure 4.1*, the position of the centroid from mid length of the trapezium length, h, and having end ordinates y_1 and y_2, is

$$
\frac{y_2 - y_1}{y_2 + y_1} \times \frac{h}{6}
$$

so that its position from $x = 0$ is

$$\frac{h}{2} + \frac{y_2 - y_1}{y_2 + y_1} \times \frac{h}{6}$$

Hence:

Moment of area about $x = 0 = \dfrac{y_1 + y_2}{2} \times h \left(\dfrac{h}{2} + \dfrac{y_2 - y_1}{y_2 + y_1} \times \dfrac{h}{6} \right)$

$$= \frac{h^2}{12} \left(2y_1 + 4y_2 \right)$$

Similarly the moment of the trapezium between y_2 and y_3 about $x = 0$ is

$$\frac{h^2}{12} (8y_2 + 10y_3)$$

Therefore:

$$\text{Total moment} = \frac{h^2}{12} (2y_1 + 12y_2 + 10y_3) \tag{4.4}$$

If the same parabolic curve is considered as was used previously then:

$$\text{Moment} = \frac{1^2}{12} (2 \times 10 + 12 \times 13 + 10 \times 14)$$

$$= 26.33 \text{ units}$$

The moment by integration is given by

$$\int_0^2 x(10 + 4x - x^2)\,\mathrm{d}x = \left(5x^2 + \frac{4}{3}x^3 - \frac{x^4}{4} \right)_0^2 = 26.66 \text{ units}$$

The error in this case is about $1\frac{1}{4}\%$.

Simpson's rule

The replacement of a curve by a series of straight lines as in the trapezoidal rule has severe limitations, and the rules mostly used in naval architecture calculations replace the actual curve by a mathematical curve of higher order. One rule which is used extensively is Simpson's rule, which assumes that the actual curve can be replaced by a curve of the form

$$y = a_0 + a_1 x + a_2 x^2$$

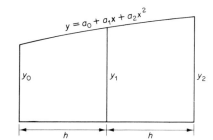

$$y = a_0 + a_1 x + a_2 x^2$$

y_0 y_1 y_2

Figure 4.3

h h

The origin of the rule apparently goes back a long way and it appears that it can be traced to Newton.

Figure 4.3 shows a curve of this form. It is expressed by three evenly spaced ordinates y_0, y_1 and y_2. The values of these ordinates are a_0, $a_0 + a_1 + a_2$ and $a_0 + 2a_1 + 4a_2$ for $x = 0$, 1 and 2.

The area under the curve is

$$\int_0^2 (a_0 + a_1 x + a_2 x^2)\,dx = \left(a_0 x + \frac{a_1 x^2}{2} + \frac{a_2 x^3}{3} \right)_0^2 = 2a_0 + 2a_1 + \frac{8}{3} a_2$$

Now $a_0 = y_0$ and $y_1 = y_0 + a_1 + a_2$, and $y_2 = y_0 + 2a_1 + 4a_2$.

It follows that

$$y_0 + 2a_1 + 4a_2 - 2y_0 - 2a_1 - 2a_2 = y_2 - 2y_1 = -y_0 + 2a_2$$

or
$$a_2 = (y_2 - 2y_1 + y_0)/2$$

Now

$$a_1 = y_1 - y_0 - a_2 = y_1 - y_0 - (y_2 - 2y_1 + y_0)/2 = -\tfrac{3}{2}y_0 - y_2/2 + 2y_1$$

Hence:

$$\text{Area} = 2y_0 - 3y_0 - y_2 + 4y_1 + \frac{8}{3}\left(\frac{y_2 - 2y_1 + y_0}{2} \right) = \frac{1}{3}(y_0 + 4y_1 + y_2)$$

This is for a spacing of ordinates of unity. If the spacing was h then

$$\text{Area} = \frac{h}{3}(y_0 + 4y_1 + y_2) \tag{4.5}$$

This formula calculates correctly the area underneath a curve of the second order. It will approximate the area underneath any curve which passes through the same three points, the degree of accuracy depending upon the extent to which the actual curve is represented by the parabola which replaces it. The rule is called 'Simpson's rule' and is the one most widely used in ship calculations.

The rule can be extended to calculate the area under a long curve such as a waterline. In this case it would be unreasonable to imagine that such a curve could be represented by a parabola. If, however, the length of the curve is divided into n equal parts, then it would not be unreasonable to consider that the part of the waterline extending over two adjacent divisions is of parabolic form. If, for example, the curve in *Figure 4.4* were divided into ten divisions, then the area from 0 to 2 would be given by $\frac{1}{3}(L/10)(y_0+4y_1+y_2)$, where $L/10$ is the length of the divisions into which the length of the curve has been divided and is usually known as the 'common

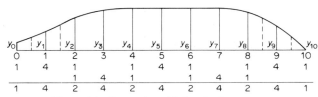

Figure 4.4   *Simpson's multipliers for long curve*

interval'. Similarly the area from 2 to 4 would be $\frac{1}{3}(L/10)$ $(y_2+4y_3+y_4)$ and so on. It will be seen that the multipliers for the complete curve are as shown at the bottom of *Figure 4.4*, so that calling $L/10$ C.I.:

$$\text{Total area} = \frac{\text{C.I.}}{3}(y_0+4y_1+2y_2+4y_3+2y_4+4y_5+2y_6$$

$$+4y_7+2y_8+4y_9+y_{10}) \qquad (4.6)$$

One further point is that in the calculation of areas of waterplanes it is often necessary to close up the spacing towards the ends, as there is often considerable curvature at the ends of the waterlines. Thus, ordinates are often taken at $\frac{1}{2}$, $1\frac{1}{2}$, $8\frac{1}{2}$ and $9\frac{1}{2}$. The area from 0 to 1 would then be given by

$$\frac{\text{C.I.}/2}{3}(y_0+4y_{\frac{1}{2}}+y_1)=\frac{\text{C.I.}}{3}(\tfrac{1}{2}y_0+2y_{\frac{1}{2}}+\tfrac{1}{2}y_1)$$

and from 1 to 2 by

$$\frac{\text{C.I.}/2}{3}(y_1+4y_{\frac{1}{2}}+y_2)=\frac{\text{C.I.}}{3}(\tfrac{1}{2}y_1+2y_{1\frac{1}{2}}+\tfrac{1}{2}y_2)$$

and similarly at the other end.

The additional ordinates at the half-stations are shown dotted in *Figure 4.4* and it will be seen that the combined multipliers for the whole curve are as shown in *Figure 4.5*. It follows that

$$\text{Total area} = \frac{\text{C.I.}}{3}(\tfrac{1}{2}y_0 + 2y_{\frac{1}{4}} + y_1 + 2y_{1\frac{1}{4}} + 1\tfrac{1}{2}y_2 + 4y_3 + 2y_4 + 4y_5$$

$$+ 2y_6 + 4y_7 + 1\tfrac{1}{2}y_8 + 2y_{8\frac{1}{4}} + y_9 + 2y_{9\frac{1}{4}} + \tfrac{1}{2}y_{10}) \qquad (4.7)$$

In the past it has usually been found sufficiently accurate to divide the length of the ship up into ten equal parts for calculations of this sort and to introduce ordinates at half-intervals at the ends as

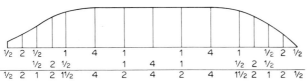

Figure 4.5   Simpson's multipliers for long curve with intermediate ordinates at the ends

described. If it was desired to avoid the awkward multipliers arising from the reduced spacing at the ends then the necessary accuracy could be achieved by halving the spacing throughout, in which case the length of the ship would be divided into 20 equal parts. This might be more convenient when programming the calculation for the computer.

Calculation of moments of area and centroids by Simpson's rule

Consider the curve shown in *Figure 4.6* and suppose that the moment of the area under the curve about $x = 0$ is required. Then it will be seen that for an elementary area $y\,dx$ moment $= xy\,dx$.

$$\text{Total moment} = \int_0^L xy\,dx$$

The integral represents the summation of the products of the ordinates y and their distances x from the axis about which the moments are required. This summation can be carried out by using Simpson's rule and if a long curve is considered which is divided into n equal parts, the distances of the ordinates from $x = 0$ being x_1, x_2, etc., then

$$\text{Moment} = \frac{\text{C.I.}}{3}(x_0 y_0 + 4x_1 y_1 + 2x_2 y_2$$

$$+ 4x_3 y_3 + \ldots + 4x_{n-1} y_{n-1} + x_n y_n) \qquad (4.8)$$

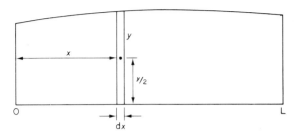

Figure 4.6

It is often convenient to express the various distances from $x = 0$ as multiples of the common interval C.I., so that $x_0 = 0 \times$ C.I., $x_1 = 1 \times$ C.I., $x_2 = 2 \times$ C.I. etc. It follows that the moment is then given by:

$$M = \frac{\text{C.I.}^2}{3}(y_0 + 1 \times 4y_1 + 2 \times 2y_2 + 3 \times 4y_3$$

$$\dots + (n-1)4y_{n-1} + ny_n) \qquad (4.9)$$

This saved a great deal of arithmetic when the calculation had to be carried out manually but is not so important when the calculations are done by computer.

In carrying out calculations of this sort for long curves such as the waterlines of a ship it is customary, although not necessary, to take moments about the mid length (amidships). There would then be two moments of different sign, the net moment being the difference between them. Hence

$$\text{Moment of area about amidships} = \int_0^{L/2} xy\,dx - \int_{L/2}^{L} xy\,dx$$

where in this case the distances x are measured from amidships.

When the moment of the area has been obtained in this way the centroid of the area from the axis about which moments have been taken is obtained by dividing the moment by the area.

For areas such as the waterplane of a ship the position of the centroid in the transverse direction is automatically found, because by symmetry the centroid must lie on the centre line of the ship. It is sometimes, however, necessary to find the centroid of a figure bounded by a curve on one side and by a straight line on the other side. Such a figure is the half waterplane of a ship.

Referring to *Figure 4.6* again, it will be seen that the centroid of the elementary area $y\,dx$ is $y/2$ from the line OL.

Hence

$$\text{Moment} = y\,dx \times y/2 = \tfrac{1}{2}y^2\,dx$$

and

$$\text{Total moment} = \int_0^L \tfrac{1}{2}y^2\,dx$$

In this case the summation represented by $\int y^2 dx$ can be carried out by the use of Simpson's rule and the result would be

Moment of area about OL

$$= \frac{1}{2} \times \frac{\text{C.I.}}{3}\,(y_0^2 + 4y_1^2 + 2y_2^2 + 4y_3^2 \dots 4y_{n-1}^2 + y_n^2) \quad (4.10)$$

Second moments of area or moments of inertia

Some ship calculations require the determination of second moments of area or moments of inertia and these calculations can also be performed by the use of Simpson's rule. Referring once more to *Figure 4.6*, the second moment of area about a transverse axis of the elementary ship $y\,dx$ is $x^2 y\,dx$ and the total second moment is $\int_0^L x^2 y\,dx$.

Here the summation $\int_0^L x^2 y\,dx$ is carried out using Simpson's rule and for a curve divided into n equal parts:

$$\text{S.M.A.} = \frac{\text{C.I.}}{3}(x_0^2 y_0 + 4x_1^2 y_1 + 2x_2^2 y_2 + 4x_3^2 y_3$$

$$\dots + 4x_{n-1}^2 y_{n-1} + x_n^2 y_n) \quad (4.11)$$

If as in the case of moments the distances x_0, etc., are expressed as multiples of the common interval, then:

$$\text{S.M.A.} = \frac{\text{C.I.}^3}{3}(0 \times y_0 + 1^2 \times 4y_1 + 2^2 \times 2y_2 + 3^2 \times 4y_3$$

$$\dots (n-1)^2\,4xy_{n-1} + n^2 y_n) \quad (4.12)$$

As was the case with moments, it is customary in ship calculations to calculate the second moments about amidships when dealing with long curves as represented by the waterlines of a ship. Thus:

$$\text{S.M.A.} = \int_0^{L/2} x^2 y\,dx + \int_{L/2}^L x^2 y\,dx$$

The second moment of an area is always least about an axis passing through its centroid. If this is denoted by I_{NA} and the second moment about an axis xx by I_{xx} then:

$$I_{NA} = I_{xx} - A\bar{y}^2$$

where A is the area and $\bar{y}$ is the distance from its centroid to the axis xx. This correction has to be made when calculating the second moment of the area of a waterplane, since in general its centroid is not at amidships.

In ship calculations the second moment of area is also required about the centre line of the waterplane. In *Figure 4.6*, y is the half-ordinate of a waterplane measured from the centre line. The second moment of area of the elementary strip about the centre line is

$$y\,dx \times (y/2)^2 + \tfrac{1}{12}y^3\,dx = \tfrac{1}{3}y^3\,dx$$

$$\text{Total second moment of half waterplane area} = \int_0^L \tfrac{1}{2}y^3\,dx$$

Now the second moment of the total area will be twice this amount, and this will be the second moment about an axis passing through the centroid, since the waterplane is symmetrical about the centre line. The summation $\int y^3\,dx$ can be performed by Simpson's rule and if the length were divided into n equal parts where n is an even number the result would be as follows:

$$I_{\text{C}} = 2 \times \frac{\text{C.I.}}{3} \times \frac{1}{3}(y_0^3 + 4y_1^3 + 2y_2^3 \ldots + 4y_{n-1}^3 + y_n^3) \qquad (4.13)$$

Example of use of Simpson's rule in calculating properties of water-planes

The following example will illustrate how the various properties of a waterplane can be calculated by the use of Simpson's rule.

A ship 180 m long floats at a waterline whose half-ordinates at evenly spaced sections starting from the after end are as follows: 0.0, 4.1, 9.6, 11.9, 12.0, 12.0, 12.0, 12.0, 11.1, 7.2 and 0.0 m. Calculate the area, the position of the centroid forward or aft of amidships and the second moments of area about a transverse axis passing through the centroid and about the centre line.

When these calculations were done manually or by calculating machine they were set out in tabular form. This is not necessary when the calculations are programmed for the computer, but for the purposes of illustrating the use of Simpson's rule the tabular method will be followed here. The area, the position of the centroid in the fore and aft direction, and the second moment about a transverse axis passing through the centroid can all be calculated from one table. *Table 4.1* shows how this calculation is set out.

In this table the half-ordinates are multiplied by the Simpson's multiplier appropriate to the position of the ordinate, i.e. 1, 4, 2, etc. The result is put in the column marked 'Area product', and the products when summed and multiplied by 1/3 of the common interval (in this case 18 m), and by two for both sides, give the area of the waterplane. The moment is calculated about amidships and as mentioned above the distances of the various ordinates from amidships are given as the number of divisions of the length. The area products are multiplied by these distances and the results set down in the column marked 'Moment product'. The sum of these products represents the moment of the area. There are two sums, one for the after part of the waterplane and one for the forward part. It will be seen that the difference of these sums divided by the sum of the area products and multiplied by the common interval gives the distance of the centroid from amidships.

The second moment of area product is obtained by multiplying the moment product by the lever. The sum of these products multiplied by the cube of the common interval and by two for both sides and divided by three gives the second moment of area of the complete waterplane.

Table 4.1 CALCULATION OF PROPERTIES OF A WATERPLANE USING SIMPSON'S RULE

Section	Half-ordinate, m	Simpson's multiplier	Area product	Lever	Moment product	Second moment product
0	0.0	1	—	5	—	—
1	4.1	4	16.4	4	65.6	262.4
2	9.6	2	19.2	3	57.6	172.8
3	11.9	4	47.6	2	95.2	190.4
4	12.0	2	24.0	1	24.0	24.0
5	12.0	4	48.0	0	242.4	—
6	12.0	2	24.0	1	24.0	24.0
7	12.0	4	48.0	2	96.0	192.0
8	11.1	2	22.2	3	66.6	199.8
9	7.2	4	28.8	4	115.2	460.8
10	0.0	1	—	5	—	—
			278.2		301.8	1526.2

Area $A = \frac{2}{3} \times 18 \times 278.2 = 3338.4 \, \text{m}^2$

Centroid from amidships $\bar{y} = (301.8 - 242.4)/278.2 \times 18 = 3.84$ m for'd

$I_{\otimes} = \frac{2}{3} \times 18^3 \times 1526.2 = 5\,933\,865 \, \text{m}^4$

less $3338.4 \times 3.84^2 \quad = \quad \underline{49\,227 \, \text{m}^4}$

$I_{CG} \qquad\qquad\qquad = 5\,884\,638 \, \text{m}^4$

The calculation for the second moment about the centre line is shown in *Table 4.2*. The ordinates are cubed and then multiplied by the appropriate Simpson's multiplier. The sum of the products multiplied by 2/9 of the common interval gives the second moment of area.

Table 4.2   CALCULATION OF SECOND MOMENT ABOUT THE CENTRE LINE

Section	Half-ordinate, m	Half-ordinates cubed, m³	Simpson's multi-plier	Second moment product
0	0.0	—	1	—
1	4.1	68.9	4	275.7
2	9.6	884.7	2	1 769.5
3	11.9	1685.2	4	6 740.6
4	12.0	1728.0	2	3 456.0
5	12.0	1728.0	4	6 912.0
6	12.0	1728.0	2	3 456.0
7	12.0	1728.0	4	6 912.0
8	11.1	1367.6	2	2 735.3
9	7.2	373.2	4	1 493.0
10	0.0	—	1	—
				33 750.1

$$I_{\bar{q}} = \tfrac{2}{9} \times 18 \times 33\,750.1 = 135\,000\,\text{m}^4$$

Calculation of volumes and centroids

It is frequently necessary to calculate volumes of spaces such as the holds of a ship or the under-water volume of the ship itself. In addition the centroids of these volumes are often required. These calculations can also be done by Simpson's rule. The procedure is similar to that used in calculating areas and their centroids, except that in calculating a volume the area underneath a curve of areas is being determined. *Figure 4.7* shows a curve which represents the cross-sectional areas of a space. The volume of the space is then the area underneath this curve. If the length is divided into n equal divisions, then:

$$\text{Volume} = \frac{\text{C.I.}}{3}(A_0 + 4A_1 + 2A_2 \ldots + 4A_{n-1} + A_n) \quad (4.14)$$

The centroid is obtained by taking moments of these areas about some axis and summing by Simpson's rule, as was done for finding the centroid of an area.

Notes on Simpson's rule

It was stated that the basis of Simpson's rule was that it calculates the area under a curve of parabolic form which is assumed to replace the actual curve for which the area is required. It is exact for a curve

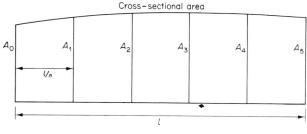

Figure 4.7

which contains powers of x up to the second. It can be shown that the rule would calculate accurately the area of a curve containing powers up to the third. It would therefore calculate the moment $\int xy\,dx$ correctly for a curve of the form $y = a_0 + a_1x + a_2x^2$. If, however, the second moment of area about say $x = 0$ of a curve of this form were required, it would be necessary to evaluate the integral $\int x^2y\,dx = \int(a_0x^2 + a_1x^3 + a_2x^4)\,dx$, i.e. a curve containing a fourth power in x. Error would therefore be introduced even for the parabolic curve. Additional error may arise for an arbitrary curve.

When the second moment about a horizontal axis is considered then the integral $\int y^3\,dx$ has to be evaluated. Thus $\int(a_0 + a_1x + a_2x^2)^3\,dx$ has to be determined and it will be seen that this will involve powers of x up to the sixth, so that greater error is likely to occur. However, the evidence is that Simpson's rule calculates with sufficient accuracy areas, moments and second moments of the curves met with in ships provided that a sufficiently close spacing of ordinates is adopted.

When dealing with the cross sections of a ship care needs to be taken in calculating the area from the base to the lowest waterline. *Figure 4.8* shows a section of a ship near amidships where it will be seen that the curve runs into a straight line (the rise of floor line) at the point A. The shaded area OABC is a trapezoid and this area can be calculated accurately when the position of A and the rise of floor are known. The area of the portion ADB can be obtained by

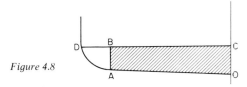

Figure 4.8

using Simpson's rule, either with horizontal ordinates measured from AB to the section, or with vertical ordinates measured from BD to the section.

If the area of the section below CD was calculated using horizontal ordinates measured from OC, error would arise unless a very close spacing of ordinates were adopted near the base because of the rapid change in ordinate with increasing height above base. In calculating the areas of sections, often in the past the area below the 3 ft or 4 ft waterline was determined by means of instruments such as the planimeter or the integrator, but this is not now normally done when the calculation is programmed for the computer. Some such method as that described here is adopted to obtain accurate values for the area of this part of the section.

Other rules for calculation

Attention has been concentrated on the derivation and use of Simpson's rule for ship calculations. There are, however, many other rules which can be used for this purpose. They are all based on the assumption that the curve to be integrated can be replaced by a curve of known mathematical form. One such rule assumes that the actual curve can be replaced by a curve of the form $y = a_0 + a_1 x + a_2 x^2 + a_3 x^3$. This leads to the following formula for the area:

$$A = \tfrac{3}{8} \times \text{C.I.} (y_0 + 3y_1 + 3y_2 + y_3) \tag{4.15}$$

and is thus applicable to a curve represented by four ordinates spaced distances C.I. apart. It can be extended to long curves as in the case of Simpson's rule and would in this case yield the multipliers 1 3 3 2 3 3 ... 3 3 1. It can in this form be applied to a curve represented by $4 + 3n$ evenly spaced ordinates where n is any whole number.

Another useful rule is that known as the 5, 8 -1 rule. It assumes that the curve is of the form $y = a_0 + a_1 x + a_2 x^2$ and enables the area under the curve between y_1 and y_2 to be determined in terms of the three ordinates y_1, y_2, and y_3 (see *Figure 4.9*). This area is given by:

$$A = \frac{l}{12}(5y_1 + 8y_2 - y_3) \tag{4.16}$$

It is very useful in some calculations to be able to find the area under a curve by simply summing the ordinates of the curve and multiplying by the length and some appropriate constant. A rule

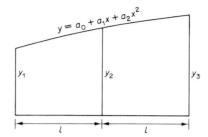

Figure 4.9

which will accomplish this is Tchebycheff's rule. If n ordinates are taken at various positions in the length l of the curve the area is given by

$$A = \frac{l}{n}(y_1 + y_2 + y_3 \ldots + y_n) \qquad (4.17)$$

when using Tchebycheff's rule. In this rule the ordinates are not equally spaced and their position in the length depends upon the number of ordinates. *Table 4.3* shows where the various ordinates have to be positioned.

Table 4.3 POSITIONING OF ORDINATES FOR TCHEBYCHEFF'S RULE

Number of ordinates	Position of ordinates from centre of length expressed as fraction of half length				
2	0.5773				
3	0	0.7071			
4	0.1876	0.7947			
5	0	0.3745	0.8325		
6	0.2666	0.4225	0.8662		
7	0	0.3239	0.5297	0.8839	
8	0.1026	0.4062	0.5938	0.8974	
9	0	0.1679	0.5288	0.6010	0.9116
10	0.0838	0.3127	0.5000	0.6873	0.9162

The rules mentioned are a few of those which are available for calculating areas, volumes, centroids, etc. One feature which is common to them all is that the area underneath a curve is given by

$$A = \text{Length} \times \frac{\text{Sum of products of ordinates and multipliers}}{\text{Sum of multipliers}}$$

This means that the mean ordinate of the curve is equal to the sum of the products divided by the sum of the multipliers.

5

Buoyancy, stability and trim

Conditions for equilibrium of body floating in still water

When a body is floating in equilibrium in still water there is a force acting downwards which is due to gravity, so that if the body is of mass m, this force called the weight of the body is equal to mg. Since the body is in equilibrium it is correct to conclude that there must be a force of the same magnitude acting upwards. This force is generated by the hydrostatic pressures which act normally to the body as shown in *Figure 5.1*. The forces normal to the surface have

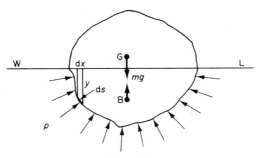

Figure 5.1

vertical and horizontal components. The sum of the vertical components must add up to give a force equal to the weight mg. This force is called the 'buoyancy'. The horizontal components of the hydrostatic pressures cancel out, giving a zero horizontal force. If p is the normal pressure, p_v and p_h the vertical and horizontal components, and da the element of area on which the pressure acts, it follows that

$$\sum p_v \, da = mg \tag{5.1}$$

$$\sum p_h \, da = 0 \tag{5.2}$$

The gravitational force mg can be imagined to be concentrated at a point G which is the centre of mass or is more commonly known as the centre of gravity. Similarly the buoyant force can be imagined to be concentrated at a point B called the centre of buoyancy, which can be considered to be the centroid of the underwater volume. For equilibrium then G and B must lie in the same vertical line.

Consider now the hydrostatic force acting on a small element of the surface length ds, a distance y below the free surface:

Pressure = Head × Density × Gravitational acceleration = $\rho g y$
Normal force per unit length = $\rho g y \, ds$

If θ is the angle of inclination of the surface to the horizontal, then the projection of ds on the horizontal plane is dx, and $dx/ds = \cos \theta$. Hence:

$$\text{Normal force} = \rho g y \, \frac{dx}{\cos \theta}$$

Vertical component of force = Normal force × $\cos \theta$

$$= \rho g y \, \frac{dx}{\cos \theta} \times \cos \theta = \rho g y \, dx$$

Thus the vertical force is equal to the density of the fluid multiplied by the immersed area of the element. It follows that

$$\text{Total vertical force} = \sum \rho g y \, dx = \rho g \sum y \, dx = \rho g A \text{ per unit length}$$

where A is the immersed cross-sectional area of the section. By integrating this along the length of the body the total buoyant force is obtained and is therefore $\rho g \sum A \, dz$, if the longitudinal co-ordinate is z. Hence:

$$\text{Total buoyant force} = \rho g V \tag{5.3}$$

where V is the immersed volume of the body. Since this force is equal to the gravitational force on the mass m of the body, then

$$\rho g V = mg$$

or

$$m = \rho V \tag{5.4}$$

In other words the mass of the body is equal to the mass of the fluid displaced by the body. This is an important result as far as ship calculations are concerned. It is possible to calculate the mass of the ship by calculating the mass of the structure fittings, machinery, etc., and adding them all. It is a long, tedious and often inaccurate procedure. However, by making use of equation 5.4 it is possible to calculate the mass of a ship by determining its underwater volume, observing the density of the fluid in which it is floating and multiplying the two together. Both the underwater volume and the density of the fluid can be determined very accurately, so that a very close estimate of the mass of the ship can be obtained in this way. As will be seen later a calculation like this is carried out for a ship when it is near completion and thus enables the light mass of the ship to be calculated.

Calculation of underwater volume

The underwater volume can be calculated in many ways and the 'rules' which were discussed in Chapter 4 can be used for the purpose. For example, if the immersed areas of a number of sections throughout the length of a ship are calculated by Simpson's rule it is possible to plot a sectional area curve as shown in *Figure 5.2*. It then follows that if A is the cross-sectional area at a particular section, the underwater volume would be $V = \int_0^L A \, dz$.

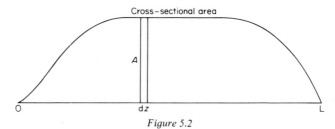

Figure 5.2

If immersed cross-sectional areas are calculated to a number of waterlines parallel to the base of the ship, then the underwater volumes to each of these waterlines can be determined and a curve of underwater volume against draught can be plotted as in *Figure 5.3*. Suppose now that it is observed that a ship is floating at a draught d: by drawing a horizontal line at d above base it is possible to pick off the volume V. The density of the water is measured by some form of hydrometer, measuring the densities of samples taken at different positions round the ship. Hence it is possible to determine the mass of the ship.

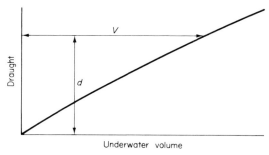

Figure 5.3

Another method of finding the underwater volume is to make use of what are called 'Bonjean curves'. These are curves of immersed cross-sectional areas plotted against draught and are often drawn on the profile of the ship as shown in *Figure 5.4*. Suppose the ship is floating at some waterline such as WL in the figure. By drawing this waterline on the profile it is possible to read off the immersed areas by drawing horizontal lines (shown dotted) from the intercept of the waterline with the section to the Bonjean curve for that section. Having the areas for all the sections, one can calculate the underwater volume and also the longitudinal position of its centroid (called the 'longitudinal centre of buoyancy'). The use of Bonjean curves enables the volume to be obtained to waterlines which are not parallel to the base.

When the calculation of underwater volume or the 'displacement' of a ship was done manually it was customary to use what was called a 'displacement sheet'. A typical layout for a displacement sheet is shown in *Figure 5.5*. The displacement from the base up to in this case the 5 m waterline was determined by the use of Simpson's rule applied to half-ordinates measured at waterlines spaced 1 m apart and at sections taken at every one tenth of the length. The calculations were done in two ways. First of all, areas of sections were calculated and then integrated in the fore and aft direction to give volume. Then areas of waterplanes were calculated and

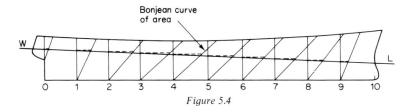

Figure 5.4

Displacement sheet

Section	Simpson's multipliers	Levers from amidships	0.0 Half-ordinate	0.0 Area product	0.5 Half-ordinate	0.5 Area product	1.0 Half-ordinate	1.0 Area product	2.0 Half-ordinate	2.0 Area product	3.0 Half-ordinate	3.0 Area product	4.0 Half-ordinate	4.0 Area product	5.0 Half-ordinate	5.0 Area product	Sum of products of sectional areas	Simpson's multipliers	Volume products	Levers from amidships	Moments about amidships
0	1	5																1		5	
1	4	4																4		4	
2	2	3																2		3	
3	4	2																4		2	
4	2	1																2		1	
5	4	0																4		0	$M_A \otimes$
6	2	1																2		1	
7	4	2																4		2	
8	2	3																2		3	
9	4	4																4		4	
10	1	5																1		5	
Total products of water-plane areas																			A		$M_F \otimes$
Simpson's multipliers			$\frac{1}{2}$		2		$1\frac{1}{2}$		4		2		4		1						
Volume products																					
Levers above base			0		$\frac{1}{2}$		1		2		3		4		5		B				
Moments about base																	M_Y				

Note: "Waterlines above base (metres)" spans the columns 0.0, 0.5, 1.0, 2.0, 3.0, 4.0, 5.0, each divided into Half-ordinate and Area product.

Figure 5.5 Displacement sheet

integrated vertically to give volume. The two volume products denoted by A and B in *Figure 5.5* should of course be exactly the same if the arithmetic is correct, so a check was thus obtained. The need for a check on the calculation arises because in designing the form of the ship it is most important that sufficient displacement is obtained at the designed draught to carry all the weights which the ship is intended to carry. Any error in this displacement could have serious consequences for the final design of the ship. It should be noted that what this system could not check was the correctness of the ordinates lifted from the lines plan and used in the calculation.

Included in the displacement sheet was the calculation of the vertical and fore and aft positions of the centroid of the underwater volume, i.e. the position of the centre of buoyancy.

Cross checking the displacement by calculating the volume in two ways has become unnecessary since the use of the digital computer became common for ship calculations. So long as the input data (in this case the values of the ordinates lifted from the lines) are correct, the computer program, once tested, should give the correct answer. Such programs make use of Simpson's or some other rule, and the calculations are carried out in the way described in Chapter 4.

Stability

The concept of the stability of a floating body such as a ship can be explained by considering what happens when it is inclined from the vertical by some external force and then the force is removed. *Figure 5.6* shows a ship floating originally at a waterline W_0L_0 and then inclined through a small angle θ so that it floats at a waterline W_1L_1. Both waterlines are shown on the same diagram.

The inclination does not affect the position of G, the centre of

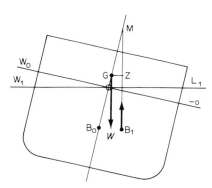

Figure 5.6

gravity, unless there is some weight which is free to move. The inclination does, however, affect the shape of the underwater form and this in turn causes the position of the centre of buoyancy to move from B_0 to B_1. This arises because a volume W_0OW_1 has come out of the water and an equal volume L_0OL_1 has gone into the water to maintain equilibrium.

If g_e and g_i are the centroids of the two wedges each of volume v and $g_e g_i = h$ then:

$$B_0B_1 = \frac{v \times g_e g_i}{V} = \frac{v \times h}{V}$$

where V is the total volume.

It will now be seen that the buoyancy acts upwards through B_1, and if the vertical through B_1 is continued upwards it intersects the original vertical at a point which is denoted by M in the diagram. This point is called the 'metacentre' and for small angles of inclination is fixed in position. This is true for a ship for the first few degrees of inclination from the vertical. It will be seen that the weight $W = mg$ acting downwards and the buoyant force of equal magnitude acting upwards are not in the same straight line but form a couple of magnitude $W \times GZ$, where GZ is the perpendicular on B_1M drawn from G. The couple will restore the body to its original position and in this condition the body is said to be in stable equilibrium. It will be seen that $GZ = GM \sin \theta$. GZ is called the 'righting arm' or 'lever' and GM is called the 'metacentre height'. Since G is fixed in position and M can be considered to be fixed for small inclinations, GM is constant for a body or ship floating at a particular waterline. From *Figure 5.6* it will be noted that M is above G for stable equilibrium and in this case GM is regarded as being positive.

Figure 5.7 illustrates a second condition which may exist when a

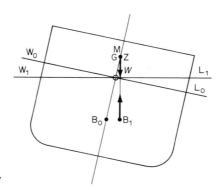

Figure 5.7

ship is inclined from the vertical. In this case the new position of the centre of buoyancy B_1 is directly under G and the three points M, G, and Z are coincident. It follows that since the two forces of weight and buoyancy are acting in the same straight line there is no moment acting on the ship to bring it back to its original position, and consequently when the external disturbing force is removed it would remain in the inclined position. The ship is then in neutral equilibrium and would not return to the vertical, nor would it move further from it when the external force is removed. When this situation exists both *GM* and *GZ* are zero.

A third condition exists, shown in *Figure 5.8*. Here, after inclination

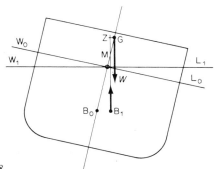

Figure 5.8

the new position of the centre of buoyancy B_1 lies nearer to the centre line than G, and consequently there is a moment $W \times GZ$ which will take the ship further away from the original vertical. This is the condition of unstable equilibrium and the inference is that the ship will capsize. In actual fact this will not necessarily happen, but the ship will move to a position of stable equilibrium, which may mean that it rolls to a considerable angle from the vertical. When a ship is in unstable equilibrium M is below G and both *GM* and *GZ* are considered to be negative.

The above considerations refer to what is called the 'initial stability' of the ship, i.e. the stability in the upright condition, and the criterion of initial stability is the metacentric height *GM*. The three conditions may be summarised as follows:

M above G	*GM* and *GZ* positive	stable
M at G	*GM* and *GZ* zero	neutral
M below G	*GM* and *GZ* negative	unstable

Applying these results to a ship it is clear that the ship should be stable in all normal conditions of loading and it is one of the conditions that the designer must ensure. Also, ship personnel must be sufficiently informed to make sure that the ship is never loaded in such a way that it becomes unstable, i.e. the metacentre height becomes negative.

Having established the basic principles for the stability of a surface vessel the next stage is to study the factors which affect the two quantities involved, i.e. the position of the centre of gravity G, and the position of the metacentre M.

Position of centre of gravity

The position of the centre of gravity of a ship depends only on the distribution of masses in the ship. If M is the total mass and KG the distance of its centre of gravity above base, then if there are a number of masses m with centres of gravity y from the base:

$$M \times KG = \sum my$$

or
$$KG = \frac{\sum my}{M} \tag{5.5}$$

Thus the moments about the base of all the masses in the ship in any particular condition of loading are summed and divided by the total mass to determine the height of the centre of gravity above base. It is usual at the same time to consider the moments of these masses about amidships in order to find the fore and aft position of the centre of gravity. The importance of the position of the centre of gravity in this direction will be discussed later.

Although the position of the centre of gravity can be calculated, it will be shown later that it can be determined experimentally, which avoids much tedious calculation which may be liable to considerable error. For the present, however, it will be assumed that by some means the position of the centre of gravity can be obtained. It remains then to investigate how the position of the metacentre can be obtained.

Position of transverse metacentre

The position of the metacentre can be studied by considering small inclination of a ship from the vertical. *Figure 5.9* shows a ship floating originally at a waterline $W_0 L_0$ and then inclined to an angle θ so that it floats at a waterline $W_1 L_1$. For small angles of inclination of the order of $2°$ or $3°$ these two waterlines will intersect at a point

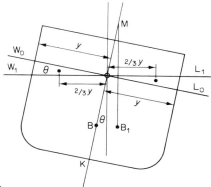

Figure 5.9

O on the centre line of the ship. The position of intersection must satisfy the condition that the volume coming out of the water W_0OW_1 must equal the volume entering the water L_0OL_1 to maintain the same volume of displacement, V.

For the small angles of inclination being considered, the emerged and immersed wedges at any section (W_0OW_1 and L_0OL_1), are approximately triangular, and if y is the half-ordinate of the original waterline at any section the emerged or immersed area is $\frac{1}{2}y \times y \tan\theta = \frac{1}{2}y^2\theta$ for small values of θ. Hence

$$\text{Total volume} = \int_0^L \tfrac{1}{2}y^2\theta\,dx$$

where the integral is taken along the length of the ship.

This volume is moved from one side to the other and if the sections are considered to be triangular it will be seen that the centre of gravity moves from a position $\frac{2}{3}y$ from the centre line on one side to a position $\frac{2}{3}y$ from the centre line on the other side. It follows then that if moments of volume about the centre line are considered, the excess moment is given by:

$$\int_0^L \tfrac{1}{2}y^2\theta\,dx \times \tfrac{4}{3}y = \theta\int_0^L \tfrac{2}{3}y^3\,dx$$

since θ is constant along the length of the ship. Now it was shown when dealing with rules for calculations that $\int_0^L \tfrac{2}{3}y^3\,dx$ was the second moment of area or the moment of inertia of a waterplane about its centre line and may be denoted by I. Hence

$$\text{Excess moment} = I\theta$$

The transverse movement of the centre of buoyancy BB_1 is then given by

$$BB_1 = \frac{I\theta}{V}$$

where V is the volume of displacement.

Referring to *Figure 5.9* for the small angles of inclination being considered, $BB_1/BM = \theta$ or $BM = BB_1/\theta$,

hence: $$BM = \frac{I\theta/V}{\theta} = \frac{I}{V} \qquad (5.6)$$

Thus the height of the metacentre above the centre of buoyancy BM is found by dividing the second moment of area or the moment of inertia of the waterplane about its centre line by the volume of displacement. The height of the centre of buoyancy above base KB is the height of the centroid of the underwater volume above base, hence the height of the metacentre above base can be calculated, since

$$KM = KB + BM \qquad (5.7)$$

Knowing KM and KG, the difference of these two quantities gives the metacentric height *GM*.

Position of transverse metacentre for simple geometrical forms

Vessel of constant rectangular cross section
Consider the rectangular form shown in *Figure 5.10* of breadth B and length L floating at some draught d. If the section is constant throughout the length

Volume of displacement $= LBd$

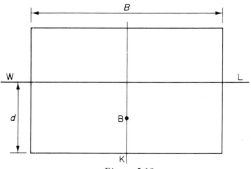

Figure 5.10

Second moment of area of waterplane about centre line $= \frac{1}{12}LB^3$

$$\therefore \quad BM = \frac{\frac{1}{12}LB^2}{LBd} = \frac{B^2}{12d}$$

Height of centre of buoyancy above base $KB = d/2$

$$\therefore \quad \text{Height of metacentre above base } KM = \frac{d}{2} + \frac{B^2}{12d} \qquad (5.8)$$

The two factors here which affect the height of the metacentre above base are the draught and the beam of the ship. At low draughts in relation to the beam the second term $B^2/12d$ is predominant and at zero draught KM would be infinite. At deep draughts the influence of the second term becomes small and the first term is the important one. It should be noted, further, that the length of the ship does not enter into the formula for the height of the transverse metacentre. The only ship dimension which is important for this simple form is the breadth and the other factor is the draught at which the vessel floats.

It is instructive to evaluate equation 5.8 for a ship of some given dimensions. For example, calculate the height of the metacentre for a vessel of constant rectangular cross section 15 m broad for draughts from 1 to 6 m. The formula for KM would be in this case

$$KM = 0.5d + \frac{15^2}{12d} = 0.5d + \frac{18.75}{d}$$

The values of KM for various draughts are shown in *Table 5.1* and are plotted in *Figure 5.11*.

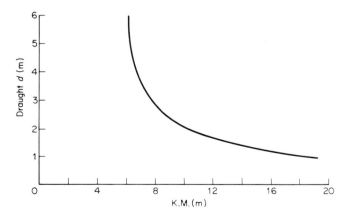

Figure 5.11

Table 5.1   VALUES OF KM FOR VARIOUS DRAUGHTS

d, m	0.5 d, m	18.75 d, m	KM, m
1	0.5	18.75	19.25
2	1.0	9.37	10.37
3	1.5	6.25	7.75
4	2.0	4.69	6.69
5	2.5	3.75	6.25
6	3.0	3.12	6.12

It will be seen that the height of the metacentre above base is large at low draughts and then falls rapidly with increasing draught. If the calculation were carried far enough KM would reach a minimum value and then start to increase again. It is possible to find the draught at which KM is minimum by differentiating equation 5.8 with respect to d and equating to zero. $dKM/dd = \frac{1}{2} - B^2/12d^2 = 0$ for minimum

$$\therefore \quad d^2 = \frac{B^2}{6} \quad \text{or} \quad d = \frac{B}{\sqrt{6}}$$

For the example quoted $d = 15/\sqrt{6} = 6.12$ m.

Vessel of constant triangular section
Figure 5.12 shows a vessel of triangular cross section, the breadth at the top being B and the depth D. The breadth of the waterline at any draught d is given by

$$b = (d/D) \times B$$

$$I = \tfrac{1}{12}L \times \{(d/D) \times B\}^3$$

$$V = \tfrac{1}{2}L \times (d/D) \times B \times d$$

$$\therefore \quad BM = \frac{\tfrac{1}{12}L \times \{(d/D) \times B\}^3}{\tfrac{1}{2}L \times \{(d/D) \times B\}d} = \frac{1}{6} \times \frac{B^2}{D^2} \times d$$

$$KB = \tfrac{2}{3}d$$

$$\therefore \quad KM = \tfrac{2}{3}d + \frac{1}{6}\frac{B^2}{D^2} \times d \tag{5.9}$$

In this case the curve of KM on draught is a straight line starting from zero at zero draught.

Curve of KM for ship forms

The cross sections of a ship can be considered to have the character partly of the rectangular section and partly of the triangular section.

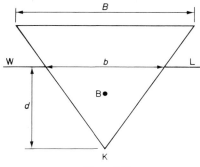

Figure 5.12

As far as the metacentre curve is concerned, the rectangular type of section is dominant and the KM for a ship shows very much the same character as that for a form of constant rectangular cross section. Usually the metacentre is a minimum somewhere about the load draught for a ship. The initial stability for a ship in the loaded condition may therefore be critical and may be the deciding factor in determining the ratio of breadth in relation to draught.

Longitudinal inclination

So far stability has been considered from the point of view of transverse inclination of a ship, and it has been seen that a measure of the stability is the distance between the vertical position of the centre of gravity and the metacentre. If longitudinal inclinations are considered the same principles will apply as for transverse stability. There is, however, one difference: the distance between the centre of buoyancy and the metacentre will be dependent on the second moment of area or moment of inertia of the waterplane about a transverse axis passing through its centroid. This quantity is many times the value for the second moment of area about the centre line and since this is divided by the same volume as for transverse stability then BM_L is a very large quantity, often of the order of the length of the ship. For this reason the longitudinal metacentric height is a large quantity and it is therefore impossible for a surface vessel to be unstable when inclinations in the fore and after direction are considered. The formula for the height of the metacentre above base for longitudinal inclination is

$$KM_L = KB + BM_L = KB + \frac{I_L}{V} \qquad (5.10)$$

where I_L is the second moment of the waterplane area about a transverse axis passing through the centroid.

Consider a ship floating at a waterline W_0L_0 as in *Figure 5.13*. For equilibrium the centre of gravity G and the centre of buoyancy B must be in the same vertical line. Suppose now that a moment is applied to the ship by moving a weight w (which is already on board) through a distance h forward. The applied moment is wh and

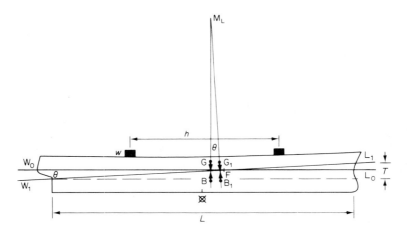

Figure 5.13

if W is the weight of the ship the centre of gravity will move forward to G_1 and the shift of the centre of gravity is given by

$$GG_1 = \frac{wh}{W}$$

The ship is now no longer in equilibrium, since there is now a moment WGG_1 tending to rotate it. It therefore trims forward until it floats at a waterline W_1L_1 which cuts off the same displacement and brings the centre of buoyancy to B_1 directly underneath G_1.

Because of the non-symmetry of the ship form about amidships the waterlines do not intersect at amidships but at a point F called the 'centre of flotation'. The position of F is such that volume $W_0FW_1 = $ volume L_0FL_1 and for small trims can be considered to be the centroid of the waterplane area.

For longitudinal inclinations of the ship the angle of inclination is best measured by the difference of the end draughts called the trim,

T, divided by the length of the ship. From *Figure 5.13* it follows that

$$\tan \theta = \frac{T}{L} = \frac{GG_1}{GM_L} = \frac{wh}{WGM_L}$$

$$wh = \frac{TWGM_L}{L}$$

The quantity wh is the moment which will cause a trim of T. Hence

$$\text{Moment causing unit trim} = \frac{WGM_1}{L} \qquad (5.11)$$

If imperial units are used then W would be in tons, GM_L and L in feet, so that equation 5.11 would give a moment to change trim one foot. This is a very large quantity and it was customary to divide it by 12 to give a 'moment to change trim 1 in'. Hence

$$\text{M.C.T. 1 in} = \frac{WGM_L}{12L} \qquad (5.12)$$

and the units of M.C.T. 1 in would be ton ft.

If SI units are employed then W would be in meganewtons, MN, GM_L and L in metres, so that equation 5.11 would in this case give moment to cause or change trim of one metre, and this moment would be in MN m. Once again this would be a large quantity and although the centimetre is not allowed in the SI system, it would be much more convenient to have a 'moment to change trim one centimetre', which would be $(WGM_L)/100L$.

The quantity moment to cause or change trim is exceedingly useful in ship calculations for the determination of the draughts at which a ship will float in a particular condition of loading.

Suppose the weight of a ship has been worked out as W and the position of the centre of gravity as x forward of amidships. If the ship floated on an even keel it would float at a draught d with the centre of buoyancy y forward of amidships. It will be seen that in this case there is a moment $W(y-x)$ causing the ship to trim by the stern so that if M.C.T. is the moment causing unit trim then

$$\text{Trim } T = \frac{W(y-x)}{\text{M.C.T.}}$$

The ship trims about the centre of flotation F and if this is a distance λ forward of amidships

$$\text{Draught aft } d_a = d + \frac{(L/2)+\lambda}{L} \times T$$

$$\text{Draught for'd } d_f = d - \frac{(L/2)-\lambda}{L} \times T$$

   The draughts have here been worked out for the case where the trim is by the stern and the centre of flotation is forward of amidships. The signs will be different for other cases. In all there are four cases: the trim may be either by the stern or the bow, and the centre of flotation may be either forward or aft of amidships. Formulae such as those above can easily be worked out for all these cases.

   The determination of the end draughts of the ship is of great practical importance, particularly where depth of water is limited. It will be clear that if there is a restriction on draught the ship will be able to carry its maximum load when floating on an even keel, so that it is advisable to distribute the load to bring this about.

Hydrostatic curves

In the above sections it has been shown how the displacement of a ship and the position of the centre of buoyancy can be calculated and also how the position of the metacentres and the centre of flotation can be determined. It is customary to calculate all these quantities for about six or seven waterlines parallel to the base and spaced one metre (3 or 4 ft) apart. The results so obtained are plotted in a diagram with draught measured vertically. The curves drawn in this way are called 'hydrostatic curves'. A typical set of hydrostatic curves is shown in *Figure 5.14*.

   Two curves of displacement are shown. One is called the 'moulded

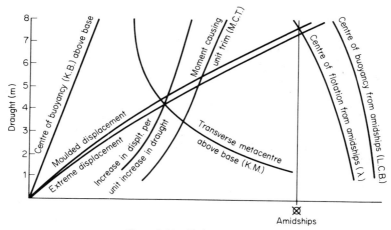

Figure 5.14   Hydrostatic curves

displacement' and it is the displacement obtained to the moulded lines of the ship between perpendiculars. To obtain the extreme displacement it is necessary to add on to this the shell displacement, the displacement of the cruiser stern and bulb forward, if fitted, and in the case of multiple screw ships the displacement of the bossing enclosing the shafting. Sometimes the displacement of the rudder and propeller and shafting are included in the extreme displacement.

It is also important to correct the position of the centre of buoyancy for these items, and this would apply particularly to the longitudinal position of the centre of buoyancy since the volume of such items as bossing can have a major effect.

With regard to the displacement of the shell, this is determined by first of all calculating the wetted surface area. This area when multiplied by the mean thickness of the shell plating will give the volume displaced by the shell. The wetted surface area is not easy to calculate since the outside surface of a ship has double curvature. It can be approximated to by taking girths round the various sections and then applying Simpson's rule to find the area. The procedure ignores the curvature of the hull surface in the fore and aft direction, (the 'obliquity effect' as it is sometimes called) but this is often not of great magnitude.

Various approximate formulae have been developed for the calculation of wetted surface area and they will be referred to later when dealing with ship resistance.

Shell displacement represents only a small percentage of the total displacement of a ship but is of sufficient magnitude to justify its inclusion in the calculation of the displacement. In a large modern vessel it could amount to many hundreds of tons.

Included in *Figure 5.14* is a curve which gives the increase in displacement for unit increase in draught. If A is the area of the waterplane at which the ship is floating, then for unit increase in draught the volume added is $A \times 1$ assuming the ship to be wall sided in the neighbourhood of the waterline. It follows that increase in displacement $= \rho g A$. When imperial units are used the weight per unit volume of sea water is given as $\frac{1}{35}$ ton/ft^3, so that increase in displacement $= A/35$, with A in square feet, which may be called the 'tons per foot immersion'. As this is quite a large quantity it was usually divided by 12 to give 'tons per inch immersion'. Therefore:

$$\text{T.P.I.} = A/420 \quad \text{for sea water}$$

When using SI units it is probably more convenient to leave this quantity in the form given above, i.e. $\rho g A$ where ρ is the density in kg/m^3, g is the acceleration due to gravity and A is the waterplane area in m^2. For $\rho = 1025\,\text{kg/m}^3$ and $g = 9.81\,\text{m/s}^2$:

Increase in displacement per metre increase in draught
$$= 1025 \times 9.81 \times 1 \times A = 10\,055A\ \text{N}$$
$$= 0.010\,055A\ \text{MN}$$

For 1 cm immersion this would become $0.000\,100\,55A$ MN.

The increase in displacement per unit increase in draught is useful in approximate calculations when weights are added to the ship. The weight added divided by this quantity gives the parallel sinkage of the ship. The calculation is only reasonably correct for the addition of relatively small weights, since the increase in displacement per unit increase of draught varies with the draught.

Hydrostatic curves are most useful in working out the end draughts and the stability of a ship as represented by metacentric height in various conditions of loading. This is done for all the normal working conditions of the ship and the information so obtained is supplied to the master.

Hydrostatic calculations are nowadays normally done by computer. Computer programs exist which will carry out all the calculations which have been discussed in this chapter. The input data required consist of ordinates at various waterlines defining the form of the ship. When this is put into the computer the program calculates all the quantities necessary for plotting hydrostatic curves. It can be done in a very short space of time, whereas in the days of hand calculations the production of a set of hydrostatic curves required about two man weeks.

Problems in trim and stability

In the following sections several examples will be discussed of calculations which can be carried out making use of hydrostatic curves.

Determination of displacement and longitudinal position of centre of gravity from observed draughts

It is often necessary to calculate the displacement of a ship and the position of the centre of gravity when the end draughts are available. Suppose these draughts are d_a and d_f as shown in *Figure 5.15*. Then mean draught $d = (d_a + d_f)/2$ and a first approximation to the displacement would be to set off d on the hydrostatic curves and lift off the corresponding value of the displacement. Remembering that the displacement curve is drawn for a series of even keel waterlines, this

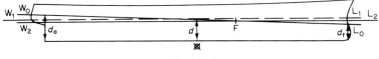

Figure 5.15

procedure assumes that the original waterline W_0L_0 corresponding
to draughts d_a and d_f will cut off the same displacement as the level
keel waterline W_1L_1, distance d from the keel and intersecting W_0L_0
at amidships. It has, however, been shown that because of the non-
symmetry of the form about amidships the ship trims about the
centre of flotation so that the level keel waterline cutting off the
same displacement as W_0L_0 is actually W_2L_2 in *Figure 5.15*. Consequently the displacement to W_1L_1 from the hydrostatic curves
would overestimate the displacement by a small layer. If T is the
trim (i.e. $d_a - d_f$) and λ is the distance of the centre of flotation from
amidships, then

$$\text{Thickness of layer} = \frac{\lambda}{L} \times T$$

and if i is the increase in displacement per unit increase in draught:

$$\text{Displacement of layer} = \frac{\lambda}{L} \times T \times i$$

Consequently the actual displacement equals $\Delta - (\lambda/L)T \times i$ where
Δ is the displacement from the hydrostatic curves at draught d. The
displacement obtained in this way will require to be corrected for
the density of the water in which the ship is floating. It is usual to
draw the hydrostatic curves for sea water which has a density of
1025 kg/m^3 ($1/35 \text{ ton/ft}^3$) and if ρ_a is the actual density, in metre units
the density correction factor becomes $\rho_a/1025$.

The correction to the displacement for trim has been worked out
here for a particular case in which the correction was a deduction
from the displacement obtained from the hydrostatic curves. There
are four possible cases, i.e. trim either by the stern or by the head,
each associated with position of centre of flotation either forward or
aft of amidships. It can easily be shown that when trim and position
of centre of flotation are both forward or both aft the correction is
an addition to the displacement, and in the other two cases the
correction is a deduction.

The calculation for displacement has assumed that the keel of the
ship remains a straight line. In large modern ships there is likely to
be some bending even in still water, so that a draught taken at
amidships may not have the value of the mean draught, i.e. $(d_a + d_f)/2$,

but may have some value d_m giving a deflection of the hull δ. If the ship sags then the calculation described above would underestimate the volume of displacement and if it hogs the volume would be over-estimated. It would be a reasonable assumption to consider that the deflected profile of the ship was parabolic, so that the deflection at any point distance x from amidships would be given by $\delta(1-(x/\tfrac{1}{2}L)^2)$ where L is the length of the ship. Hence

$$\text{Volume} = \int_{-L/2}^{+L/2} b\delta\left(1-\left(\frac{x}{L/2}\right)^2\right)dx$$

where b is the waterline breadth.

Unless an expression is available for b in terms of x then this integral cannot be evaluated mathematically. If a waterline of constant breadth B is considered, i.e. a rectangular waterplane, then

$$\text{Volume} = \int_{-L/2}^{+L/2} B\delta\left(1-\left(\frac{x}{L/2}\right)^2\right)dx$$
$$= \tfrac{2}{3}A\delta$$

where A is the waterplane area.

For a real ship the volume will differ from this because b is not constant. In the analysis which follows it will be assumed that $b = B(1-(x/\tfrac{1}{2}L)^n)$ where n will depend upon the fullness of the waterline and x is measured positive from either side of amidships. This would give a triangular waterplane for $n = 1$ and would approach the rectangular waterplane shape for very large values of n. The volume to be added or subtracted from the basic displacement calculation is then given by

$$\int_0^{+L/2} B\left(1-\left(\frac{x}{L/2}\right)^n\right)\delta\left(1-\left(\frac{x}{L/2}\right)^2\right)dx$$

for the half-length of the ship.

$$= \int_0^{L/2} B\delta\left(1-\left(\frac{x}{L/2}\right)^2 - \frac{x^n}{(L/2)^n} + \frac{x^{n+2}}{(L/2)^{n+2}}\right)dx$$
$$= B\delta\left(x - \frac{x^3}{3L^2/4} - \frac{x^{n+1}}{(n+1)(L/2)^n} + \frac{x^{n+3}}{(n+3)(L/2)^{n+2}}\right)_0^{L/2}$$
$$= B\delta L\left(\frac{1}{2} - \frac{L}{6} - \frac{1}{2(n+1)} + \frac{1}{2(n+3)}\right)$$
$$= B\delta L\left(\frac{1}{3} - \frac{1}{(n+1)(n+3)}\right) = B\delta L\left(\frac{(n+1)(n+3)-3}{3(n+1)(n+3)}\right) \quad (5.13)$$

Now area of waterplane from $\otimes$ to $L/2$

$$= \int_0^{L/2} B\left(1 - \left(\frac{x}{L/2}\right)^n\right)dx = \frac{BL}{2}\left(\frac{n}{n+1}\right)$$

Waterplane area coefficient $C_A = \dfrac{\text{Area}}{\text{Length} \times \text{Breadth}}$

$$= \frac{n}{n+1} \quad \text{or} \quad n = \frac{C_A}{1 - C_A}$$

If this value of n is substituted in equation 5.13 then

$$\text{Volume} = B\delta L\frac{4C_A - 3C_A^2}{9 - 6C_A}$$

A similar expression would be obtained for the other half of the waterplane, so that the volume for the whole waterplane would be given by

$$V = B\delta L\frac{8C_A - 6C_A^2}{9 - 6C_A} \tag{5.14}$$

where in this case C_A is the waterplane area coefficient for the whole waterplane. The value of the coefficient has been determined for a range of values for C_A and the results are given in *Table 5.2*.

Table 5.2 VALUE OF COEFFICIENT IN EQUATION 5.14 FOR A RANGE OF VALUES FOR C_A

C_A	$(8C_A - 6C_A^2)/(9 - 6C_A)$
0.5	0.416
0.6	0.488
0.7	0.554
0.8	0.609
0.9	0.650
1.0	0.666

The variation in the value of the coefficient is not very great for a considerable variation in the waterplane area coefficient. An average value over the working range of values of C_A from 0.7 to 0.9 would be about 0.60 so that a good approximation to the volume would be 0.60 $B\delta L$.

The longitudinal position of the centre of gravity of a ship floating in still water can be obtained as follows. Suppose the ship is floating in equilibrium at a waterline W_0L_0 as in *Figure 5.16*, with the centre of gravity a distance x from amidships, which distance is yet to be determined. The centre of buoyancy B_0 must be directly underneath G for equilibrium. If the ship is imagined to be brought to an even

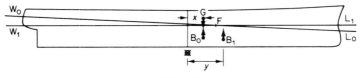

Figure 5.16

keel and floating at a waterline W_1L_1 which cuts off the correct displacement, then the position of the centre of buoyancy will be at B_1, distance y from amidships, which distance will be obtained from the hydrostatic curves for the waterline W_1L_1. It follows then that if T was the trim when the ship was at W_0L_0

$$\Delta(y-x) = T \times \text{M.C.T.}$$

where M.C.T. is the moment to cause unit trim.

$$\therefore \quad y-x = \frac{T \times \text{M.C.T.}}{\Delta}$$

and x is given by

$$x = y - \frac{T \times \text{M.C.T.}}{\Delta} \tag{5.15}$$

Hence the longitudinal position of the centre of gravity has been determined.

Direct method for finding displacement and position of centre of gravity

The methods described in the previous section for finding the displacement and the longitudinal position of the centre of gravity are usually sufficiently accurate when trim is small. When large trims are involved the accuracy may not be so great. To obtain accurate results the Bonjean curves can be used. If the end draughts of the ship have been observed then the draught at any particular section can be calculated, since

$$d_x = d_a - \frac{d_a - d_f}{L} \times x \tag{5.16}$$

where x is the distance from where d_a is measured, and d_a and d_f are measured at a distance L apart. The draughts so obtained can be corrected for hog or sag of the ship if this is seen to be of considerable magnitude. If the deflection is assumed to be parabolic the correction to the draughts obtained above will be $(4\delta/L)(x - x^2/L)$

from 0 to $L/2$ and will be an addition for sagging and a deduction for hogging. This formula will also calculate the correction for the forward sections if x is measured from the forward end towards amidships.

Having calculated the draughts on each section they can be set up on the Bonjean curves and the immersed area at each cross section thus obtained. The immersed volume and position of centre of buoyancy can then be calculated using Simpson's rule, and since for equilibrium the centre of gravity and the centre of buoyancy must be in the same vertical line, this will give the position of the centre of gravity. It is of course necessary to ensure that the position of the centre of buoyancy in this calculation includes cruiser stern bossing, bulb, etc., and they would normally have to be added on as separate items. Knowing the volume of displacement and having observed the density of water in which the ship is floating, one can determine the displacement and hence the mass and weight of the ship.

Effect of shifting a weight transversely

In *Figure 5.17* a ship is shown originally upright and at rest in still water. A weight w is shifted transversely through a distance h. The centre of gravity of the ship is originally at G but due to the shift of the weight it moves to G_1, so that $GG_1 = wh/W$ where W is the weight of the ship. The upright position is now no longer the equilibrium position and the ship heels through an angle θ until the centre of buoyancy moves to a position B_1 vertically underneath G_1, so that equilibrium is again restored. From *Figure 5.17* it will be seen that $GG_1/GM = \tan \theta$

Hence
$$\tan \theta = \frac{wh/W}{GM} = \frac{wh}{W\,GM} \tag{5.17}$$

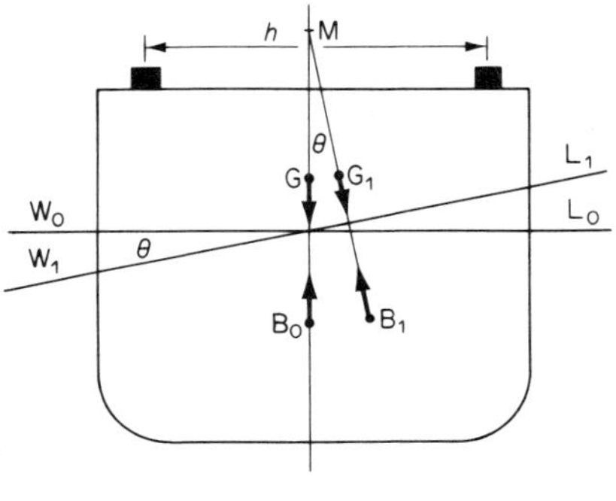

Figure 5.17

It is thus possible to calculate the angle to which the ship will incline when the weight is moved across the deck, always assuming that W and GM are known.

Influence of loose fluids in tanks on stability

A ship in service will often have tanks which are only partially filled with fluid; for example fuel tanks will have a free surface when part of the fuel has been consumed. The same will apply to fresh water tanks and may apply to ballast tanks. In ships carrying liquid cargoes such as oil tankers the cargo tanks may not be filled to the top.

When a ship in such a condition is heeled to the vertical, fluid will move to the inclined side so as to maintain the fluid surface horizontal, as shown in *Figure 5.18(a)*. In what follows the problem

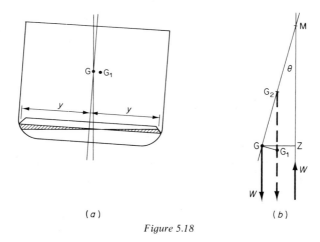

(a)   (b)

Figure 5.18

is treated from a purely statical point of view and the results would not necessarily apply to a rolling ship, in which the loose fluid would itself be moving about. Referring to *Figure 5.18(a)* it will be seen that in the static case a wedge of fluid is transferred to the inclined side.

$$\text{Volume of wedge} = \int_0^l \tfrac{1}{2} y^2 \theta \, dx$$

for small angles where l is the length of the tank. Similarly

$$\text{Moment of volume} = \int_0^l \tfrac{1}{2} y^2 \theta \, dx \times \tfrac{4}{3} y$$

on the assumption that the wedges can be treated as triangles.

Hence   Moment of volume $= \theta \int_0^l \tfrac{2}{3} y^3 \, \mathrm{d}x = \theta I$

where I is the second moment of area or moment of inertia of the free surface.

Moment of mass moved $= \rho_f \theta I$, where ρ_f is the density of the fluid in the tank.

The centre of gravity of the ship will move transversely because of the shift of mass to a position G_1 and

$$GG_1 = \frac{\rho_f g \theta I}{W} = \frac{\rho_f g \theta I}{\rho g V} = \frac{\rho_f}{\rho} \times \frac{\theta I}{V}$$

where ρ is the density of the water in which the ship is floating and V is the volume of displacement.

The effect of the transverse movement of the centre of gravity is to reduce the righting arm GZ by the amount GG_1 (see *Figure 5.18(b)*). Hence righting moment $= W(GZ - GG_1)$, so that there is a reduction in the moment tending to bring the ship back to the vertical, or in other words there is a loss of stability. Now $GZ = GM \sin \theta$ for small angles and it follows that the influence of the shift of G to G_1 is equivalent to raising G to G_2 on the centre line so that $GG_1 = GG_2 \sin \theta$ and the righting moment $= W(GM \sin \theta - GG_2 \sin \theta) = W(GM - GG_2) \sin \theta$.

It will be seen that the effect of the movement of the liquid is equivalent to a rise of GG_2 in the height of the centre of gravity so that there is a loss of GM of this amount. This is the way in which the effect is usually regarded and it is generally referred to as a loss of metacentric height due to free surface. It is clear that if this loss was sufficiently large, the metacentric height could become zero or negative and in the latter case would result in the ship lolling over to some angle from the vertical. It is most important, therefore, that the free surface effect in tanks carrying liquids should be reduced to a minimum. One way of doing this is to subdivide wide tanks into two or more narrower ones. In *Figure 5.19* a double-bottom tank is shown in which a water- or oiltight centre division has been fitted. Suppose the breadth of the tank from one side of the ship to

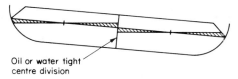

Oil or water tight
centre division

Figure 5.19

the other is B, then when the centre division is fitted the breadth is reduced to $B/2$ and there are now two tanks. Assuming the tank to be of constant section and of length l then

Second moment of area without centre division

$$= \frac{1}{12} lB^3$$

and

Second moment of area with centre division

$$= \frac{1}{12} l \left(\frac{B}{2}\right)^3 \times 2 = \frac{1}{48} lB^3$$

It will be seen that the introduction of a centre division has had the effect of reducing the free surface effect to one quarter of its value without the division.

It is the usual practice to fit a water- or oiltight centre division in the double-bottom tanks of ships, at least over the midship portion, and this means separate suction, air and filling pipes for the tank on each side of the centre line. Towards the ends of the ship the centre division is often not made fluid-tight, since the free surface effect is not nearly so great because of the reduced breadth.

The free surface effect is very important in ships carrying fluid cargoes and the oil tanker is one ship in particular where this effect must be carefully investigated. As in the case of double-bottom tanks the breadth of the tanks is reduced by fitting either an oiltight centre line bulkhead and/or oiltight wing bulkheads.

As an example, consider an oil tanker of 310 000 kN displacement carrying oil of relative density 0.92. The length of the cargo tanks is 110 m and the breadth of the tanks (considered constant throughout the length) is 24.6 m.

Loss of GM if no longitudinal bulkheads were fitted

$$= \frac{\frac{1}{12} \times 110 \times 24.6^3 \times 0.92 \times 1000 \times 9.81}{310\,000 \times 10^3} = 3.97 \,\text{m}$$

Loss of GM with centre line oiltight bulkhead

$$= 2 \times \frac{(24.6/2)^3}{24.6^3} \times 3.97 = 1.98 \,\text{m}$$

Loss of GM with two wing bulkheads evenly spaced

$$= 3 \times (\tfrac{1}{3})^3 \times 3.97 = 0.44 \,\text{m}$$

In the first case the loss of GM is unacceptable, in the second it is almost acceptable and in the third is of a magnitude which could be catered for in the design.

Free surface effects in tanks should be avoided if possible, but where this cannot be done the effect must be taken account of in design. This generally means that the ship should be designed to have sufficient metacentric height to permit the loss due to free surface without the ship becoming unstable.

Influence on stability of weight free to move transversely

In *Figure 5.20* a weight w is suspended from a point A, the distance from the weight to A being h. When the ship heels the weight moves transversely and considering the problem as a static one the weight

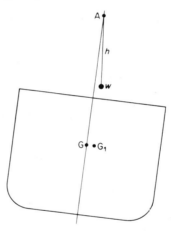

Figure 5.20

will take up a position vertically underneath the point of suspension A. If the angle of inclination is θ then

Transverse movement of weight $= h \sin \theta$

Transverse movement of centre of gravity of ship $= \dfrac{wh \sin \theta}{W} = GG_1$

Loss of righting moment $= W \times GG_1 = wh \sin \theta$

As in the case of free surface in a tank, the effect is the same as though G were raised to G_2 vertically above G_1. Hence:

$$GG_2 = \frac{GG_1}{\sin \theta} = \frac{wh \sin \theta / W}{\sin \theta} = \frac{wh}{W}$$

It follows that this can then be regarded as a loss of metacentric height of GG_2 or in other words the weight w behaves as though it was fixed at the point A. It is important to avoid having weights on board the ship which are capable of moving in this way. Sometimes, however, it cannot be avoided and one instance in particular is when a weight is being lifted by a derrick on board the ship. As soon as the weight is lifted from, say, the deck or the quayside it is as though it was concentrated at the derrick head and there is a consequent raising of the centre of gravity, which if the weight is of sufficient magnitude could cause the ship to become unstable. This is a problem which has to be taken into account in heavy-lift ships.

The inclining experiment

The vertical position of the centre of gravity of a ship has been seen to be important from the point of view of initial stability. It is therefore necessary to know this position with considerable accuracy. The position of the centre of gravity of the lightship could be determined by considering all the weights such as steel, outfit and fittings and machinery, and ascertaining their individual centres of gravity, from which could be calculated the position of the centre of gravity for the complete ship. For a particular condition of loading the weights of all the items which the ship carries would then have to be added on at their appropriate centres of gravity. In this way the position of the centre of gravity of the loaded ship could be determined.

Whilst it is customary to calculate the volumes and centres of gravity of all the hold, fuel, fresh water and ballast spaces, etc., from the drawings (and in fact there is no other satisfactory way of obtaining these quantities) the calculation of the lightship weight and the position of its centre of gravity would be a long and tedious process and the accuracy of the result would be in some doubt. For this reason the lightship weight and the position of its centre of gravity are determined experimentally, in what is known as an 'inclining experiment'.

The experiment consists very simply of causing the ship to heel to a small angle to the vertical by moving known weights known distances across the deck and observing the angle of inclination. At the same time the draughts at which the ship is floating are observed, from which the weight can be determined. It is thus possible to calculate the metacentric height of the ship in the inclining condition, and since the height of the metacentre above base is known from the hydrostatic curves for the mean draught at which the ship is floating, it is possible to determine the position of the centre of gravity above base.

Ideally the experiment should be carried out when the ship is absolutely finished but obviously it is not always possible to achieve this condition exactly. There will usually be a number of items of relatively small amount both to come off the ship (e.g. staging, tools, men, etc.) and to go on to complete the ship. The weights and centres of gravity of these items have to be ascertained or estimated and the condition of the ship as inclined corrected.

The theory underlying the inclining experiment is quite simple and the procedure adopted is illustrated in *Figure 5.21*. Two sets of weights w, each of mass m, are situated on each side of the ship approximately at amidships, the port and starboard pairs being a

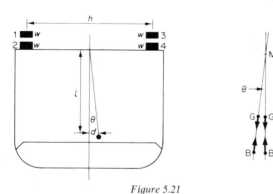

Figure 5.21

distance h apart. One set is moved transversely across the deck, e.g. No. 1 is moved through a distance h and put alongside Nos. 3 and 4. As described, G moves to G_1 and the ship will incline to an angle θ until the centre of buoyancy moves to B_1 directly underneath G_1. It follows that

$$GG_1 = \frac{mh}{M} \quad \text{or} \quad \frac{wh}{W}$$

where M and W are respectively the mass and the weight of the ship.

Also
$$\frac{GG_1}{GM} = \tan \theta$$

$$\therefore \quad GM = \frac{GG_1}{\tan \theta} = \frac{wh}{M} \cot \theta \qquad (5.18)$$

The value of θ can be obtained in several ways. One of the commonest methods is to make use of a pendulum suspended from the deck down into the hold. The deflection d is measured for each shift of weight and it will be seen that if l is the length of the pendulum, $\tan \theta = d/l$. An alternative method is to make use of an instrument called a stabilograph, which records on paper attached to a rotating drum the angle of heel to a base of time.

Because the angle of inclination is difficult to obtain accurately, especially with a pendulum, several shifts of weight are made and the mean value of the angles recorded for each shift is taken in calculating GM. Thus after weight No. 1 has been moved over to the same side as 3 and 4, No. 2 is moved and then 1 and 2 are moved back in turn, the deflection of the pendulum being measured for each shift. The process is then repeated on the other side of the ship, by moving Nos. 3 and 4 over and then returning them to their original positions. It is also customary to use more than one pendulum so that if two are used there would be in all 16 readings of deflection for a shift of weight w through a distance h.

The reason why the angle is difficult to obtain with a pendulum is that it is very lightly damped, and once set in motion takes a long time to come to rest. Also, any slight motion of the ship which may be caused by wind, for example, will set the pendulum in motion again. However, if the process described is adopted it is possible to smooth out the inaccuracies in the readings and thus arrive at a reasonable value for the angle of inclination.

If use is made of the stabilograph, then because this instrument gives a record of angle of inclination to a base of time it is possible to find accurately the mean value of the angle for a given shift of weight.

When the metacentric height has been determined, the height of the centre of gravity above base is obtained by subtracting this from the value of KM obtained from the hydrostatic curves for the mean draught at which the ship is floating. Thus

$$KG = \mathrm{KM} - GM$$

This KG as inclined must now be corrected for the items to come off and those to go on to make up the lightweight.

The inclining experiment is concerned mainly with the determination of the vertical position of the centre of gravity of the ship, but at the same time it is customary to calculate the longitudinal position of the centre of gravity as well. The calculation for this from observed draughts has already been described in an earlier section of this chapter and need not be repeated here.

In order that satisfactory results are obtained from the inclining experiment there are a few practical points which should be borne in mind. In the first place the experiment should be carried out in calm weather when there is little or no wind and for the best results should be done in dock. This eliminates the influence of tides and currents on the ship. It is essential that the ship be free to incline when the weights are moved across the deck, and for this reason all mooring ropes should be as slack as possible and gangways lifted clear. Ideally the mooring ropes should be removed altogether, but this is undesirable as under these circumstances the ship may drift even when in dock, which may lead to damage.

Another important point is that all weights capable of moving transversely should be locked in position and there should be no loose fluids in tanks, which should be absolutely full or completely empty. The influence of free surface on stability has already been demonstrated and if tanks are left partially full in the inclining condition a false value of the metacentric height *GM* will be obtained.

To avoid having to make large corrections to the condition of the ship as inclined in order to obtain the lightship weight, the ship in the inclining condition should be as near completion as possible. There should also be as few people on board the ship as possible, generally only sufficient to carry out the experiment. Movement of people during the experiment may not be significant in large ships but in smaller ships with small metacentre heights transverse movements of people can produce measurable angles of inclination and should therefore be avoided.

The draughts of the ship must be measured accurately at stem and stern and also at amidships if the ship is suspected of hogging or sagging. The density of the water in which the ship is floating should be measured carefully and this is usually done by taking samples of water at several positions round the ship and using some form of hydrometer to find the density.

One final point is that in ships having a large trim in the inclining condition it may not be sufficiently accurate to use the hydrostatic curves to determine displacement and the horizontal and vertical positions of the centre of buoyancy and the metacentre. In such cases it is advisable to carry out a detailed calculation for all these quantities for the inclined waterplane at which the ship is floating.

Stability when docking

When a ship is supported partially by the ground and partially by water, the conditions for stability differ somewhat from those for the ship floating freely. A typical example of this situation, apart from that of the ship grounding at some position along its length, is the case of the ship going into dry dock. Usually when docking a ship has a small trim by the stern so that as the water is pumped out of the dock grounding will take place first at the after end. As the level of the water is lowered the ship will gradually rise in the water and the trim will also reduce until grounding occurs all fore and aft and the trim is reduced to zero, or at least to the rake of the blocks in the dock. At the instant of grounding fore and aft the load on the sternframe will reach its greatest value and it is usually in this situation that the stability of the ship may become critical, since after grounding along the length of the keel it is possible to put in side shores to support the ship.

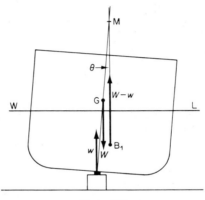

Figure 5.22

Suppose that w is the greatest load on the after end of the ship when the ship is about to ground all fore and aft. If W is the weight of the ship, it will be seen from *Figure 5.22* that the buoyant force provided by the water is $W - w$, and taking moments about the centre of gravity G

$$\text{Righting moment} = (W - w)GM \sin \theta - wKG \sin \theta$$
$$= (WGM - wGM - wKG) \sin \theta$$
$$= (WGM - wKM) \sin \theta$$
$$= \left(GM - \frac{w}{W}KM\right)W \sin \theta \qquad (5.19)$$

for a small inclination θ.

Should the quantity inside the bracket become negative the ship will be unstable, so that $(w/W)KM$ should never exceed GM. The problem then is first of all to find the value of w.

If $\bar{x}$ is the horizontal distance from the point where the ship grounds at the after end to the centre of gravity G then at the instant of grounding all fore and aft the moment applied to the ship is $w\bar{x}$ and if T is the amount of trim which has to be taken off the ship to contact the keel blocks all along its length then

$$w \times \bar{x} = T \times \text{M.C.T.}$$

where M.C.T. is the moment to cause unit trim, which can be obtained from the hydrostatic curves.

It follows that

$$w = \frac{T \times \text{M.C.T.}}{\bar{x}}$$

This value of w when inserted in equation 5.19 will indicate if the ship is stable or not. The value of KM used in this expression should be that at a waterline corresponding to a displacement $W - w$.

The method described here is only an approximate one for calculating w, since M.C.T. should also be at the new waterline for the ship, or at least at a mean line between the original and final draughts, but since the final draught depends upon the value of w this is not known initially. A second approximation, however, by repeating the calculation with the first calculated value of w to approximate to the 'grounding' draught should give a sufficiently accurate answer.

The problem of the stability of a ship when docking which has been examined here may not be so important when side blocks are provided in the dock, since in this case the ship would ground in the transverse direction at the same instant as it grounds fore and aft. There may, however, be an intermediate condition where the ship could be unstable before complete grounding takes place, in which case the ship would loll over to one side or the other.

Stability at large angles of inclination

Introduction and derivation of Atwood's formula

Hitherto the stability of a ship has been considered for very small angles of inclination from the vertical and it has been shown that the metacentric height is a measure of the stability under these conditions (often called 'initial stability'). When the angle of inclination exceeds four or five degrees for a normal ship the intersection of the vertical through the centre of buoyancy in the inclined condition with the centre line of the ship (the point M) can no longer be regarded as a fixed point relative to the ship. The quantity GM then is no longer a suitable criterion for measuring the stability of the ship and it is usual to use the value of the righting arm GZ for this purpose.

In what follows it will still be assumed that the ship is in static equilibrium under the action of its own weight and its buoyancy when inclined to some angle θ from the vertical. *Figure 5.23* shows a ship in such a condition and if W_0L_0 represents the waterline at which it floated when upright and W_1L_1 the waterline in the inclined condition, these two waterlines will cut off the same volume of displacement. If the ship were vertically sided the two waterlines would intersect on the centre line so long as the deck edge was not awash or the bilge did not come out of the water. In general, however, ship sections are not vertically sided, at least throughout the entire length of the ship, so that the two waterlines W_0L_0 and W_1L_1 cutting off the same volume do not intersect on the centre line but at some position S as shown in *Figure 5.23*.

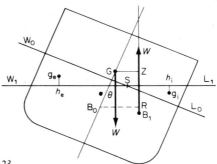

Figure 5.23

It will be seen that a volume W_0SW_1 has come out of the water and an equal volume L_0SL_1 has gone into the water. Calling this volume v and if g_e and g_i are the centroids of the two wedges, h_e and h_i being the feet of the perpendiculars from g_e and g_i on W_1L_1, the horizontal shift of the centre of buoyancy B_0 is given by

$$B_0R = \frac{v \times h_e h_i}{V}$$

where V is the volume of the displacement of the ship. It will be seen further that

$$GZ = B_0R - B_0G \sin \theta$$

$$= \frac{v \times h_e h_i}{V} - B_0G \sin \theta \qquad (5.20)$$

By evaluating v and $h_e h_i$ in equation 5.20 for various angles of inclination it is possible to plot a curve GZ to a base of θ. A typical

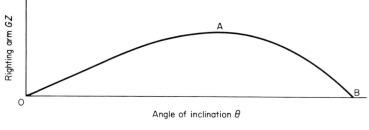

Figure 5.24

curve is shown in *Figure 5.24*. The righting arm *GZ* increases from zero when the ship is vertical, reaches a maximum value at A, declines slowly at first and then more rapidly until *GZ* becomes zero at B. If external forces were applied which carried the ship beyond B it would not return to the vertical even if the forces were removed. Thus the ship is unstable beyond B, and the range of angles from 0 to B represent the range of stability. It is important to know for all normal conditions of loading of a ship what the maximum value of *GZ* is and also the range of stability, and to this end stability calculations are normally carried out for a ship.

Equation 5.20 is often called 'Atwood's formula' and was developed in the eighteenth century. As presented it is difficult to evaluate. Because of the non-symmetry of ship forms above and below water it is not easy to assess the position of S in *Figure 5.23* and thus determine the volume and position of centroids of the immersed and emerged wedges. One method is shown in *Figure 5.25*. Assume in the first place that when the ship heels the inclined waterline intersects the original waterline at O on the centre line. This will give unequal volumes for the emerged and immersed

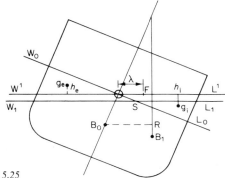

Figure 5.25

wedges. Calling these v_e and v_i respectively, the volume of displacement will have been overestimated by an amount $v_e - v_i$ so that the ship should rise by a small amount until it floats at $W_1 L_1$, so that volume $W'L'L_1W_1 = v_e - v_i$. It can be assumed that the centroid of the volume of this layer is at the transverse centre of flotation of the waterplane $W'L'$ distance λ from O. It will be seen then that the horizontal shift of the centre of buoyancy can be written

$$B_0 R = \frac{v_e\, h_e\, O + v_i\, h_i\, O - \lambda(v_i - v_e)}{V}$$

and the righting arm is given by

$$GZ = \frac{v_e\, h_e\, O + v_i\, h_i\, O - \lambda(v_i - v_e)}{V} - B_0 G \sin\theta \qquad (5.21)$$

In the nineteenth century a method was developed by F. K. Barnes[1] for evaluating the volumes of the wedges and the positions for the centroids as well as the value of λ by radial integration making use of Simpson's rule. The method involves a great deal of tedious calculation although it represented one of the first methods, if not *the* first method, of determining the stability information for a ship at large angles of inclination. The method has long been outmoded by advances in calculation methods and by different approaches to the problem. Before making some brief reference to these modern methods it is instructive to examine the application of Atwood's formula to a simple geometrical form.

Formula for righting arm of a ship vertically sided in neighbourhood of waterline

Consider a vessel of rectangular cross section as shown in *Figure 5.26*. What follows would apply equally well to a ship of varying section throughout its length which is vertically sided above and below the waterline, so long as the bilge does not come out of the water or the deck edge does not become immersed. Because the vessel is vertically sided the emerged and immersed wedges are of right angled triangular section and the volume $W_0 O W_1$ = volume $L_0 O L_1$. Let the co-ordinates of the new position of the centre of buoyancy B_1 after inclination relative to B_0 in the upright condition be ξ and η. These co-ordinates can then be calculated by considering the moment of transfer of the volume of the wedges in both the horizontal and vertical directions.

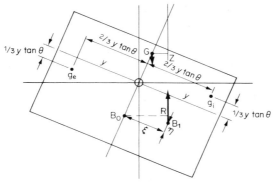

Figure 5.26

It will be seen that:

Transverse moment of volume shifted $= \displaystyle\int_0^L \frac{1}{2}y \times y\tan\theta\,\mathrm{d}x \times \frac{4}{3}y$

$$= \int_0^L \frac{2}{3}y^3 \tan\theta\,\mathrm{d}x = \tan\theta \int_0^L \frac{2}{3}y^3\,\mathrm{d}x = I\tan\theta$$

where I is the second moment of area of the waterplane about the centre line. Therefore $\xi = I\tan\theta/V = \mathrm{B_0 M}\tan\theta$ since $\mathrm{B_0 M} = I/V$.
Similarly:

Vertical moment of volume shifted $= \displaystyle\int_0^L \frac{1}{2}y^2\tan\theta \times \frac{2}{3}y\tan\theta\,\mathrm{d}x$

$$= \tfrac{1}{2}I\tan^2\theta$$

Hence $\eta = \frac{1}{2}I\tan^2\theta/V = \frac{1}{2}\mathrm{B_0 M}\tan^2\theta$.

From *Figure 5.26* it will be found that

$$\mathrm{B_0 R} = \xi\cos\theta + \eta\sin\theta$$

$$= \mathrm{B_0 M}\tan\theta\cos\theta + \tfrac{1}{2}\mathrm{B_0 M}\tan^2\theta\sin\theta$$

$$= \sin\theta\,(\mathrm{B_0 M} + \tfrac{1}{2}\mathrm{B_0 M}\tan^2\theta)$$

Now $\qquad GZ = \mathrm{B_0 R} - \mathrm{B_0 G}\sin\theta$

$$= \sin\theta\,(\mathrm{B_0 M} - \mathrm{B_0 G} + \tfrac{1}{2}\mathrm{B_0 M}\tan^2\theta)$$

$$= \sin\theta(GM + \tfrac{1}{2}\mathrm{B_0 M}\tan^2\theta) \qquad (5.22)$$

This is often called the 'walled sided formula' or the 'box formula'. It is applicable to a great many ships for angles of inclination up to as much as 10°, but will of course apply only to any ship so long as the deck edge is not awash or the bilge does not come out of the water. It may be regarded as a refinement of the simple metacentric formula for GZ which was $GZ = GM \sin \theta$, or for very small angles $GZ = GM\theta$. The formula is useful in examining conditions of equilibrium for a ship.

Taking first the case where the ship has a positive metacentric height. The ship will be in equilibrium when $GZ = 0$. Therefore, using equation 5.22:

$$0 = \sin \theta (GM + \tfrac{1}{2}B_0M \tan^2 \theta)$$

This can be satisfied by either $\sin \theta = 0$ or $GM + \tfrac{1}{2}B_0M \tan^2 \theta = 0$. The former gives $\theta = 0$ and the ship is upright, as would be expected from simple metacentric theory. The second possibility gives

$$\tan^2 \theta = \pm \sqrt{\frac{-2\,GM}{B_0M}} \qquad (5.23)$$

Now since GM and B_0M are both positive, the quantity under the root is negative so that there is no solution corresponding to this second alternative. It is reasonable to conclude, therefore, that for a ship with a positive metacentre height the only equilibrium position is the vertical.

If the case of a ship with a zero metacentre height is considered, then it will be seen once again that the only equilibrium position is when $\theta = 0$, i.e. the ship is upright. The wall sided formula does show, however, in this case that the ship is in stable equilibrium and is not in neutral equilibrium as would be suggested by the simple metacentric approach, since a positive value of GZ would be generated for an inclination from the vertical given by

$$GZ = \tfrac{1}{2}B_0M \sin \theta \tan^2 \theta \qquad (5.24)$$

The most interesting case which can be investigated by the wall sided formula is that of a ship with a negative metacentric height. In equation 5.23, if GM is put as a negative quantity it will be apparent that $\tan \theta = \pm \sqrt{\{2GM/B_0M\}}$. There are two positions of equilibrium at equal angles on either side of the vertical, given by

$$\theta = \tan^{-1} \pm \sqrt{\frac{2GM}{B_0M}} \qquad (5.25)$$

This is called the angle of loll and the ship with a negative metacentric height would incline indifferently to one side or the other. If it were inclined to say the port side and a force were applied which carried it past the upright the ship would loll to an equal angle on the starboard side even when the force was removed.

The use of the wall sided formula shows that an initially unstable ship will not necessarily capsize as would be suggested by meta-centric theory, but would take up an indifferent permanent list. On the basis of the walled sided formula the GZ curve would be given by

$$GZ = \sin\theta(-GM + \tfrac{1}{2}B_0M\tan^2\theta) \qquad (5.26)$$

This would give the type of curve shown in *Figure 5.27*. The curve would go below the base line and then gradually pick up as the second term in equation 5.26 becomes more dominant and would cross the base at the angle of loll. In this position the ship would be stable.

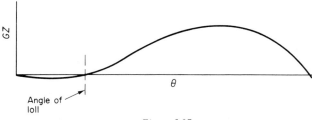

Figure 5.27

In order to obtain some idea of the magnitude of the angle of loll, consider a ship with a negative metacentric height of 0.08 m and with $B_0M = 5$ m. Then

$$\tan\theta = \pm\sqrt{\frac{2\times.08}{5}} = \pm0.18$$

$$\theta = \pm10.2°$$

An appreciable list can thus be obtained with a relatively small negative metacentric height.

The metacentric height of an initially unstable vessel

A vessel which is unstable in the upright condition will loll to some angle θ until it arrives at a position of stable equilibrium. *Figure 5.28* shows this situation where the metacentre M was originally below G in the upright condition. At the angle of loll θ the centre of buoyancy will be in some position B_1 vertically under G and the line B_1G will be the new vertical. If it is imagined that the ship is

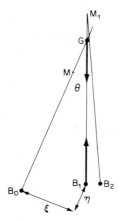

Figure 5.28

inclined through a small angle from this position the centre of buoyancy will move to B_2 and a line drawn through B_2 will intersect B_1G produced in M_1, where M_1 is the new metacentre when considering inclinations from B_1G. Now it has been seen that the height of the metacentre about the centre of buoyancy is given by I/V where I is the second moment of area of the waterplane about its centre line, so that

$$B_1M_1 = \frac{I \text{ of inclined waterplane}}{V}$$

On the assumption of vertical sidedness in the neighbourhood of the original waterline, the half-ordinates of this waterline will be increased to $y/\cos \theta$ so that

$$I_1 = \sum \frac{2}{3} \frac{y^3}{\cos^3 \theta} \mathrm{d}x = \frac{I}{\cos^3 \theta}$$

where I refers to the waterplane of the vessel in the upright condition. Therefore

$$B_1M_1 = \frac{I/\cos^3 \theta}{V} = \frac{B_0M}{\cos^3 \theta}$$

It has been seen that $\tan \theta = \pm \sqrt{\{2GM/B_0M\}}$ so that

$$\cos \theta = \frac{1}{\sqrt{\{1+(2GM/B_0M)\}}} \quad \text{and} \quad B_1M_1 = B_0M\{1+(2GM/B_0M)\}^{\frac{3}{2}}$$

Referring to *Figure 5.28* it will be seen that

$$B_0G = B_1G \cos\theta + \eta$$
$$= B_1G \cos\theta + \tfrac{1}{2}B_0M \tan^2\theta$$
$$= B_1G \cos\theta + \tfrac{1}{2}B_0M(2GM/B_0M)$$
$$= \frac{B_1G}{\sqrt{\{1 + (2GM/B_0M)\}}} + GM$$

$$GM_1 = B_1M_1 - B_1G = -(B_0G - GM)\sqrt{\{1 + (2GM/B_0M)\}}$$
$$+ B_0M\{1 + (2GM/B_0M)\}^{\frac{3}{2}}$$
$$= B_0M[\{1 + (2GM/B_0M)\}^{\frac{3}{2}} - \sqrt{\{1 + (2GM/B_0M)\}}]$$
$$= B_0M\sqrt{\{1 + (2GM/B_0M)\}} \{1 + (2GM/B_0M) - 1\}$$
$$\therefore \quad GM_1 = 2GM\sqrt{\{1 + (2GM/B_0M)\}} \tag{5.27}$$

In the example quoted $GM = 0.05$ m, $B_0M = 5$ m.

$$\therefore \quad GM_1 = 2 \times 0.05 \sqrt{\left/ \left(1 + \frac{2 \times 0.05}{5}\right)\right.} = 0.10 \times 1.0099$$
$$= 0.101 \text{ m}$$

For all practical purposes it can be taken that the value of the positive metacentric height in the lolled condition is twice the numerical value of the negative metacentric height in the initial condition.

Cross curves of stability

It was shown above that it is difficult to ascertain the exact waterline at which a ship would float in the inclined condition in order to obtain the same displacement as in the upright condition. This difficulty can be overcome by approaching the problem in another way. *Figure 5.29* shows a ship inclined to some angle θ. If the displacements to a number of waterlines I II III IV and V and the perpendicular distances of the centroids of those displacements from the line YY passing through G are calculated, it is possible to plot a curve to a base of displacement for a particular angle of inclination of the righting arm GZ as shown in *Figure 5.30*. The value of GZ for any particular displacement can then be obtained by setting off the required displacement Δ on the base and lifting from the curve. If the calculation is repeated for a series of different angles of inclination a set of curves can be drawn, each for a different angle. These curves are shown in *Figure 5.30* and are known as 'cross

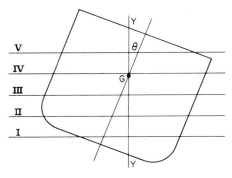

Figure 5.29

curves of stability'. A cross curve of stability may be defined as a curve of *GZ* to a base of displacement for a constant angle of inclination.

Cross curves of stability used to be calculated for ships by the use of an instrument called an 'integrator'. The instrument is capable of calculating the area of a figure and its moment about an axis by tracing a stylus round the perimeter of the figure. These areas and moments of areas when integrated all fore and aft by Simpson's rule give volumes and moments of volumes from which the values of the righting arm *GZ* could be calculated. With the advent of the digital computer the integrator has been superseded and much more accurate results have been obtained. The computer does of course make use of one or other of the rules for integration but works from offsets lifted from the body plan of the ship. Whichever method

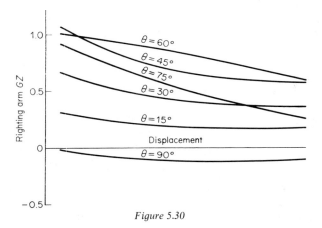

Figure 5.30

is used it is customary to carry out calculations for cross curves at angular intervals of 15°, with perhaps one or two additional calculations between 0° and 15°.

Determination of curve of statical stability from cross curves

A curve of righting arm GZ to a base of angle of inclination for a fixed displacement is called a 'curve of statical stability' and such a curve is quite readily obtained from a set of cross curves. Suppose that the displacement of the ship, say Δ, is known and the position of its centre of gravity is at G_1. The cross curves will have been

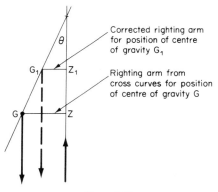

Corrected righting arm for position of centre of gravity G_1

Righting arm from cross curves for position of centre of gravity G

Figure 5.31

determined for some assumed position of the centre of gravity G, so that the procedure is first of all to set Δ off on the base of the cross curves and lift off the values of GZ at the various angles. The right arms for the actual ship are then obtained from $G_1Z_1 = GZ$ from cross curves $\pm GG_1 \sin\theta$, as shown in *Figure 5.31*. The + or − sign will be used depending upon whether G_1 is below or above G.

Factors affecting a curve of statical stability

In general a curve of statical stability rises steadily from the origin and for the first few degrees is practically a straight line, after which it becomes a curve which may have a point of inflexion changing from concave upwards to concave downwards. It eventually reaches a maximum value, after which it declines and eventually crosses the base line. The important features are the slope at the origin, the maximum value of *GZ* and the point where *GZ* becomes zero. The slope at the origin can be readily determined by considering the simple metacentric formula for *GZ*, i.e. $GZ = GM \sin \theta$. By differentiating this the slope is obtained. Hence

$$\frac{\mathrm{d}GZ}{\mathrm{d}\theta} = GM \cos \theta$$

so that when $\theta = 0$ the slope is *GM*. Thus, if the value of the metacentric height was set up at $\theta = 1$ rad (57.3°) and a straight line drawn from this point to the origin, the actual *GZ* curve should be tangent to this at the origin, as shown in *Figure 5.32*. It will be seen from

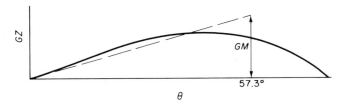

Figure 5.32

this that the factor affecting the slope of the curve at the origin is the metacentric height. A ship with a zero metacentric height would have a stability curve which is tangent to the base line and one with a negative metacentric height will have a curve which starts by going below the base and then gradually picking up until it becomes positive.

The other important features of a stability curve, the maximum righting arm and the range of stability, are to a large extent a function of the freeboard. The higher the freeboard for a given position of the centre of gravity the greater will be the maximum value of *GZ* and the larger will be the range before the ship becomes unstable. This is understandable, because the higher the freeboard the greater will be the angle at which the deck edge comes awash, and so the curve will continue to rise, as will be seen by considering the wall sided formula, so long as the bilge does not come out of the water.

Another feature of the ship which would increase both the maximum value of *GZ* and the range is the presence of watertight superstructures on the weather deck. They have the effect of increasing the freeboard locally if they are only partial superstructures, i.e. if they do not cover the whole length of the deck. Doubt often exists as to whether superstructures are watertight and if therefore they should be included in the calculations for stability, because they may have sidelights or windows in their sides and doors which are not watertight in their ends. If there is any doubt about the watertightness of these superstructures they are usually left out of the calculations.

It has been seen in Chapter 3 that the draught to which a ship is allowed to load is limited by regulation, and this is governed by ensuring that the ship has a certain minimum freeboard. The minimum freeboard is usually sufficient to provide an adequate range of stability. What is an adequate range of stability for a particular ship is very difficult to decide and there appears to be no unanimity on this subject.

Although in dealing with stability it is usual to plot GZ against angle of inclination, the important quantity as far as stability is concerned is the righting moment. Thus, very often much larger values of GZ are shown in the light condition than in the fully loaded condition, but the former are associated with a very much smaller displacement so that in assessing stability one should really consider the product $\Delta \times GZ$.

Influence on stability of transverse movement of a weight

When dealing with initial stability the problem of a 'slung' weight capable of moving transversely was dealt with and it was seen that the influence of this was to produce a virtual reduction in the metacentric height. Another problem is that of a weight which moves permanently some distance transversely across the ship. This may be a piece of cargo which has not been properly secured and has become detached when the ship is rolling at sea. Suppose that the weight of the item is w and it moves horizontally a distance h, the total weight of the ship being W. Then

$$\text{Transverse movement of centre of gravity } GG_1 = \frac{wh}{W}$$

as shown in *Figure 5.33*. The value of GZ is reduced by $GG_1 \cos \theta$, so that

$$\text{Modified righting arm} = GZ - \frac{wh}{W}\cos \theta$$

Unlike the slung weight problem, the weight w will not necessarily come back to its original position when the ship rolls in the opposite direction but may remain at the position to which it has moved. The righting lever and the righting moment are therefore reduced on one

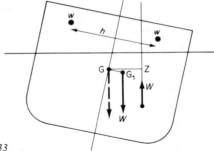

Figure 5.33

side of the ship but of course will be increased for the other side. The situation is shown in *Figure 5.34*. If $GG_1 \cos \theta$ is plotted on the stability curve for the particular condition of loading of the ship the two curves will intersect at B and C. The point B will represent the equilibrium position when the ship is in still water and C will now give the angle of vanishing stability when the ship becomes unstable.

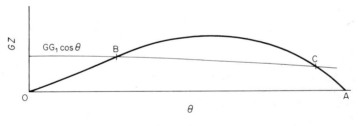

Figure 5.34

The range of stability is very greatly reduced on the side to which the ship lists, as also is the maximum value of *GZ*. On the other side of the vertical, stability would be increased, both the range and the maximum value of *GZ* being greater than for the ship before the cargo shifted. It is, however, the worst condition which must be considered so that every precaution should be taken to avoid cargo moving transversely.

Bulk cargoes

In the previous section the general problem of a weight moving transversely across a ship has been considered and it has been shown that this can have a major effect on stability. A particular case of very great importance occurs in the carriage of dry bulk cargoes. There are many cargoes of this type, such as grain, ore and coal. One of the problems with a bulk cargo such as grain is that although the holds may be completely filled after loading, the cargo settles down when the ship goes to sea and thus leaves spaces at the tops of the holds. All substances of this type have an angle of repose, so that if the ship rolls to a greater angle there could be movement of cargo to one side of the ship and the cargo will not necessarily move back to its original position when the ship comes back to the upright. Consequently there is a permanent movement of weight to one side of the ship resulting in a permanent list, as has been seen, and a reduction of the range of stability on the inclined side. Every effort should be made to prevent this movement of cargo; in the past there have been many losses of ships due to this cause.

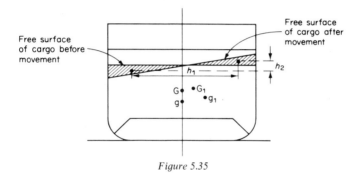

Figure 5.35

Figure 5.35 shows the section through the hold of a ship which is carrying a bulk cargo. When the cargo settles down after loading its centre of gravity is at g. If the ship rolls at sea the free surface of the cargo could take up some position as shown by the inclined line. It will be seen that there is a horizontal movement of weight h_1 and a vertical movement h_2. If the weight moved is w and the total weight of cargo is W then the centre of gravity of the cargo will move to g_1 and the co-ordinates of g_1 relative to g are

$$x = wh_1/W$$
$$y = wh_2/W$$

Similarly the co-ordinates of the centre of gravity of the ship after the movement of cargo (G_1) relative to the original centre of gravity G are

$$X = \frac{Wx}{\Delta} \quad \text{and} \quad Y = \frac{Wy}{\Delta}$$

It follows in this case that the modified righting arm

$$G_1Z_1 = GZ - X\cos\theta - Y\sin\theta$$

where GZ is the righting arm at the angle θ before the cargo moved. In this case there is a small correction due to the rise of the centre of gravity of the cargo moved, but it should also be remembered that a small improvement in stability will have occurred because the cargo centre of gravity will be slightly lower due to the cargo settling down.

Prevention of movement of bulk cargoes

Regulations have been in existence for a long time to minimise the movement of bulk cargoes and in particular grain. In the first place when a hold is filled with grain in bulk it must be trimmed so as to fill all the spaces between beams and at the ends and sides of the hold. Similar precautions in regard to trimming were always taken in the carriage of coal in ordinary dry cargo ships but in the case of coal, when loaded into specially designed self-trimming colliers, manual trimming became practically unnecessary.

Another important feature which would reduce the movement of grain cargoes is the provision of centre line bulkheads and shifting boards in the holds. They have a similar effect to watertight centre divisions in tanks carrying liquids in that they reduce the free surface effect and hence reduce the movement of the cargo. Centre line bulkheads and shifting boards were at one time required to extend from the tank top to the lowest deck in holds and from deck to deck in 'tween deck spaces. The present regulations in the International Convention for the Safety of Life at Sea 1960 require that these bulkheads and shifting boards shall extend for a distance of one-third the depth of the hold or 2.44 m (8 ft), whichever is the greater, downwards from the deck. In the 'tween deck spaces they are required to extend from deck to deck.

The centre line bulkheads are fitted clear of hatches, and are usually of steel. They have a dual purpose in that they can be regarded as a line of pillars supporting the beams if they extend from tank top to deck. Shifting boards are of wood and are placed on the centre line in way of hatches. They are portable and can be removed if required when bulk cargoes are not being carried.

Although the provision of centre line bulkheads and shifting boards reduces the amount of transverse movement of the cargo it is possible for spaces to be left at the tops of holds and 'tween decks when the grain settles down. In an attempt to fill these spaces feeders are fitted to provide a head of grain which will feed into the empty spaces. Hold feeders are usually formed by trunking in part of the hatch in the 'tween decks above. The capacity of a feeder must be 2% of the volume of the space which it feeds.

The above gives a general outline of the precautions which are taken to prevent the movement of grain cargoes. They have enabled such cargoes to be carried with a high degree of safety.

Dynamical stability

So far the stability of a ship has been considered as a purely statical problem. A true measure of stability can only be made by considering the problem as a dynamical one. The dynamical stability of a ship is defined as the work which must be done upon it to heel it to a particular angle. In *Figure 5.36* the shaded area is the work done in

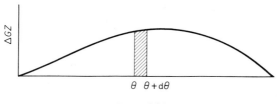

Figure 5.36

heeling the ship from an angle θ to $\theta + d\theta$ and it is equal to $\Delta GZ\, d\theta$. It follows that the total work done in heeling the ship to any angle α say is

$$\int_0^\alpha \Delta GZ\, d\theta$$

The work done in heeling a vessel to a given angle represents the potential energy which it possesses at that angle if the upright position is regarded as the zero of potential energy. The value of $\int \Delta GZ\, d\theta$ up to the angle of vanishing stability represents the maximum amount of potential energy which the ship possesses. If a force were applied which gives the ship more energy than this then it will move beyond the point of vanishing stability and capsize.

Influence of wind on stability

If a beam wind blows on a ship then the force generated on the exposed above water surface is resisted by the hydrodynamic force

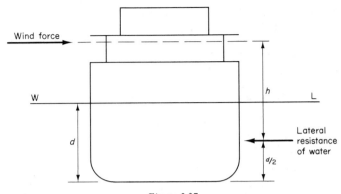

Figure 5.37

produced by the slow sideways motion of the ship through the water. The wind force may be taken to act at the centroid of the above water area whilst the hydrodynamic force can be considered with sufficient accuracy to act at half-draught (see *Figure 5.37*). Let the distance between the lines of action of these two forces be h and let the area of the above water form be A. As the ship heels the projected area will reduce to $A \cos \theta$ approximately and the lever h will also reduce to $h \cos \theta$. This is strictly speaking true only for a flat surface, and in the case of the ship because of its breadth it will not be quite accurate. The wind force is also proportional to the square of the velocity V, so that this force can be written

$$P = kA \times V^2 \cos \theta$$

and the moment will be

$$M = kAhV^2 \cos^2 \theta$$

where k is some empirical coefficient.

If the curve of wind moment is plotted on a diagram giving the righting moment ΔGZ, then as shown in *Figure 5.38* when the wind moment is applied gradually the ship would take up the angle of steady heel show at A and the range of stability from this position would be from A to B. The problem would then be similar to the case of the movement of a weight transversely. On the other hand if the wind moment were applied suddenly the amount of energy applied to the ship while it heeled to A would be represented by the area DACO. The ship would only absorb energy represented by the area OAC, so that the energy equal to the area DAO would be sufficient to carry it beyond A to some angle F, so that area AEF = area DAO. Should F be beyond B then the ship would capsize.

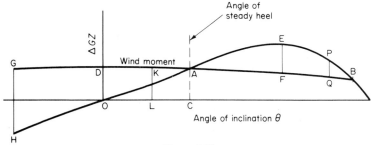

Figure 5.38

The worst case would probably be when the ship is rolling and is inclined to its maximum angle to windward and about to return to the vertical. Suppose this position is represented by GH in *Figure 5.38*. The ship would already have sufficient energy to carry it to some angle denoted by KL in the diagram. In the absence of damping this angle would be the same as the angle to windward, but because of the energy absorbed by damping forces then the angle would be somewhat less than the windward angle. The amount of energy put into the ship by the wind up to the angle L is now represented by the area GDKLOH. The ship must therefore continue to heel until this energy is absorbed so that it may reach some angle represented by PQ. Once again if this position should be beyond B the ship will capsize.

The influence of wind on stability and the possibility of capsizing due to dynamical considerations were more important in sailing ships than in modern power driven vessels. In the former the area exposed to the wind (the sails) was much greater than the area of the above water form presented to the wind in the latter, and also the distance *h* between the centre of pressure of the sails and the centre of lateral resistance of the under water form was much greater. Hence larger wind moments had to be catered for. Nevertheless, in ships with small metacentric heights large angles of heel might be experienced with sudden gusts of wind on the beam, since the metacentric height controls the form of the stability curve near the origin.

Angle of heel due to turning

A problem akin to that of the influence of a wind moment acting on a ship is the heel which arises when a ship turns. If a ship is imagined to be turning in a steady circle of radius r there is an outward force of magnitude $(MV^2)/r$ where M is the mass of the ship and V is its speed. This force acts at the centre of gravity and the lateral resistance R to the sideways motion of the ship through the water can be imagined to act at half-draught as in the wind problem. There will thus be a moment $(MV^2h)/r$ heeling the ship outwards. The angle of heel can then be obtained by plotting this moment on a ΔGZ curve as in the case of the wind. The intersection of the two curves gives the steady heel. If the moment is applied suddenly then the instantaneous heel would be obtained by considering the dynamical stability as in the previous case. It is unlikely, however, that the centrifugal force due to turning would be applied very suddenly. In any case the motion of the ship is very complicated during the initial stages of turning. Turning is accomplished by the force generated by the rudder, which may act quite low down and probably below the centre of lateral resistance so that initially the ship is likely to heel inwards due to the moment generated.

If the angle of heel due to turning is likely to be small then the angle can be determined by using the metacentric height GM to determine the righting moment. On this assumption it follows that

$$\frac{MV^2h}{r} = MgGM \sin \theta$$

from which

$$\sin \theta = \frac{V^2h}{rgGM}$$

Loading conditions

It has long been customary for loading conditions of a ship to be worked out and information supplied to the master in some suitable form. The information supplied consists of a drawing of the profile of the ship indicating the positions of all the loads on board, a statement of the end draughts and trim of the ship and also the value of the metacentric height. Stability information in the form of curves of statical stability is often also supplied. The usual loading conditions which are worked out are as follows:

(1) Lightship
(2) Fully loaded departure condition with homogeneous cargo
(3) Fully loaded arrival condition with homogeneous cargo
(4) Ballast condition
(5) Any other conditions in which the ship is likely to be in service.

A trim and stability booklet is prepared for the ship showing all these conditions of loading.

Nowadays the supply of much of this information is compulsory under the Load Line Regulations. It is in fact one of the conditions of assignment of a freeboard that such information shall be supplied to the ship.

Other data supplied include hydrostatic particulars, cross curves of stability and plans showing the position, capacity and positions of the centroids of all spaces on board. They enable the master to work out the trim and stability of the ship for what might be called non-standard conditions which may not be included in the stability and trim information supplied.

Flooding and damaged stability

In the consideration of trim and stability dealt with so far in this chapter it has been assumed that the ship is intact. In the event of collision or grounding or even springing a leak of the hull water may enter the ship and if unrestricted flooding were permitted the ship would eventually sink. To prevent this, or at least reduce the probability, the hull is divided into a series of watertight compartments by means of watertight bulkheads, which are transverse sheets of stiffened plating extending from side to side of the ship. It must be pointed out that the provision of such bulkheads will not ensure 100% safety in the event of damage. It is clear, for example, that if the damage is 'horizontal' then although there are bulkheads several compartments may be opened to the sea and the ship may not then be able to sustain loss of buoyancy of this magnitude. This was well illustrated in the loss of the *Titanic* in 1912. The ship was well subdivided but the extent of the damage after collision with an iceberg rendered much of the subdivision ineffective, with the result that the ship sank with considerable loss of life.

Another effect of flooding and one which may be just as important as the ability of the ship to float after damage is the question of loss of stability. If the damage is sufficiently large it may cause the ship to list and if it does not capsize it may be at least difficult or impossible to launch the lifeboats. These two aspects of flooding will be considered here.

Trim when a compartment is open to the sea

Suppose in *Figure 5.39* the compartment shown shaded is open to the sea. The buoyancy of the ship between the bulkheads bounding

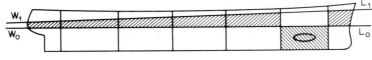

Figure 5.39

this compartment is lost. The ship must therefore sink in the water until it picks up buoyancy on the above water form to maintain equilibrium. At the same time because of the loss of buoyancy the longitudinal position of the centre of buoyancy is altered so that in order to restore it to its original position the ship must trim. The ship which was floating originally at the waterline $W_0 L_0$ will therefore now float at $W_1 L_1$. Should the waterline $W_1 L_1$ be higher at any point than the deck at which the bulkheads stop (the bulkhead deck) it is usually considered that the ship would be lost because the pressure of water in the damaged compartments could force off the hatches and unrestricted flooding would occur all fore and aft. In practice, however, the ship may still be capable of floating for a considerable time after such damage, especially if the bulkhead deck is not the weather deck.

To calculate the position of the waterline $W_1 L_1$ is not an easy matter and usually has to be done by successive approximation. There are two approaches: the 'lost buoyancy method' and the 'added weight method'.

Lost buoyancy method

In the lost buoyancy method the area of the damaged waterplane is first calculated. Suppose that the area of the intact waterplane was A and that an area a has been lost in way of the damage; then the area of the damaged waterplane is $A - a$ and if v is the volume of buoyancy lost up to $W_0 L_0$ the parallel sinkage is given by $y = v/(A - a)$ and the ship would float at $W_0' L_0'$ (see *Figure 5.40*). A

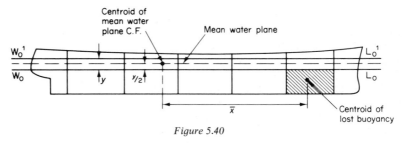

Figure 5.40

second approximation to the sinkage is required, however, as there will be a change in the area of the waterplane due to the increase in draught. This second approximation can be made by considering a mean waterplane midway between d and $d + y$. For this mean waterplane the area A_m, position of centroid CF and second moment of area I_m about a transverse axis through its centroid are calculated. The position of the centroid can be taken as the centre of buoyancy of the added layer which replaces the lost buoyancy. It then follows that

$$\text{Sinkage} = \frac{v}{A_m} \qquad (5.28)$$

If M.C.T.$_m$ is the moment causing unit trim for the mean waterplane then

$$\text{Trim} = \frac{\rho g v \bar{x}}{\text{M.C.T.}_m} \qquad (5.29)$$

If λ is the distance of the centre of flotation of the final waterline from amidships then

$$\text{Draught aft} \quad = d + \frac{v}{A_m} - \frac{(L/2) - \lambda}{L} \times \frac{\rho g v \bar{x}}{\text{M.C.T.}_m} \qquad (5.30)$$

$$\text{Draught forward} = d + \frac{v}{A_m} + \frac{(L/2) + \lambda}{L} \times \frac{\rho g v \bar{x}}{\text{M.C.T.}_m} \qquad (5.31)$$

This assumes that the centre of flotation is aft of amidships and damage takes place forward. Similar expressions for the draught would be obtained for other cases. Another approximation could now be made by determining the properties of a second mean waterplane at $d + \frac{1}{2}(v/A_m)$ above base, but usually the results given by equations 5.30 and 5.31 will be sufficiently accurate.

One point requires further consideration and this concerns the permeability of the compartment damaged. When a compartment contains cargo, fuel, etc., then the amount of water which can enter on damage is less than the volume of the empty compartment. The ratio of the volume entering to the volume of the empty compartment is called the 'permeability' and is denoted by μ. For cargo spaces it is taken as 60% and for passenger and crew spaces as 95%. Empty spaces are taken as 95% also, because the structure within the compartment occupies some volume. It follows then that the volume v used in determining sinkage should be replaced by μv and similarly $\rho g v$ used in calculating trim will become $\mu \rho g v$. In calculating the properties of the damaged waterplane the area in way of the damaged compartment is not totally ineffective because of the permeability of the cargo, so that the effective area of the waterplane becomes $A - \mu a$. The centroid and second moment of area from which the moment to cause trim is calculated should be determined by considering this reduced area μa to be deducted from the intact area. Some have suggested that different values of the permeability should be used for areas and volumes but for practical purposes the two can be taken as having the same value.

Added weight method

In this method the water entering the damaged compartment up to the original waterline can be regarded as an added weight. The modification to the position of the centre of gravity of the ship can then easily be calculated and the increase in draught and the trim determined making use of the hydrostatic information for the ship for the intact condition. This calculation is similar to those which would be carried out for finding the draughts of the ship for particular conditions of loading. It will be seen, however, that when the draughts have been calculated in this way more water would enter the compartment because the ship is now deeper in the water in way of the damage. A second calculation is therefore necessary to take this additional weight of water into account so that a second approximation is obtained to the waterline in the damaged condition. This new waterline will now involve a further addition of weight in the damaged compartment so that a third calculation would be necessary to obtain yet a closer approximation to the correct waterline. It will be seen that this is an iterative process and if the calculation were programmed for computer any number of iterations could be made to obtain any desired degree of accuracy. The method has the advantage that it makes use of existing hydrostatic data for the intact ship.

Stability in the damaged condition

Considering first stability from the point of view of metacentric height, the effect of loss of buoyancy in the damaged compartment is to remove buoyancy from a position below the original waterline to some position above this waterline so that the centre of buoyancy will rise. If b is the position of the centroid of the lost buoyancy and b_1 is the position of the centroid of the buoyancy recovered on the above water form, then

$$\text{Rise of centre of buoyancy} = \frac{\mu v \, bb_1}{V}$$

where V is the total volume. The value of BM, the distance between the centre of buoyancy and the metacentre, will decrease because of the loss of waterplane area in way of the damage, so that if the second moment of area of the damaged waterplane about a transverse axis is I_d, then $BM = I_d/V$.

The height of the centre of gravity above base KG will remain unchanged. Hence:

$$\text{Modified metacentric height} = GM \text{ (intact)} + \frac{\mu v \, bb_1}{V} - \frac{I_d}{V}$$

This may or may not be a reduction in the metacentric height but generally the metacentric height will be reduced.

If the added weight method is used then the modified vertical position of the centre of gravity should be determined for the water added in the damaged compartment and the metacentric height calculated, making use of the value of the height of the metacentre above base (KM) for the intact ship and for the increased draught. In this case it will be necessary to correct the metacentric height for the free surface effect of the water in the damaged compartment.

Unsymmetrical flooding

When there are longitudinal bulkheads in a ship damage in a particular compartment may not pierce a longitudinal bulkhead, so that the possibility of flooding on one side of the ship only must be considered. It has been considered that in the event of collision damage the penetration is not likely to extend beyond 20% of the breadth of the ship. This is only a very rough guide but it could be expected that a longitudinal bulkhead further away from the ship's side than this distance would remain undamaged. Under these circumstances water enters one side of the ship only and an angle of heel will develop. Consider the ship shown in *Figure 5.41* where a volume of water v has entered the ship with its centroid distance z from the centre line. This is equivalent to a weight $\rho g v$ added at distance z from the centre of gravity. The heeling moment is $\rho g v z$ in the upright condition so that the ship will heel until a righting moment of this magnitude is produced. Within the metacentric range of stability the righting moment is $\Delta GM \sin \theta$ or if V is the volume of displacement $\rho g \times V GM \sin \theta$. Hence:

$$\rho g V GM \sin \theta = \rho g v z$$

or

$$\sin \theta = \frac{vz}{V GM} \tag{5.32}$$

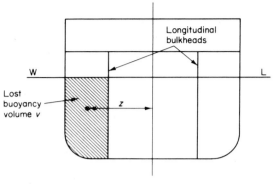

Figure 5.41

This angle would require correction for the additional water which would enter the compartment as the vessel heeled, just as the draughts required correction when the added weight method was used in the flooding calculation so that a second approximation would be obtained to the angle of heel.

It is important to limit the angle of heel due to unsymmetrical flooding and usually, if this becomes excessive, means are provided to flood the corresponding compartment on the opposite side of the ship. It is done by pipes extending from one side to the other. While this cross flooding would make the ship sink deeper in the water it would eliminate the angle of heel, which is considered less dangerous.

Floodable length

In the previous sections the consequence of flooding a particular compartment has been considered. The flooding problem can be looked at in another way by determining what length of the ship could be flooded without loss. It is difficult to define what condition would correspond to loss of the ship but it is generally accepted that when the waterline at which the ship would float in the damaged condition is tangent to the bulkhead deck line at side the limit has been reached. It is usual to have a small margin on this condition and to say that the limit is when the waterline is tangent to a line drawn 75 mm below the bulkhead deck at side. This line is called the 'margin line'.

The problem can be approached as follows. Suppose as in *Figure 5.42* a ship is floating in the intact condition at a waterline W_0L_0

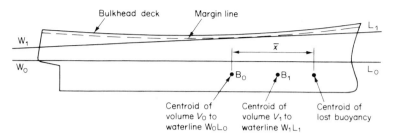

Figure 5.42

and its volume of displacement is V_0 with its centre of buoyancy at B_0. Imagine now that the ship is depressed by some external force so that it floats in the intact condition at a waterline W_1L_1 which is tangent to the margin line. The volume of displacement in this condition will be V_1 and the centre of buoyancy will be at B_1. For the ship to float freely at this waterline it would have to lose a volume of buoyancy $= V_1 - V_0$ at such a position that the centre of buoyancy moves vack to B_0.

Taking moments about the original centre of buoyancy $V_1 \times B_0B_1 = (V_1 - V_0) \times \bar{x}$ where $\bar{x}$ is the distance of the centroid of the lost buoyancy from B_0. Hence

$$\bar{x} = \frac{V_1 \times B_0B_1}{V_1 - V_0} \tag{5.33}$$

From this equation it is possible to find the position of the centroid of the lost buoyancy and knowing the value of the lost buoyancy $V_1 - V_0$ it is then possible to convert this into a length of ship which can be flooded. A first approximation can be obtained as follows.

In *Figure 5.43* is shown a curve of cross-sectional areas of the ship to the margin line. If the distance $\bar{x}$ is set off along the base and an

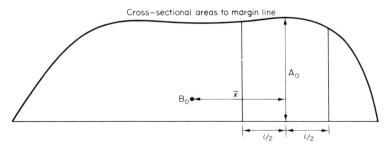

Figure 5.43

ordinate A_0 lifted from the curve a first shot at the length which can be flooded is $v/A_0 = (V_1 - V_0)/A_0 = l$. Setting half this length on either side of the position where the ordinate is A_0 would give the approximate positions of two bulkheads which would give a length of compartment of the correct volume and position of centroid. For positions near the ends of the ship where the area is changing rapidly this would not be very accurate. The compartment would probably have to be moved bodily forward or aft and the length increased. This would have to be done by a process of trial and error until the correct volume and position of centroid was obtained.

Calculations of this sort can be done for a series of waterlines tangent to the margin line at different positions in the length of the ship, from which a series of values of the floodable length can be obtained for different positions and hence a series of values of the floodable length can be obtained for different positions in the length. It is thus possible to draw a curve of floodable length as shown in *Figure 5.44*. This curve is drawn so that the ordinate at any point represents the length of the ship which can be flooded with the centre of that length at the point considered. Thus if l is the floodable length at some point then by setting off $l/2$ on each side of that point the position of two bulkheads which would give a compartment of the required length and position would be obtained. Proceeding in this way throughout the length of the ship it is possible to obtain the positions of all the bulkheads so that any one compartment can be flooded without the ship sinking.

It should be noted that the permeability of the compartments will affect the floodable length and it is usual to work out the average permeability for the machinery space, the part forward of the machinery space and the part aft, so that three different curves could be obtained for the floodable length of the complete ship, as shown in *Figure 5.45*.

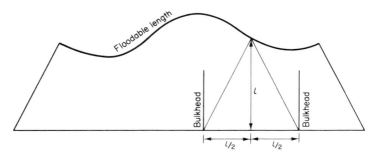

Figure 5.44

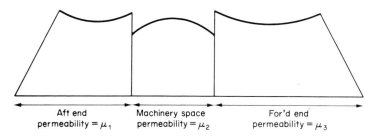

Aft end	Machinery space	For'd end
permeability = μ_1	permeability = μ_2	permeability = μ_3

Figure 5.45

The condition that a ship should be able to float with any one compartment open to the sea is the minimum requirement for ocean going passenger ships. Existing regulations require that as the length of the ship and the extent to which it is a passenger ship increase a higher standard of subdivision will be required. It is achieved by introducing the idea of a permissible length which is some fraction of the floodable length, the fraction being called the 'factor of subdivision'. In some ships the factor of subdivision is 0.5, in which case it would be possible to flood two adjacent compartments without the ship sinking.

Regulations for the watertight subdivision of passenger ships have for many years been the subject of international agreement, as was pointed out in Chapter 3. The whole question of survivability after damage has been under review for many years and it is expected that there will be extensive modifications to existing regulations in the near future.

REFERENCE

1. Barnes, F. K., 'On a method of calculating the statical and dynamical stabilities of a ship', *Trans. Instn. of Naval Architects*, 1861

6

The sea and ship motions

The sea is the natural environment of the ship and to understand the behaviour of a ship it is important to study the sea itself. The study of the behaviour of the ship in still water is a relatively simple task and has been dealt with in the previous chapter. The sea is, however, rarely if ever absolutely still because of the presence of waves and in this chapter the properties of waves will be examined.

Waves arise on the surface of the sea because of wind and the sizes of waves generated are related to the wind speed. The mechanism of the generation of waves by wind will not be discussed here but some of the theory covering the movement of waves over the water surface will be dealt with. As a starting point it is fitting to discuss the properties of regular waves on the surface of a fluid.

Regular waves

Regular waves may be considered to be waves which do not change their pattern with time, i.e. if a wave train is imagined to be passing a fixed point each succeeding wave has exactly the same size and shape. It will be shown later that this does not give a true picture of ocean waves but it is a suitable starting point.

Several different types of waves can be recognised and in the first place they will all be considered to be long crested, which simply means that the motion is being considered in two dimensions only, in which case the waves are of infinite breadth measured along the crests.

The different types of waves may be enumerated as follows:

(1) Deep sea waves
(2) Shallow water waves
(3) Waves of translation
(4) Ripples or capillary waves.

Types (1), (2) and (3) are what may be called gravity waves, gravity being the major force acting. In type (4) the important force is surface tension and waves of this type are necessarily of very short length and of small height. Deep sea waves are waves in which the depth of water from the surface to the bottom does not have any effect on the wave motion, whereas in the second type the depth of water is important. Classical theory does not distinguish between these two types but shows that (1) is the limiting case of (2). Waves of these types are oscillating waves in which the water particles move in closed paths without bodily movement of fluid. The wave form moves over the surface and energy is transmitted. Type (3), the wave of translation, involves bodily movement of fluid in the direction of motion of the wave.

Figure 6.1 shows the profile of a regular wave which may be considered to be a deep sea wave. The distance between successive crests is the wave length and will be noted by λ. The distance from the trough to the crest is the wave height h. If such a wave system moves over a fluid it will be observed that the crests move forward with some velocity v, but it can easily be demonstrated that there is not any forward motion of the fluid, for if a light floating object such as a cork is thrown into the fluid it will be carried around by the motion of the fluid particles but the centre about which it rotates will not move forward.

The classical theory of waves shows that there is a relation between the velocity v of the waves, the length λ, and the depth of water d. This relation is

$$v = \sqrt{\left(\frac{g\lambda}{2\pi} \tanh \frac{2\pi d}{\lambda} \right)} \qquad (6.1)$$

Suppose the depth of water $d = \lambda/2$, then

$$v = \sqrt{\left(\frac{g\lambda}{2\pi} \tanh \pi \right)}$$

$\tanh \pi$ is practically equal to unity, so that beyond depths of approximately half a wave length the influence of depth is unimportant. At the other end of the scale, if d becomes very small

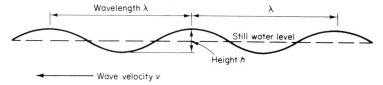

Wavelength λ λ

Still water level

Height h

Wave velocity v

Figure 6.1

then $\tanh 2\pi d/\lambda$ tends to $2\pi d/\lambda$ and the expression for velocity becomes

$$v = \sqrt{(gd)} \qquad (6.2)$$

This can be shown to be the speed of a wave of translation.

It will be noted that the height of the wave does not affect the velocity so that all waves of a given length would have the same velocity regardless of their heights. Since the velocity of waves in a fluid of finite depth is dependent on the factor $\tanh 2\pi d/\lambda$, as ocean waves approach shallow water they are slowed up so that the crests become crowded together; this is particularly noticeable as waves approach a beach.

The trochoidal theory

Observations on ocean waves have shown that the crests are sharper than the troughs. This has led to the theory of Gertsner,[1] which assumes that the wave profile is a trochoid. The trochoid is shown in *Figure 6.2* and is the path swept cut by a point P at radius r from the centre of a circle, radius R, which rolls underneath a base line. This produces a stationary wave and could be imagined to be the surface of the sea if it was frozen. If the angular velocity of the circle radius R (the rolling circle) is ω then the horizontal velocity v of the centre will be ωR and R is $\lambda/2\pi$. Imposing a velocity v in the opposite direction would bring the rolling circle to rest and the point P would rotate in a circle with angular velocity ω and the wave form would move forward with velocity v, thus producing a progressive wave. The co-ordinates of a point on the trochoid are given by

$$\left.\begin{array}{l} x = R\theta - r\sin\theta \\ y = r\cos\theta \end{array}\right\} \qquad (6.3)$$

when the origin is at the centre of the rolling circle at $\theta = 0$.

The trochoidal theory yields the same relation between speed and length for deep sea waves as does the classical theory, although it does not fulfil all the conditions for what is called irrotational motion. It would not in the form quoted enable the influence of depth of water to be taken into account. It shows that the paths

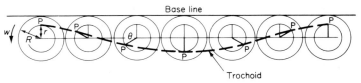

Base line

Trochoid

Figure 6.2

of the water particles are circles, whereas the classical theory shows that the paths are ellipses which tend to circles as depth of water increases.

Both theories show that for a deep sea wave, if conditions are considered some distance below the surface, then the radius of the orbit circles of the particles diminishes. If $r_0 = h/2$ is the radius of the surface particles and r is their radius at some sub-surface distance y below the free surface, then

$$r = r_0 \exp(-y/R) = r_0 \exp\{-(2\pi y)/\lambda\} \qquad (6.4)$$

where y is considered positive downwards.

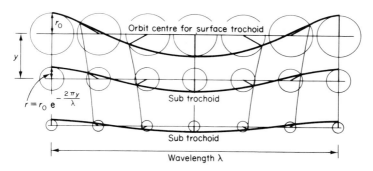

Figure 6.3

It will be seen then that the disturbance below the surface diminishes rapidly with depth, the situation being as shown in *Figure 6.3*, where the sub-surface trochoids as they may be called rapidly flatten out. This is an important point since it will be seen that a fully submerged ship such as a submarine would not be subject to the same disturbance as a surface vessel and in fact at a depth little over half a wave length there would be negligible movement of the water.

In shallow water where the influence of the depth is important the elliptical orbits of the particles flatten out with depth below the free surface and at the bottom the vertical movement is prevented altogether and the particles move horizontally only.

Pressure in a wave

In still water the pressure at any point is proportional to the distance below the free surface, but it is not so in the case of a wave.

It can be shown that the still water level corresponding to any point in a trochoidal wave lies at a distance $r^2/(2R) = (\pi r^2)/\lambda$ below the line drawn through the orbit centre of the particles corresponding to that point. Thus at the surface the still water level would be $(\pi r_0^2)/\lambda$ below the orbit centres, and at some point distance y below the surface the still water level would be $(\pi r^2)/\lambda$ below the orbit centres.

The distance between these two still water levels is

$$y - \frac{\pi r_0^2}{\lambda} + \frac{\pi r^2}{\lambda}$$

and the pressure in the wave can be shown to be proportional to this distance. Hence pressure in wave $= \rho g\{y-(\pi/\lambda)(r_0^2-r^2)\}$. This is shown in *Figure 6.4*. Its effect is that pressure is reduced below the static value at the crest of a wave and is increased beyond the

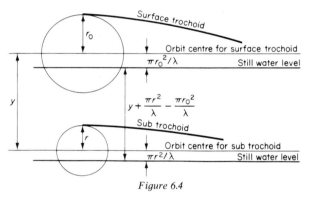

Figure 6.4

static value at the trough, which influences the buoyancy of a ship amongst waves. If for example a ship were imagined to be poised instantaneously on a wave with crests at the ends the buoyancy distribution would be as shown in *Figure 6.5(a)* compared with the buoyancy based on static conditions only. On the other hand, if there was a crest at amidships the buoyancy distribution would be as shown in *Figure 6.5(b)*.

The influence of dynamic considerations on the pressure in a wave is often referred to as the 'Smith effect'.[2]

Wave period

The time interval between two crests passing the same point is the wave period and its value can be deduced quite easily from the

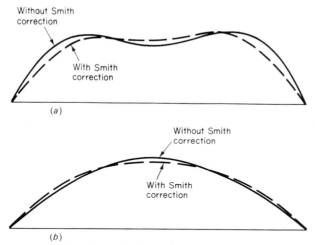

Figure 6.5 Buoyancy distribution of ship in waves (a) *Wave crests at ends* (b) *Troughs at ends*

expression for the wave velocity. The distance which the wave travels in a time T_w which is the wave period is one wavelength λ. Hence:

$$T_w = \frac{\lambda}{v} = \frac{\lambda}{\sqrt{\{(g\lambda)/(2\pi)\}}} = \sqrt{\frac{2\pi\lambda}{g}} \qquad (6.5)$$

This gives the period relative to some fixed position. When a ship moving through the sea is considered it is the period relative to the ship which is required. This will differ from the period T_W given by equation 6.5 because of the ship speed. The period relative to the ship is called the 'apparent' or 'effective period' or the 'period of encounter'. Consider a ship moving through long-crested waves as shown in *Figure 6.6*, the inclination of the course of the ship to the wave direction being α. The distance which a crest has to travel in time T_E which is the period of encounter is now $T_E(v - V\cos\alpha)$ where V is the ship speed. This distance must be equal to one wavelength, so that

$$T_E(v + V\cos\alpha) = \lambda$$

or

$$T_E = \frac{\lambda}{v - V\cos\alpha} = \frac{\lambda/v}{1 - (V/v)\cos\alpha} = \frac{T_W}{1 - (V/v)\cos\alpha} \qquad (6.6)$$

where T_W is the actual wave period.

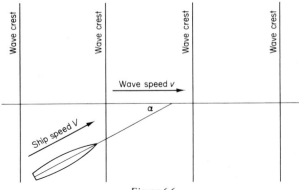

Figure 6.6

It will be seen that if the ship is in a following sea, i.e. if α is between $0°$ and $90°$, then the period of encounter is greater than the wave period, and that when α is greater than $90°$ the period of encounter is less than the wave period.

Wave energy

It can be shown that the energy in a deep sea wave is half potential and half kinetic. A formula which can be derived from the trochoidal theory for the total energy of a single wave per unit breadth is

$$E = \frac{1}{8}\rho g\lambda h^2 \left(1 - \frac{\pi}{2}\frac{h^2}{\lambda^2}\right)$$ (6.7)

where ρ is the density of the fluid.

The second term in the bracket is quite small and for a wave in which $h/\lambda = \frac{1}{20}$ the error involved by neglecting this term would be only a little over 1%. It is therefore sufficiently accurate to take $E = \frac{1}{8}\rho g\lambda h^2$, from which it will be seen that the energy is proportional to the wavelength and the square of the height, or alternatively by dividing by λ it can be said that the energy per unit area of sea surface is proportional to the square of the height only.

Sea surface

The idea of long-crested waves of regular form does not convey a true picture of the sea conditions to which a ship is subjected in service. Observations of actual sea conditions show that the pattern is quite confused and is not even uni-directional. At a storm centre waves of different heights, lengths and directions are generated which all combine to give the sea state as it is observed. It is only at some distance from the storm centre that any appearance of regularity might be noticeable due to the longer and hence faster waves having outrun the shorter waves.

Consider first a uni-directional sea in which a typical record of wave height to a base of time might be as shown in *Figure 6.7*.

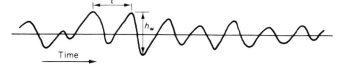

Figure 6.7   Wave record

This is the wave pattern observed at a fixed point. The apparent wave heights observed are the vertical distances between a crest and an adjacent trough. They are denoted by h_w in the figure. Several ways are available for measuring the apparent wave period. This may be the time interval τ between two crests, or the interval between two adjacent up or down crossings of the mean line. It will be found that in the analysis of wave records what is known as the 'significant wave height' is used. This is usually defined as the mean of the $\frac{1}{3}$ highest waves.

The data from a record such as that shown can be plotted in histogram form. If wave height is divided into a number of divisions, say 0–0.5 m, 0.5–1.0 m, etc., and the number of waves in the record in each of these divisions is determined, then a histogram of the type shown in *Figure 6.8* can be plotted. Here the number of waves

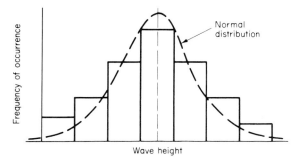

Figure 6.8   Histogram of wave heights

in each division is expressed as a percentage of the total number in the record. Alternatively, the record can be analysed in terms of wave periods (see *Figure 6.9*).

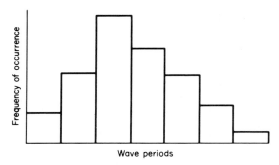

Figure 6.9   *Histogram of wave periods*

If the deviation of the wave height from the mean height of a particular record is plotted it will be found that the resulting histogram follows the normal or Gaussian distribution. The mathematical curve corresponding to this distribution is shown plotted in *Figure 6.8*. The curve is expressed by

$$f(h) = \frac{1}{\sigma\sqrt{2\pi}} \exp\{-(h-\bar{h})^2/2\sigma^2\} \tag{6.8}$$

where

$f(h)$, which is the height of the curve, is the frequency of occurrence

h = wave height

$\bar{h}$ = mean wave height from the record

σ = standard deviation = $\sum\limits_{1}^{N}\sqrt{\{(h-\bar{h})^2/N\}}$

where N = total number of observations in the record.

If, on the other hand, the actual wave heights are plotted then it has been found that they follow quite a different distribution. It is known as the 'Rayleigh distribution' and is given by

$$f(h) = \frac{2h}{E} \exp(-h^2/E) \tag{6.9}$$

where in this case the parameter $E = (1/N) \sum_1^N h^2$, N being the total number of observations in the record as before.

The Rayleigh distribution is only strictly speaking applicable to short term records of 20–30 min duration during which weather conditions may be reasonably assumed to be constant. It applies quite well to observations of wave heights and also to the response of the ship, i.e. amplitudes of heaving, pitching, rolling, etc.

Energy spectrum

One of the most useful concepts in the consideration of the sea is the energy spectrum. It has been seen that wave height records present a very irregular pattern. It is nevertheless possible to represent this pattern by a series of regular components, the total height being the sum of all these components. The components can be obtained by Fourier analysis and the elevation of the sea surface at any point becomes

$$h = \sum_{n=1}^{n=\infty} h_n \cos\left(\omega_n t + \varepsilon(\omega_n)\right) \qquad (6.10)$$

where h_n is the height of the individual component and ω_n is the circular frequency of that component, $\varepsilon(\omega_n)$ being an arbitrary phase angle. The analysis would yield a whole series of components, as shown in *Figure 6.10*, and while theoretically the number should be infinite, sufficient accuracy could be achieved in practice by taking a limited number of these.

Now it has been shown that the energy per unit area of sea surface is proportional to the square of the height of the wave, so that the energy corresponding to a particular component will be proportional to h_n^2. Consequently the total energy of the sea will be

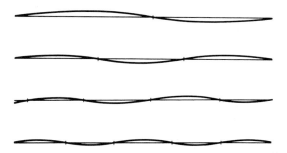

Figure 6.10   Regular wave components in irregular sea

obtained by summing up the energies corresponding to these components. Hence:

$$\text{Total energy} \propto \sum_{n=1}^{n=\infty} h_n^2 \qquad (6.11)$$

since

$$h^2 = \sum_{n=1}^{n=\infty} h_n^2$$

By considering the mean wave height over a small range of frequency $\delta\omega$ at various values of ω it is possible to plot a histogram of wave height squared (which represents energy) to a base of ω. From this,

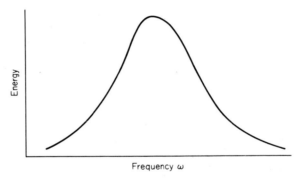

Figure 6.11

the type of curve shown in *Figure 6.11* can be derived. This is the energy spectrum. However, in order that the area under the curve shall be proportional to the total energy of the sea it is usual to plot an ordinate $S(\omega)$ which has the unit's height squared × seconds and which is really a measure of the energy per unit frequency. This is called the spectral density. Hence the total energy of the sea is proportional to $\int_0^\infty S(\omega)\,d\omega$.

Sea spectra

While it is not intended to discuss here how sea spectra are derived, the factors upon which they depend should be considered.

It has already been stated that wave heights are dependent upon wind speed. Other factors upon which wave heights would depend are the duration for which the wind has been blowing and the length of 'fetch', i.e. the distance over which the wind blows. For instance in the neighbourhood of land the wave heights generated would be lower than they would be in the open sea. It is possible therefore to distinguish between partially developed and fully developed spectra.

Considering the influence of wind duration time on the spectrum, *Figure 6.12* shows the shape and size of the spectrum for increasing

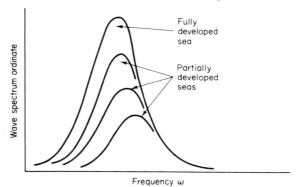

Figure 6.12   Partially and fully developed spectra

duration. At first only the higher frequency components (short wave length) are generated, but as the wind continues to blow the spectrum grows in height and the lower frequency components (long wave lengths) gradually appear. A similar effect can be observed if spectra are drawn for increasing wind speeds, the longer waves only appearing at the higher speeds.

Formulae for one-dimensional wave spectra

Many formulae exist for the determination of wave spectra and in some way they must include a factor which assesses the weather intensity. The spectra are often therefore related to the wind speed, although sometimes significant wave height is used and a separate formula employed to relate significant wave height to wind speed. A formula of the former type is that due to Pierson. This gives for the energy spectrum ordinate

$$S(\omega) = \frac{8.1 \times 10^{-3} g^2}{\omega^5} \exp\left\{-0.74(g/V\omega)^4\right\} \qquad (6.12)$$

where g = acceleration due to gravity in cm/s^2
ω = frequency in rad/s, and
V = wind speed in cm/s.

As an example of the other type of formula the Bretscheinder spectrum may be quoted, where the spectral density ordinate is given as

$$2h^2(\omega) = 4200H_S^2/T_S^4\omega^5 \exp(-1050/T_S^4\omega^4) \qquad (6.13)$$

where H_S is the significant wave height and T_S is the significant period and is actually the average period of the significant waves. The values of H_S and T_S could be taken from sea data for the particular area of the ocean for which the information is required. However, formulae relating both of these quantities to wind speed have been given and are:

$$H_S = 0.025V_K^2 \text{ (ft)}$$

$$T_S = 0.64V_K \text{ (s)}$$

In its 1970 Report the Committee on Environmental Conditions of the International Ships' Structures Congress recommended the following formula for one-dimensional energy spectra:

$$E(f) = \frac{0.11H_v^2}{(T_n f)^5} \exp\{-0.44/(T_n f)^4\} \qquad (6.14)$$

where $E(f)$ = wave energy density
H_v = visual wave height, but could be taken as the significant wave height
T_v = visual estimate of wave period
f = wave frequency in hertz.

Many other formulae exist for wave spectra but those quoted give some idea of the form they take.

Directional spread

In the spectra discussed in the previous section it is assumed that all the energy contained in the sea state is concentrated in waves travelling in the same direction. Under actual sea conditions this is very rare and to take into account the fact that waves come from all directions it is necessary to introduce a spreading function. Two such functions have been proposed, namely:

$$f(\theta) = \frac{2}{\pi}\cos^2\theta$$

and
$$\qquad (6.15)$$

$$f(\theta) = \frac{4}{3\pi}\cos^4\tfrac{1}{2}\theta$$

in which θ is the inclination of the direction of a wave component to the predominant direction of the waves. It follows that the wave energy is now a function of ω and θ, so that

$$S(\omega_1\theta) = S(\omega) \times f(\theta)$$

and the total energy in the sea will be

$$\int_0^\infty \int_{-\theta}^{+\theta} S(\omega) \times f(\theta)\,d\omega\,d\theta \qquad (6.16)$$

Response of ship to the sea

The ship has six degrees of freedom as a rigid body, i.e. heaving, swaying, surging, rolling, pitching and yawing. The first three are linear motions, heaving being the vertical motion of the ship, whilst swaying and surging are respectively the athwartship motion and the fore and aft motion. Rolling is rotation about a longitudinal axis, while pitching is rotation above a transverse axis and yawing is rotation about a vertical axis. The ship being an elastic body has other degrees of freedom. These will not be dealt with at present but will be discussed in Chapters 7 and 11.

In a realistic sea state all six motions can be generated, but here attention will be focused mainly on heaving, pitching and rolling. These are true oscillatory motions, that is to say if the ship is displaced from its equilibrium position by some force, when that force is removed the ship will oscillate until the motion is damped out.

In considering the response of the ship to the sea it is important to determine the natural periods of oscillation in the various modes.

Heaving

Consider first of all a ship heaving freely in still water. In *Figure 6.13* a ship is shown in the equilibrium position where the gravitational force Mg is exactly equal to the buoyancy ρgV. The dotted lines show the ship heaved into the water an amount y. The increase in buoyancy, assuming the waterplane area A to be constant, is ρgAy and this force will tend to restore the ship to its original position. The ship will have vertical acceleration d^2y/dt^2 so that the equation of motion is in the absence of damping forces

$$(M + M')\frac{d^2y}{dt^2} + \rho gAy = 0$$

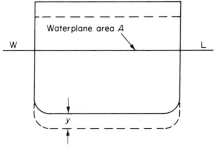

Waterplane area A

W   L

y

Figure 6.13

or
$$\frac{d^2 y}{dt^2} + \frac{\rho g A}{M + M'} \times y = 0$$

in which M' is the added virtual mass of the entrained water. Since the mass of the ship can be written ρV, where V is the volume of displacement in still water, and M' can also be written $\rho V'$, where V' is the volume of the entrained water, the equation becomes

$$\frac{d^2 y}{dt^2} + \frac{g A}{V + V'} \times y = 0$$

The solution to this equation is

$$y = C \sin \left(t \sqrt{\left(\frac{g A}{V + V'} \right)} + \delta \right)$$

where C is an arbitrary constant and δ is a phase angle. The period of the motion, i.e. the time for one complete oscillation, is

$$T_{\mathrm{H}} = 2\pi \sqrt{\frac{V + V'}{g A}} \qquad (6.17)$$

The added virtual mass of the entrained water can be shown to be at least equal to the mass of the ship (see Chapter 11) so that V' is approximately equal to V.

As an example, consider a vessel of constant rectangular cross section in which $V = LBd$ and $A = LB$, where d is the draught. If $V' = V$ then

$$T_{\mathrm{H}} = 2\pi \sqrt{\frac{2LBd}{gLB}} = 2\pi \sqrt{\frac{2d}{g}}$$
$$= 2.83\sqrt{d} \text{ with } d \text{ in metres}$$

In a ship in which $d = 10$ m, $T_H = 2.83 \sqrt{10} = 8.95$ s. This gives some idea of the magnitude of the heaving period.

In an actual ship $V = LBd \times C_B$ and $A = LBC_A$, so accepting that $V' = V$

$$T_H = 2\pi \sqrt{\frac{2dC_B}{gC_A}}$$

Since C_B is always less than C_A, the heaving period for a ship shape form would be rather less than that deduced for the vessel of constant rectangular cross section.

In all real problems involving oscillations damping forces exist, and if it is assumed that there is a damping force proportional to the velocity, i.e. $F = \mu\{dy/dt\}$ where μ is a damping coefficient, then

$$T_H = 2\pi \sqrt{\left[\frac{V + V'}{gA}\left(\frac{1}{1 - [\mu^2/\{4\rho^2(V + V')gA\}]}\right)\right]} \qquad (6.18)$$

In many systems the influence of damping on the period is small but this may not be so in the case of heaving, where damping is quite large. The damping force arises largely because of the creation of surface waves as the ship moves into and out of the water.

Heaving in regular waves

When a ship moves through regular waves a vertical force is created which can be shown to be a simple harmonic function of time. Heaving motion is thus generated. The magnitude of this motion is dependent upon amongst other factors the ratio of the heaving period of the ship to the period of encounter of the waves with the ship, i.e. T_H/T_E. Maximum amplitudes of heave will occur when this ratio approaches unity. This is the resonance condition and under these circumstances it is only the presence of damping which prevents the amplitude from becoming infinite, at least theoretically. To avoid these large amplitudes the resonance condition should be avoided and under given sea conditions the only way to do this is to alter course or speed so as to change the period of encounter.

Apart from the ratio of the periods or the 'tuning factor' as it is called, the amplitude of heave will depend upon the height of the waves through which the ship is passing. In the linear theory of heaving motion it is assumed that the heaving force is proportional to the wave height and as the resulting ship motion is proportional to the force then this is also proportional to the wave height. It is possible, therefore, to plot heaving amplitude ÷ wave height as a non-dimensional response amplitude operator, as it is sometimes called, against frequency of encounter or against tuning factor.

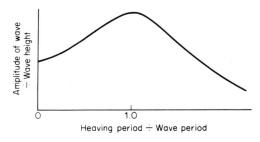

Figure 6.14

The type of curve shown in *Figure 6.14* is obtained. Results of this sort are usually obtained from experiments on models since regular waves would not generally exist at sea.

Heaving in irregular waves

It has been seen that an irregular wave system can be analysed into a series of regular components of varying frequencies and amplitudes. The response of the ship to any one of these components can thus be obtained by taking the response amplitude operator from a diagram such as *Figure 6.14* at the appropriate frequency and multiplying by the height of the component wave. The total response of the ship is then the sum of the responses to these individual components. If $S(\omega)$ is the ordinate of the sea spectrum which really represents the wave height squared, then the mean square of the response of the ship in irregular waves is given by

$$\int_0^\infty S(\omega)(\text{R.A.O.})^2 d\omega \tag{6.19}$$

This is the value of E in the Rayleigh distribution quoted in equation 6.9. Hence the probability density would be given by

$$P(y) = \frac{2y}{E}\exp(-y^2/E) \tag{6.20}$$

To find the probability that the value of y would exceed some value y_j, equation 6.20 would have to be integrated between y_j and ∞, giving

$$P(y > y_j) = \exp(-y_j^2/E) \qquad (6.21)$$

It is not intended to pursue this further here but to obtain probable maximum values of the heaving in the lifetime of the ship it would be necessary to evaluate equation 6.21 for different weather intensities, which would mean different values for the sea spectrum, and then assess the length of time that the ship is likely to be subjected to these weather intensities. In this way it is possible to ascertain the total probability of achieving any given heaving amplitude. This subject is considered in more detail in Chapter 7.

Rolling and pitching

Rolling and pitching are similar in the sense that they are both rotations, so that the approach to the pitching problem is similar to that of the rolling problem with certain modifications.

Consider first a ship heeled to an angle θ as shown in *Figure 6.15*. The moment tending to return the ship to the vertical in the absence of damping is $MgGZ$ where GZ is the righting lever and M is the mass of the ship. Now $I = Mk^2$ is the moment of inertia of the ship and $d^2\theta/dt^2$ is the angular acceleration. Hence

$$Mk^2 \frac{d^2\theta}{dt^2} + MgGZ = 0$$

$$\text{or} \quad \frac{d^2\theta}{dt^2} + \frac{g\,GZ}{k^2} = 0 \qquad (6.22)$$

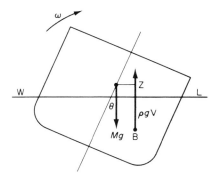

Figure 6.15

In general GZ is a complicated function of θ but if attention is restricted to small angles of inclination then GZ could be written $GZ = GM\theta$ where GM is the metacentric height. If this is substituted in equation 6.22 then

$$\frac{d^2\theta}{dt^2} + \frac{g\,GM}{k^2}\,\theta = 0 \qquad (6.23)$$

This once again is the equation of simple harmonic motion and the solution is

$$\theta = A\,\sin\left(t\,\sqrt{\left(\frac{k^2}{gGM}\right)} + \delta\right)$$

giving for the rolling period

$$T_R = 2\pi\frac{k}{\sqrt{gGM}} \qquad (6.24)$$

In the derivation of this formula, added virtual mass has been ignored as this is likely to be small in the case of rolling.

It will be seen that the rolling period is inversely proportional to the square root of the metacentric height, so that to keep the period long with small accelerations GM should be small. When GM tends to zero the formula is liable to be inaccurate and when it is actually zero, although the formula would suggest that the period would be infinite it can be shown that the ship would have a finite but long period. In this case the period is not independent of the angle of inclination.

The rolling period of a ship is much more under the control of the designer than the periods of the other motions and often in ships where large metacentric heights might exist attempts are made to reduce the value so as to give longer rolling periods.

What was said in connection with the heaving of a ship in both regular and irregular waves would apply equally well to the rolling problem. In regular waves, when the period of encounter of the waves with the ship is near or equal to the rolling period, large angles of roll may occur. It will be seen that this is more likely to occur in ships with short rolling periods, because the wave lengths corresponding to these periods are shorter.

In irregular seas, although the motion is quite complicated, if a component wave of the system has a frequency near the natural frequency of the ship and is of sufficient height, large angles of roll may develop in that frequency while the response of the ship to the other components of the wave system might be quite small.

Reduction of rolling

Many devices have been suggested for the reduction of rolling in ships, the most successful being the passive water tank or channel and activated fins. The former consists of a channel or tank placed athwartships in which water is free to oscillate in the tranverse direction. By suitably adjusting the period of the water in the tank to be approximately equal to the ship period it is possible to get the water moving in a direction which applies a moment to the ship of opposite sign to that applied by the waves, thus reducing the rolling motion. The device has proved very suitable for ships which operate for long periods at zero or near zero speeds.

The activated fin system consists in having one or sometimes two fins projecting from each side of the ship, the fins being inclined at angles of incidence to the flow so that lift forces are generated as the ship moves through the water. The angles of incidence on the two sides of the ship are of opposite sign so that a couple is generated. The attitudes of the fins are controlled by a mechanism which in turn is controlled by the motion of the ship, and the mechanism can be so arranged that the couple generated by the fins opposes the motion of the ship. The system has been widely adopted in passenger ships and naval vessels. It is more effective in fast ships, since the forces generated on the fins are proportional to the square of the speed through the water.

Pitching

A formula for pitching period can be developed similar to that for rolling, but in this case added virtual mass cannot be neglected. The formula neglecting damping is

$$T_P = \frac{2\pi K}{\sqrt{(gGM_L)}}\sqrt{\frac{V+V'}{V}} \qquad (6.25)$$

where K = longitudinal radius of gyration
GM_L = longitudinal metacentric height
V = volume of displacement
V' = volume of the entrained water.

If damping was included, the pitching period would become

$$T_P = \frac{2\pi K}{\sqrt{(gGM_L)}}\sqrt{\left(\frac{V+V'}{V}\right)} \times \frac{1}{\sqrt{(1-[\mu^2/\{4\rho^2(V+V')VgGM_LK^2\}])}}$$

$$(6.26)$$

where μ is the damping coefficient, assuming that there is a moment $\mu \mathrm{d}\theta/\mathrm{d}t$ opposing the motion.

The designer has very little control over the natural pitching period since GM_L is so large (of the order of the length of the ship) that any small change which may be made in the height of the centre of gravity G will have a negligible influence. The only other factor which could be altered is K, and this can only be done to a limited extent for a particular condition of loading. Moving loads to the ends of the ship would increase K and lengthen the period, while concentrating loads at amidships would reduce the period. Variation of load distribution for this purpose may not, however, be desirable, as it may result in large longitudinal bending moments which could produce high stresses in the structure.

The pitching motion of a ship in regular and irregular waves can be treated in the same way as that due to heaving and rolling.

If heavy pitching is likely to occur it can usually only be avoided by altering the period of encounter of the waves with the ship, which in turn means altering speed or course.

General

A very brief outline has been given here of the main motions which are likely to affect a ship in service. From several points of view it is desirable to limit the amplitudes of these motions if at all possible. All the motions involve accelerations and this can result in large forces being applied to the structure with the possibility of structural damage. With pitching, in particular, if the motion is sufficiently large the bow of the ship may leave the water altogether and when it becomes immersed again large hydrodynamic forces can be generated on the fore end of the structure. This is the phenomenon known as 'slamming' and not only has it an effect on the local structure but stress may be transmitted to other parts as well. Heavy pitching can also increase the resistance of the ship, resulting in loss of speed for a given power or alternatively the need to supply more power for a given speed.

Both pitching and heaving and to some extent rolling can affect the 'wetness' of a ship, i.e. the extent to which it takes water on board. A very wet ship could in the long run become an unseaworthy ship and may lead to foundering.

Finally, the necessity of reducing ship motions in passenger ships need hardly be stressed, since passenger comfort is of primary importance in such ships.

REFERENCES

1. Lamb, H., *Hydrodynamics*, Cambridge University Press, 1965
2. Smith, W. E., 'Hogging and sagging strains in a seaway as influenced by wave structure', *Trans. Instn. of Naval Architects, 1883*

7

Structural strength

In Chapter 1 it was stated that one of the requirements in the design of a ship was that the structure should be sufficiently strong to withstand without failure the forces imposed upon it when the ship is at sea. In this chapter the problem of structural strength will be studied in more detail.

The problem consists first of all in assessing the forces acting on the ship and secondly in determining the response of the structure to those forces, i.e. in determining the stresses which are created in the material of the structure and also the deformation of the structure. The structural strength problem is really a dynamic one. It has been seen that the ship is rarely in calm water and in consequence the motion of the sea generates motions in the ship itself. The motions generated because of the six degrees of freedom of the ship, i.e. heaving, swaying and surging, which are linear motions, and rolling, pitching and yawing, which are rotations, all involve accelerations which generate forces on the structure. It is also important to recognise that even in still water the ship is subjected to forces which distort the structure, the forces being due to hydrostatic pressure and the weight of the ship itself and all that it carries. A complete study of structural strength should take into account all these forces and in the present day development of the subject this is in fact what is done. It is fitting, however, to examine the problem from the static point of view first of all.

Static forces on a ship in still water

It has been seen that the hydrostatic forces on a floating body or ship in still water provide a vertical force B, say, which is exactly equal to the gravitated force acting on the mass M of the ship, i.e. Mg. Hence $B = Mg$.

If the distribution of these forces along the length of the ship is examined it will be found that the gravitational force per unit length is not equal to the buoyancy per unit length at every point. If the mass per unit length at every point is m and the immersed cross-sectional area at that point is a then

Buoyancy per unit length $= \rho g a$ and

Weight per unit length $\quad = mg$

so that the net force per unit length is

$$\rho g a - mg \qquad (7.1)$$

The ship under these circumstances carries a load of this magnitude which varies along the length and is therefore loaded like a beam. It follows that if this load is integrated along the length there will be a force tending to shear the structure so that

$$\text{Shearing force} = \int(\rho g a - mg)\,\mathrm{d}x \qquad (7.2)$$

On integrating a second time the bending moment causing the ship to bend in a longitudinal vertical plane can be determined. Hence

$$\text{Bending moment} = \int\int(\rho g a - mg)\,\mathrm{d}x\,\mathrm{d}x \qquad (7.3)$$

It will be seen that what is called longitudinal bending of the structure can be distinguished and this generates shear and bending stresses in the material.

Longitudinal bending is then a most important aspect of the strength of the structure of a ship and an accurate assessment of the longitudinal shearing force and bending moment is necessary in order to ensure safety of the structure.

The accurate determination of the still water shearing force and bending moment is a relatively easy task and while it does not give a complete picture of the longitudinal bending of the structure at sea it is most useful to calculate these quantities. High values of shearing force and bending moment in still water will usually indicate high values at sea, so that from still water calculations it is possible to obtain some idea of loading distributions which are likely to be undesirable.

The calculations of shear and bending stresses in the material of the structure will be dealt with later. The other result arising from these forces and moments is that there is overall deflection of the structure, i.e. the ends of the ship move vertically relative to the centre. When the ends move upwards relative to the centre the ship is said to 'sag' and the deck is in compression while the bottom is in tension. If the reverse is the case then the ship 'hogs', with the deck in tension and the bottom in compression.

The longitudinal bending of the ship due to the static forces of weight and buoyancy has been dealt with above. These forces have other effects on the structure, as will be seen from *Figure 7.1*. This represents a transverse section through the ship and it will be seen that the hydrostatic pressures are tending to push the sides of the ship inwards and the bottom upwards. The weight of the structure and the cargo, etc., which is carried are tending to pull the structure

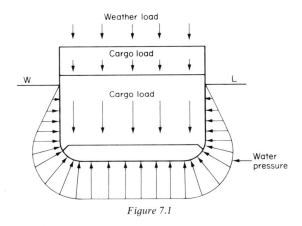

Figure 7.1

downwards. The result is that there is transverse distortion of the structure and it follows that there must be material distributed in the transverse direction to resist this type of distortion.

A third consequence of the forces acting upon the ship is local deformation of the structure. A typical example of this is the bending of plating between frames or longitudinals due to water pressure. Others are the bending of beams, longitudinals and girders under local loads such as those arising from cargo or pieces of machinery.

From the consideration of the forces acting upon the ship which have been discussed it is possible to distinguish three aspects of the strength of ships' structures. They are longitudinal strength, transverse strength and local strength. They are usually treated separately, although it is not strictly speaking correct since longitudinal and transverse bending are really interconnected. Considering, however, the complex nature of the problem of the strength of ships' structures it is a satisfactory approach, at least in the initial stages.

Function of the ship's structure

The primary requirement of the ship's structure, i.e. that it should resist longitudinal bending, necessitates that a considerable amount of material should be distributed in the fore and aft direction. This 'longitudinal' material as it may be called is provided by the plating of decks, sides and bottom shell and tank top, and any girders which extend over an appreciable portion of the length. The plating is thin relative to the principal dimensions of the transverse section of the structure and would buckle under compressive loads very easily if it was not stiffened. It is therefore necessary that there should be transverse stiffening of decks, shell and bottom, for this reason if for no other. The stiffening is provided in transversely framed ships by rings of material extending around the ship and spaced some 0.70–1 m (2–3 ft) apart. In the bottom the stiffening consists of vertical plates extending from the outer bottom to the inner bottom, the plates being called 'floors'. The sides and decks are stiffened by rolled sections such as bulb angles, inverted angles or channels, called 'side frames' and 'beams'. The transverse material so provided has the dual function of maintaining the transverse form of the structure, i.e. providing transverse strength, and preventing buckling of the longitudinal material.

The spacing of the transverse material in relation to the plating thickness is an important factor both in resisting compressive stresses and in preventing local deformation due to water pressure, so that the span thickness ratio S/t cannot be allowed to be too great. Where thin plating is employed, as would be the case in small ships, the span S between the floors, frames and beams would have to be small but may be greater in larger ships where thicker plating is employed. This will be found to be general practice, the frame spacing in small ships being less than in large ships.

Additional longitudinal strength is provided by longitudinal girders in the bottom of the ship. The centre girder is an important member in this respect. It is a continuous plate running all fore and aft and extending from the outer bottom to the tank top. Side girders are also fitted, and they are usually intercostal, i.e. cut at each floor and welded to them. The number of side girders depends upon the breadth of the ship. The double bottom egg box type of construction provided by the floors and longitudinal girders is very strong and capable of taking heavy loads such as might arise in docking and in emergencies in going aground.

The practice of stiffening ships transversely has in recent years been largely replaced by a system of longitudinal framing. This method, which actually goes back a long way to such vessels as the

Great Eastern, was initially adopted on a large scale in tankers and was known as the Isherwood System.[1] The system consists of stiffening decks, side and bottom by longitudinal members which may be either plates or rolled sections, the spacing being approximately of the same magnitude as beams, frames and floors in transversely framed ships. The longitudinals are supported by deep, widely spaced transverses consisting of plates with face flanges, the spacing being of the order of 3–4 m (10–12 ft). These transverses provide the transverse strength for the structure. Additional transverse strength is provided in all ships by watertight or oiltight bulkheads. These are transverse sheets of stiffened plate extending from one side of the ship to the other. Their main purpose is of course to divide the ship into watertight or oiltight compartments so as to limit flooding of the ship in the event of damage, but they have the additional function of providing transverse strength.

The original Isherwood System was applied to oil tankers but was not favoured in dry cargo ships, largely because of the restriction in cargo space created by the deep transverses. With the large scale development of welding in ships, however, resulting in greater distortion of the plating than would normally be found in riveted construction, longitudinal stiffening in the bottom and deck has become quite common in these vessels, while the side structure is framed transversely as formerly. It will be found that this 'combined' system of construction is now almost universally adopted in dry cargo ships. As a matter of fact the combined system was used for a time in oil tankers, but with their increasing size in the years since 1945 the complete longitudinal system has been reverted to.

Two of the advantages of the longitudinal system are that the longitudinals themselves take part in the longitudinal strength of the ship and it can be shown also that the buckling strength of the plating between longitudinals is nearly four times as great as the strength of the plating between transverse stiffeners of the same spacing.

When the decks of a ship are stiffened by transverse beams, if these were supported only at the two sides of the ship without any intermediate support, they would be required to be of very heavy scantlings to carry the loads. By introducing pillars at intermediate positions the span of the beams is reduced with the result that they can be made of lighter scantlings, thus providing a more efficient structure from the strength/weight point of view. Pillars were formerly closely spaced, being fitted on alternate beams with angle runners under the deck to transmit the load to the beams not supported by pillars. This meant that pillars were spaced about 1.5 m (5 ft) apart so that access to the sides of holds was very restricted. For this reason heavy longitudinal deck girders were introduced which had the same function as a line of pillars, the girders being supported by widely spaced pillars. Thus, in a cargo hold there would be two deck girders supported by two heavy pillars at the hatch covers. In this way access to the sides of the holds was improved.

Even these widely spaced pillars can be eliminated by fitting heavy transverse hatch end beams to support the longitudinal girders, the hatch end beams themselves being supported by longitudinal centre line bulkheads clear of the hatchways.

In ships in which the longitudinal system of framing is adopted the deep transverses take the place of the longitudinal girders and give intermediate support to the longitudinals, thus reducing their scantlings.

Nearly every part of the structure of a ship has some local strength function to fulfil. For example, the bottom and side shell plating has to resist water pressure in addition to providing overall longitudinal strength of the structure. Thus, local stresses can arise due to the bending of the plating between frames or floors. The complete state of stress in such parts of the structure is very complex because of the various functions which they have to fulfil, and even if the actual loading was known accurately it would be a difficult task to calculate the exact value of the stress.

Forces on a ship at sea

When a ship is moving through a seaway the forces acting on the ship are very different to those in still water. In the first place the static buoyancy is altered because the immersion of the ship at any point is increased or decreased compared with the still water immersion because of the presence of the waves. Secondly it has been seen that the pressure in a wave differs from the normal static pressure at any depth below the free surface. The ship also has motions which cause dynamic forces due to the accelerations involved. The two major effects are due to heaving and pitching. As was stated above, the problem then becomes a dynamic one. The traditional practice has, however, been to reduce this dynamic problem to what is considered to be an equivalent static one. The procedure adopted was to imagine that the ship was poised statically on a wave and to work out the shearing force and bending moments for this condition. Until relatively recently it was the procedure adopted in determining the longitudinal bending moments acting upon the ship at sea.

The static longitudinal strength calculation

In this calculation the wave upon which the ship is assumed to be poised statically is considered to be of trochoidal form and to have a length equal to the length of the ship. The height of the wave chosen for the calculation greatly affects the buoyancy distribution and this at one time was taken as the ship length divided by 20, i.e. $h = L/20$. More recently, however, a height given by $h = 0.607 \sqrt{L}$ m ($h = 1.1 \sqrt{L}$ ft) has been used in the calculation. This was considered to represent more closely the proportions of height to length in actual sea waves.

Two conditions were usually examined, one with wave crests at the ends of the ship and one with a wave crest at amidships. In

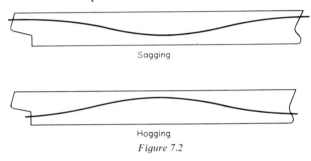

Sagging

Hogging

Figure 7.2

the former condition the bending moment due to the buoyancy provided by the wave produced sagging, and in the latter case hogging was produced. The two conditions are shown in *Figure 7.2*. Associated with these two positions of the wave it was customary to assume loading conditions for the ship which would give the greatest bending moment. Thus, with the wave crests at the ends a concentration of loading amidships would yield the greatest sagging moment, while with a crest amidships concentration of load at the ends would give the greatest hogging moment. This could possibly lead to some unrealistic conditions of loading for the ship. It is more satisfactory to consider the actual conditions in which the ship is likely to be in service and to work out the bending moments for the two positions of the wave. In this way it is possible to determine for any given loading condition the cycle of bending moment through which the ship would go as a wave of any particular dimensions passes the ship.

In the procedure described the total bending moment is obtained, including the still water moment. It is often desirable to obtain these moments separately so that the influence of the still water moment on the total can be examined. The wave moment depends only on the size of wave chosen and the ship form for any condition of loading, whereas the still water moment is dependent on the load distribution as well as the still water buoyancy distribution.

The first step in the calculation is to balance the ship on the wave, which means working out the total mass M of the ship and the longitudinal position of its centre of gravity G. The problem then is to adjust the wave on the ship to give a buoyancy equal to Mg and a position of the centre of buoyancy which is vertically below G. In doing this it is usual to ignore the Smith effect referred to in Chapter 6. To find the correct position of the wave is not an easy task since the free surface is a curve and not a straight line as in the still water position. Methods have been developed but will not be dealt with here. It will be supposed that this has been

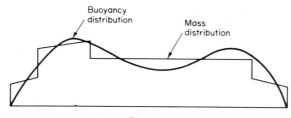

Figure 7.3

achieved, in which case the ship will be in static equilibrium under the gravitational force acting on the mass of the ship and the buoyancy provided by the wave.

The next step in the calculation is to find the distribution of buoyancy and mass along the length of the ship. The former is easy since the buoyancy per unit length is simply $\rho g A$, where A is the immersed cross-sectional area at any point in the length. The distribution of mass involves calculating the mass per unit length at a number of positions along the length and this is a tedious calculation requiring accurate estimates of the masses of the various parts of the ship. The calculation is facilitated to some extent by dividing the total mass into the lightmass of the ship and the masses of the deadweight items. The details of these calculations will not be entered into here.

Having obtained the distribution of buoyancy and mass, which in simplified form would look like the curves shown in *Figure 7.3*, it is possible to plot a load curve which is simply the difference between weight and buoyancy as in the still water calculation, i.e.

Load per unit length $= \rho g A - mg$

from which the shearing force and bending moment are given by

$$F = \int (\rho g A - mg)\,\mathrm{d}x$$

$$\text{B.M.} = \int\int (\rho g A - mg)\,\mathrm{d}x\,\mathrm{d}x$$

Because of the non-mathematical nature of the load curve these integrations have to be done graphically or can be carried out by an instrument called the 'integraph'. In recent years, largely because of the development of the computer, a tabular method has been developed. It consists of dividing the length of the ship up into a number of equal parts (say 40) each of length l, and calculating the mean buoyancy per unit length b and the mean weight per unit length w in each of these divisions. It then follows that

Shearing force $F = \Sigma(b-w)\,l = l\Sigma(b-w)$

If then the mean value of the shearing force in each of these divisions is F_m then

Bending moment B.M. $= l\Sigma F_m$

This procedure can be readily programmed for the computer and the shearing force and bending moment obtained very easily.

Characteristics of shearing force and bending moment curves

The shearing force and bending moment curves for a ship poised on a wave are shown in *Figure 7.4* and for most ships the curves follow this pattern. Both shearing force and bending moment are zero at the ends of the ship. The shearing force rises to a maximum value at a point which is roughly one quarter of the length from the end then falls to zero near amidships and changes sign, reaching a maximum value somewhere near a quarter-length from the bow. The bending moment curve rises to its maximum value at or near amidships, the exact positions occurring where the shearing force is zero.

The influence of the still water bending moment on the total moment can be seen from the curves in *Figure 7.5*, which shows the wave sagging and hogging moments superimposed on the still water moment. For a ship of any given total mass and for given still water draughts, the wave sagging and hogging moments are constant so that if the still water moment is varied by varying the loading the total moment may be altered considerably. The aim should be to keep the total moment as small as possible. If the wave sagging and hogging moments were equal then the smallest total moment would be obtained with a zero still water moment. However, the wave sagging moment is usually greater than the wave hogging moment, the proportion depending amongst other factors upon the

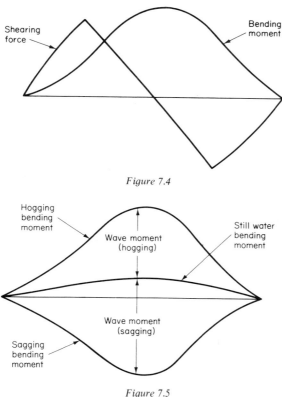

Figure 7.4

Figure 7.5

block coefficient. *Figure 7.6* shows the influence of block coefficient on these two moments.

Whether the total bending moment is a sagging or a hogging moment depends very much on the type of ship. In ships such as tankers and bulk carriers, with machinery aft, the greater moment is usually sagging whilst in large passenger ships the hogging moment is usually greater.

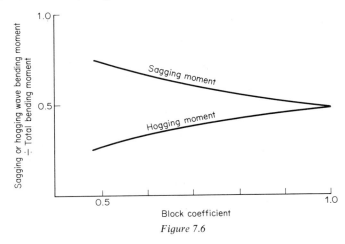

Figure 7.6

Example of calculation of bending moment for a simplified form

Consider as an example a vessel of constant rectangular cross section, 140 m long by 20 m beam and 13 m deep. The total mass is 25 830 t, 20 830 t of which is distributed uniformly over the entire length, the remaining 5000 t being distributed uniformly over a length of 10 m with its centre of gravity at amidships. The mass distribution curve would be as shown in *Figure 7.7* and the still water buoyancy distribution would be constant in this case, since the ship is of uniform section and would float on an even keel. The still water bending moment at amidships is obtained as follows:

$$\text{Buoyancy moment} = \frac{25\,830}{2} \times 35 \qquad\qquad = 452\,025 \;\text{t m}$$

$$\text{Mass moment} \;\;= \frac{20\,830}{2} \times 35 + \frac{5000}{2} \times 2.5 = \underline{370\,775} \;\text{t m}$$

$$\text{Net moment} \qquad\qquad\qquad\qquad\qquad = \;\;\;81\,250 \;\text{t m}$$

This moment requires to be multiplied by g which is 9.81 m/s^2 to convert it into units of meganewton × metre. Hence

$$\text{Still water bending moment} = 81\,250 \times 9.81 \times 10^{-3}$$
$$= 797.062 \;\text{MN m}$$

This is a sagging moment.

Suppose now that the vessel is poised on a wave equal to the length of the ship and of height $.607\sqrt{L}$. This gives a wave 140 m long × 7.2 m from trough to crest. In order to simplify the

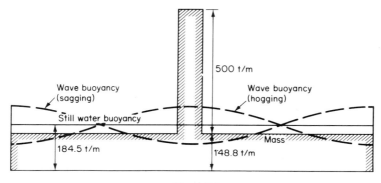

Figure 7.7

calculations it will be assumed that the wave is of cosine form
instead of being a trochoid, in which case the height at any point
above the still water level will be given by

$$h = \frac{7.2}{2} \cos \frac{2\pi x}{140}$$

for crests at the ends. Then

Buoyancy per metre $= 1.025 \times 20 \times 3.6 \times 9.81 \cos \frac{2\pi x}{140} \times 10^{-3}$ MN/m

$$= 0.724 \cos \frac{2\pi x}{140} \text{ MN/m}$$

In this the figure 1.025 is the density of sea water.

Integrating the buoyancy gives the shearing force due to the wave

$$\therefore \quad \text{Wave shearing force} = \int 0.724 \cos \frac{2\pi x}{140} \, dx$$

$$= \frac{0.724 \times 140}{2\pi} \sin \frac{2\pi x}{140} + A$$

but since the shearing force is zero at $x = 0$, $A = 0$

$$\text{Wave bending moment} = \int \frac{0.724 \times 140}{2} \sin \frac{2\pi x}{140} \, dx$$

$$= -\frac{0.724 \times 140^2}{4\pi^2} \cos \frac{2\pi x}{140} + B$$

The condition that the bending moment is zero at $x = 0$ gives

$$B = \frac{0.724 \times 140^2}{4\pi^2}$$

so that

$$\text{Bending moment} = \frac{0.724 \times 140^2}{4\pi^2}\left(1 - \cos\frac{2\pi x}{140}\right)$$

Putting $x = 70$ the wave bending moment at amidships is obtained and is 719.59 MN m sagging.

The bending moment with a crest at amidships would be numerically the same in this case but would be a hogging moment. The total bending moment can now be calculated for the two conditions:

Sagging

Still water moment	797 MN m sagging
Wave moment	719 MN m sagging
Total	1516 MN m sagging

Hogging

Still water moment	797 MN m sagging
Wave moment	− 719 MN m hogging
Total	78 MN m hogging

The influence of load distribution on the bending moment can be demonstrated by considering the load of 5000 t to be distributed uniformly over the whole length of the ship, in which case the still water bending moment would be zero. The total bending moment would then be 719 MN m in both sagging and hogging conditions.

Response of the structure

In the foregoing it has been shown how the shearing force and bending moment for a ship in any assumed loading and sea condition can be calculated. The next stage is to determine the response of the structure to these forces and moments, which simply means the calculation of the stresses in the structure and if required the overall deflection. The calculation of bending stress will be considered first.

If a beam is considered in which the bending moment at some distance x from one end is M, then the stress f in the beam at some distance z from the neutral axis is given by

$$f = \frac{M}{I/z} \tag{7.4}$$

in which I is the second moment of area about the neutral axis of the section at x. The quantity I/z when z is measured to the extreme edge of the section is called the 'section modulus' and is usually denoted by Z. If the section is unsymmetrical, i.e. if the neutral axis is not at half-depth, there will be two values of $z-z_1$ and z_2—and there will thus be two values of $Z-Z_1$ and Z_2.

The theory yielding this formula was developed for a very simple case of bending, namely a bar of uniform section bent by terminal couples in which case there was no shearing force. The theory has, however, been applied by engineers to beams subjected to other distributions of loading which involve shearing forces. Comparison of theory with experiment has shown that the simple theory can determine the stress in beams under such conditions with sufficient accuracy. It has been accepted as a suitable means for calculating the longitudinal bending stress in the complex structure of a ship. Full scale experiments on the longitudinal bending of ships have shown that the use of the theory is justified in structures where there are no discontinuities. Since the greatest bending moment occurs at or near amidships the greatest stresses are likely to occur there, so that the value of I is required for the midship section of the ship. The material which should be included in the calculation is all that which is distributed in the longitudinal direction and which extends over a considerable portion of the length of the ship. This will include side and bottom shell plating, deck plating, deck and bottom longitudinals, longitudinal bulkheads, if any, and longitudinal girders.

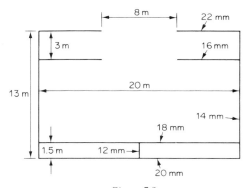

Figure 7.8

It is necessary to calculate the position of the neutral axis, i.e. the geometrical centroid of the section, so that moments of areas have to be obtained about some axis from which the position of the neutral axis can be calculated. The calculation is put down in tabular form although nowadays the actual arithmetical work would probably be done by computer. As an example, *Table 7.1* shows the calculation for the simplified section shown in *Figure 7.8*. In this table moments have been calculated about the base. Because of the fact that large portions of the deck are cut away for hatch and machinery openings the calculation is made for a section in way of the openings. The calculation is generally made for one side of the ship only.

Table 7.1 CALCULATION OF PROPERTIES OF SIMPLIFIED SECTION IN *Figure 7.8*

Item	*Scantlings*	*Area,* m^2	*Lever, y* m	*Moment,* ay m^3	*Second moment,* ay^2 m^4	*I about own N.A.,* m^4
Upper deck	6 × 0.022	0.132	13.0	1.716	22.308	
Second deck	6 × 0.016	0.096	10.0	0.960	9.600	
Side shell	13 × 0.014	0.182	6.5	1.183	7.689	2.563
Tank top	10 × 0.018	0.180	1.5	0.270	0.405	
Bottom shell	10 × 0.020	0.200	0.0	—	—	
Centre girder	1.5 × 0.012	0.009	0.75	0.007	0.005	0.001
Totals		0.799		4.136	40.007	2.564

$$\text{Neutral axis above base } z_{base} = \frac{4.136}{0.799} = 5.18 \text{ m}$$

Second moment of area of half-
section about base $= 40.007 + 2.564$ $= 42.571 \text{ m}^4$
Less area $\times z^2_{base} = 0.799 \times 5.18^2$ $= 21.439 \text{ m}^4$
I_{NA} (half-section) $= 21.132 \text{ m}^4$
I_{NA} (complete section) $= 42.264 \text{ m}^4$
$Z_{deck} = 13.00 - 5.18$ $= 7.82 \text{ m}$

$$Z_{deck} = \frac{42.264}{7.82} \qquad = 5.40 \text{ m}^3$$

$$Z_{base} = \frac{42.264}{5.18} \qquad = 8.16 \text{ m}^3$$

In calculating the second moment of area about the base in *Table 7.1* the area of each item multiplied by the square of the distance of its centroid from the base is first determined and to this must be added the second moment of area of the item about an axis passing through its centroid. Most items can be treated as rectangles, where the second moment of an item about its own neutral axis would be $ah^2/12$, where h is the vertical dimension. For 'horizontal' material, i.e. decks, tank top, bottom shell, etc., h is small, being only the thickness of the plate, and therefore the $ah^2/12$ term can be ignored. It must, however, be included for vertical material such as side shell and longitudinal bulkheads.

In the calculation no allowance has been made for deck and bottom longitudinals, but they would of course be included in an actual ship. These items have been omitted in order to simplify the calculation, which is only intended to illustrate how a calculation for the section modulus is carried out.

If the bending moments for this structure are those calculated on page 162 then the stresses in the structure can readily be calculated. Thus, the still water stresses are as follows:

$$\text{Stress in deck} \quad = \frac{797}{5.4} = 147.6 \, \text{MN/m}^2$$

$$\text{Stress in bottom} = \frac{797}{8.16} = \quad 97.7 \, \text{MN/m}^2$$

The wave bending stresses are:

$$\text{Stress in deck} \quad = \frac{719}{5.4} = 133.1 \, \text{MN/m}^2$$

$$\text{Stress in bottom} = \frac{719}{8.16} = \quad 88.1 \, \text{MN/m}^2$$

Thus:

Total stress in deck $\quad = 147.6 + 133.1 = 280.7 \, \text{MN/m}^2$ compression
or $147.6 - 133.1 = \quad 14.5 \, \text{MN/m}^2$ compression

Total stress in bottom $= \quad 97.7 + \quad 88.1 = 185.8 \, \text{MN/m}^2$ tension
or $\quad 97.7 - \quad 88.1 = \quad 9.6 \, \text{MN/m}^2$ tension

With a material such as mild steel where the ultimate strength lies between 400 and 500 MN/m^2 the sagging stresses would be much too great. Redistribution of the loading would be one way of reducing the stress and if the 5000 t load were distributed over the whole length of the ship the stresses would be reduced to 133.1 MN/m^2 in the deck and 88.1 MN/m^2 in the bottom.

The other way in which the stress could be reduced would be by adding longitudinal material to the structure. The most economical way of doing this would be by increasing the thickness of the uppermost deck and the bottom, since these parts are furthest away from the neutral axis.

Use of static longitudinal strength calculations

Because of the assumptions made in the static longitudinal strength calculation it has often been stated that the resulting stresses cannot be considered to be real but that they must be treated as comparative only. This comparison can be made between one ship and another or between different loading conditions for the same ship. As far as the former is concerned the problem can be stated as follows. Suppose that for a vessel of a given type the standard strength calculation gives a certain value for the stress and this ship has proved to be satisfactory in service. If a similar ship is being designed and the value of the stress on the basis of the standard calculation is round about the same it could be expected that the proposed ship should prove satisfactory. This procedure does not, however, tell the designer whether or not a ship is too strong for its particular purpose and will not therefore lead to economical designs. As far as comparing different conditions for the same ship is concerned the worked example shows that the calculation can be very useful for this purpose and in particular the still water strength calculation gives a good guide.

With regard to the actual values of stress used in conjunction with the standard calculation, different formulae have been proposed from time to time. One which was used at one time was as follows:

$$\left.\begin{aligned} p &= 5\left(\frac{L\text{ft}}{1000} + 1\right) \text{ton/in}^2 \\ p &= 77.2\left(\frac{L\text{m}}{304.8} + 1\right) \text{MN/m}^2 \end{aligned}\right\} \tag{7.5}$$

An alternative formula is

$$\left.\begin{aligned} p &= \sqrt[3]{L_{ft}} \text{ ton/in}^2 \\ p &= 23\sqrt[3]{L_m} \text{ MN/m}^2 \end{aligned}\right\} \tag{7.6}$$

These two formulae, although apparently widely different, give very much the same stress.

Many of the large vessels which were outside the rules of classification societies were designed on the basis of formulae of this type. In recent years, however, classification societies have developed their own standards as far as allowable stress is concerned. In their rules stress as such seldom appears, but some stress standard is implied. Most societies nowadays specify a value of the section modulus of the midship section of the ship, which is often based on the still water bending moment with some correction factor to take into account the wave bending moment. For example the current rules of Lloyd's Register of Shipping (1973) give the following rule for the determination of the section modulus:

The section modulus at deck and keel is not to be less than the greatest of the following values:

(a) M cm^3

(b) $\dfrac{M}{3} + 92$ S.W.B.M$_\text{L}$. $(C_\text{B} + 0.20)$ cm^3

(c) $\dfrac{0.85\, M_1}{3f} + 92$ S.W.B.M$_\text{B}$. $(C_{\text{B}1} + 0.20)$ cm^3

where

$M = f\,\text{KB}\,(C_\text{B} + 0.70) \times 10^5$

$M_1 = f\,\text{KB}\,(C_{\text{B}1} + 0.70) \times 10^5$

$f = 0.85$ for the class 100 A1 or 100 A1 'strengthened for heavy cargoes'

$= 0.90$ for the class 100 A1 'strengthened for heavy cargoes — specified holds may be empty'

The value of K is given in a table and is dependent upon the length of the ship

C_B is the block coefficient at the load draught or 0.045 L, whichever is the greater, but is not to be less than 0.60

$C_{\text{B}1}$ is the block coefficient at the ballast draught but is not to be less than 0.60

B is the breadth moulded

S.W.B.M$_\text{L}$. is the maximum still water bending moment in tonne metres hogging or sagging in loaded conditions

S.W.B.M$_\text{B}$. is the maximum still water bending moment hogging or sagging in ballast conditions.

Lloyd's Rules note that when the required section modulus is M the maximum associated still water bending moment in loaded conditions will be $7.25\, M/(C_\text{B} + 0.20) \times 10^{-3}$ t m, which corresponds to a still water stress of

$$7.25/(C_\text{B} + 0.20)\,\text{kg/m}^2$$

Having determined the required section modulus, how this is achieved is left largely to the ship designer, with certain limitations on scantlings such as the thickness of the deck, bottom and side plating, etc.

Other classification societies have rules similar to those described here. The scantlings determined from rules of this type are for the midship section where the bending moment is greatest. It is the practice to maintain these midship scantlings over a proportion of the length of the ship and this is usually $0.4\,L$, i.e. $0.2\,L$ on each side of amidships. From these points the thicknesses are gradually tapered down to reduced values at the ends of the ship.

Shear stress in ship's structure

Attention has been focused on the bending stress which is created in the structure due to longitudinal bending. It is, however, of importance to consider the shear stress which is generated. For this purpose the simple formula for calculating the shear stress in a beam may be used. This is

$$q = FA\bar{y}/It \qquad (7.7)$$

where

F = shearing force
A = cross-sectional area above some point distance y from the neutral axis of bending
$\bar{y}$ = the distance of the centroid of this area from the neutral axis
I = second moment of area of the complete section about the neutral axis
t = thickness of section at the position y.

Figure 7.9 illustrates for a simple I beam the distribution of shear stress over the section obtained from equation 7.7.

The formula shows that the greatest shear stress occurs at the neutral axis and at the positions in the length where the shearing force is greatest. It also shows that 'vertical' material takes the vast majority of the shear load and in the I beam, for example, it will be found that the web takes more than 90% of the shear, the contribution of the 'horizontal' material represented by the flanges being very small. The flanges of course take most of the bending stress.

In the ship the positions of the maximum shearing forces are at about the ends of the half-length amidships when the ship is at sea amongst waves. High shearing forces may, however, exist at other places in the still water condition in certain conditions of loading.

The vertical material in the ship's structure is provided by the shell plating and by longitudinal bulkheads. It could be expected, therefore, that the shear stress would be greatest in these parts of the structure and limitation of shear stress could be one of the factors which would govern the thickness of side shell and longitudinal bulkheads. The formula quoted can be used for calculating the value of the shear stress but in certain structures it must be used with reserve. It could not, for example, be used in the form quoted in structures where the shear load is shared between the shell and longitudinal bulkheads, and a more detailed study of the shear flow would be necessary in such cases.

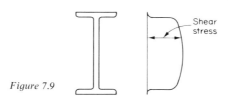

Figure 7.9

The presence of shear in the structure has two effects. In the first place shear distorts the sections so that the conditions upon which bending theory is based are no longer fulfilled, and this has the effect of altering the bending stress distribution across the section. The analysis of this problem is beyond the scope of this book but the general effect as far as the ship's structure is concerned is to increase the bending stress at the corners of the section, i.e. the deck edges and the bilge, and to reduce the stress at the centre of the deck and the bottom. The effect is only appreciable when the ratio of length to depth of the structure is small.

The second effect is an increase in the deflection of the structure due to the presence of shear stress so that the total deflection is greater than would be obtained from ordinary bending theory. This effect is relatively unimportant except in problems such as the calculation of the natural frequency of vibration of the hull. It will be seen later in Chapter 11 that the influence of shear deflection is to lower the frequency as calculated from bending theory.

Calculation of deflection

The deflection of the ship's structure can be calculated when a curve of bending moment is available. The relation between bending moment and the radius of the curve of the structure according to bending theory is given by

$$\frac{M}{I} = \frac{E}{R}$$

where E is the modulus of elasticity of the material and R is the radius of curvature. If y is the deflection of the structure measured from the chord joining the two ends of the ship at some position x from one end then it can be shown that

$$R = -\frac{\{1 + (dy/dx)^2\}^{3/2}}{(d^2y/dx^2)}$$

In ships' structures, and for that matter in structures generally, small deflections only have to be considered so that $(dy/dx)^2$ is small compared with unity and the expression for radius of curvature can be written

$$R = -\frac{1}{d^2y/dx^2} \quad \text{or} \quad -\frac{d^2y}{dx^2} = \frac{1}{R}$$

and the bending equation can be rewritten

$$-EI\frac{d^2y}{dx^2} = M \tag{7.8}$$

It follows that the deflection y can then be obtained from

$$y = -\iint \frac{M}{I}\,dxdx + Ax + B$$

where A and B are constants of integration.

In the case of the ship it is necessary to calculate I at a series of positions along the length and plot a curve of M/I which would then have to be integrated graphically. The values of the constants can be obtained by inserting the end conditions. If as stated above the deflection is measured from the chord joining the two ends of the ship then it will be zero at both of these positions. This means that B $= 0$ and

$$-\iint_L \frac{M}{EI}\,dxdx + AL = 0$$

from which

$$A = \frac{1}{L}\iint_L \frac{M}{EI}\,dxdx$$

so that

$$y = -\iint \frac{M}{EI}\,dxdx + \frac{x}{L}\iint_L \frac{M}{EI}\,dxdx \tag{7.9}$$

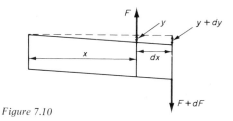

Figure 7.10

This is the deflection due to bending. The deflection due to shear is not so easily obtained. It can be approximated by assuming that the shear stress is uniformly distributed over the web area of the section. Suppose at any position in the length this area is A_w, then

$$\text{Shear stress} = \frac{F}{A_w}$$

and from *Figure 7.10* it will be seen that the shear deflection dy over a small element length dx is given by

$$dy = \frac{F}{AG}dx$$

where G is the shear modulus. Hence

$$\text{Shear deflection} = \int \frac{F}{A_w G}dx + C$$

where C is a constant which will be zero if the deflection is measured from the chord joining the two ends.

If the deflections at the centre of the ship due to bending and shear are called δ_m and δ_s respectively then the total deflection

$$\delta_t = \delta_m + \delta_s = \delta_m(1 + \delta_s/\delta_m) = \delta_m(1 + r_s)$$

where $r_s = \delta_s/\delta_m$. The importance of shear deflection can be seen from the value of r_s which varies with the square of the depth/length ratio and would generally be between 0.1 and 0.2 for normal ships.

The dynamic longitudinal strength problem

At the beginning of this chapter it was stated that the problem of longitudinal strength was essentially a dynamic one and the development of the study of the subject in recent years has been in this direction. It has of course long been recognised that ship motions such as pitching and heaving have an effect on longitudinal strength and methods have been developed whereby the ordinary static calculations can be modified to take them into account. To do this a theory of heaving and pitching was developed from which it was possible to determine the accelerations caused by the two motions so that the dynamic forces on the masses could be calculated.

Present day theory has extended this idea and the forces acting on a strip of the ship at any point in the length have been calculated. This is the so-called 'strip theory' in which forces due to the static buoyancy, weight, accelerations, damping and the hydro-dynamic forces occasioned by the motion of the strip through the fluid have been evaluated for a ship in regular waves. Knowing these forces, the shearing force and bending moment can be obtained by summing the forces and moments over the length of the ship. The theory can be extended to irregular seas by analysing the sea into a number of regular components and finding the ship response to each of these components. The total response is then the sum of these component responses so that it is possible to find the response of the ship (in this case the shearing force and bending moment) to any given sea state. The results obtained have shown good agreement with experimental results, particularly on models. The calculations involved in this approach to the longitudinal strength problem are very complex and will not be considered further here. It is, however, possible to approach the dynamic longitudinal strength problem from a practical point of view, with some use of statistical theory.

Since the end of World War II a great deal of strain and other data have been collected on ships at sea. As far as strain is concerned it has been made possible by the use of what are called 'statistical strain gauges'. They record the number of times that the strain lies in a given range, from which it is possible to deduce the stress corresponding to these strains, since stress = strain × modulus of elasticity for the material. Knowing the section modulus for the ship it is then possible to convert the stresses into bending moments. If records of this sort are made for a ship over a sufficiently long period it is possible to cover practically all the degrees of weather intensity which it is likely to meet in its lifetime.

It is usual to take records of strain as a series of short term records of 20–30 min duration, over which period it can be assumed with fair certainty that the weather intensity remains constant. *Figure 7.11* shows the results of a short term record plotted in histogram form, where the base is stress x and the ordinate is the number of reversals of stress ΔN expressed as a fraction of the total number of reversals N in the record and divided by the stress interval Δx in which the ΔN results lie. The stresses recorded are peak-to-peak values and really represent the change in stress from hog to sag as the waves pass the ship. The stresses concerned here are wave stresses and obviously do not include the still water stress, which must be calculated separately.

As was shown in Chapter 6, records of this sort can be expressed by the Rayleigh distribution, which in this case can be written as

$$f(x) = \frac{2x}{E} \exp(-x^2 E) \qquad (7.10)$$

where $E = (\Sigma x^2)/N$
which is approximately equal to $(\Sigma \bar{x}^2 \Delta N)/N$, $\bar{x}$ being the mean value of the stress in the interval of stress Δx.

Figure 7.11

From equation 7.10 the probability that the stress will exceed any value x_j can be determined by integrating from x_j to infinity. Hence it will be found that

$$Q(x > x_j) = \exp(-x_j^2/E) \qquad (7.11)$$

If logarithms of the two sides are taken then $\log Q(x > x_j) = -x_j^2/E$ so that a plot of $\log Q$ to a base of x_j^2/E would be a straight line.

It is possible to analyse any short term record of stress in this way, so that if such records are taken over a long period there will be a whole series of results all related to particular weather intensities for which different values of E will be obtained. Weather intensities can be measured in terms of the Beaufort scale, in which different intensities are given from 1 to 12, and in analysing short term records they could be divided into twelve groups. Most investigators, however, consider that this division is too fine and that it is better to use five divisions of weather intensity to cover the range. These are shown below, where their relation to the Beaufort scale and actual sea conditions is indicated.

Weather group	Beaufort number	Sea conditions
I	0–3	Calm or slight
II	4–5	Moderate
III	6–7	Rough
IV	8–9	Very rough
V	10–12	Extremely rough

Because weather intensity will vary within particular weather groups it follows that E will vary also and it has been found that this variation follows a Gaussian or normal distribution. Having all the short term records within the group it is possible to evaluate the probability that E will have a certain value within that group. Often the results are analysed in terms of $\sqrt{E}$ rather than E, where $\sqrt{E}$ then becomes the root mean square of the stresses in a short term record within a particular weather group.

The probability density of E is then given by

$$f\sqrt{E} = (2\pi S^2)^{-\frac{1}{2}} \exp\{-(\sqrt{E} - m)^2/2S^2\} \qquad (7.12)$$

where
$$S = \left\{ \left(\sum_{k=1}^{k=N} \sqrt{E_k} - m \right)^2 \middle/ N \right\}^{\frac{1}{2}}$$

is the standard derivation and $m = \sum_{k=1}^{k=N} \sqrt{E_k} \middle/ N$

and N is the number of values of $\sqrt{E}$ within the records.

The probability of $\sqrt{E}$ exceeding some particular value $\sqrt{E_j}$ is then found by integrating, i.e. by finding the area under the curve from $\sqrt{E_j}$ to infinity so that

$$F(\sqrt{E} > \sqrt{E_j}) = \int_{\sqrt{E_j}}^{\infty} f\sqrt{E}\, d\sqrt{E} \qquad (7.13)$$

Long term probablity distribution

The total probability that the stress will exceed a value x_j in a particular weather group, say i, is found by combining the Rayleigh probability and the normal probability. Thus the total probability that the stress will exceed x_j in weather group i is given by

$$Q_i(x > x_j) = \int_{-\infty}^{+\infty} (2\pi S_i)^{-\frac{1}{2}} \exp\{-(\sqrt{E_k} - m_i)^2/2S_i\} \exp\{-x_j^2/E_k\} d\sqrt{E}$$

$$(7.14)$$

If the weather remained constant, i.e. if the results could all be considered to be in the same group, then this expression would give the total probability that the stress would exceed a value x_j. However, the ship will spend part of its time in different weather groups so that it is necessary to introduce finally the probability that the ship will be in a particular weather group. This can only be determined by collecting weather data for various parts of the world and much information has been collected in recent years. Obviously for a particular ship the probability of encountering various weather conditions will depend very much on its route. As examples the following data may be quoted, in which the figures represent the fraction of time in each weather group.

	Weather group				
	I	II	III	IV	V
General routes	0.51	0.31	0.14	0.035	0.005
Tanker routes	0.71	0.23	0.055	0.00038	0.0002

It will be seen that the vast part of the life of a ship is spent in moderate weather and a very small part in extremely rough weather. If the fraction of the life of the ship spent in a particular weather group is called P_i then the final probability that the stress x will exceed x_j is given by

$$Q(x > x_j) = \sum_{i=1}^{i=V} P_i Q_i(x > x_j) \qquad (7.15)$$

Results obtained from equation 7.15 can be plotted as shown in *Figure 7.12* where the base is log $Q(x > x_j)$ and the vertical scale is x_j.

One way of interpreting these results is to say that $Q(x > x_j) = 1/n(x_j)$ where $n(x_j)$ is the number of cycles of stress during which the stress x_j is expected to be exceeded once. It is thus possible to put a second scale on the base giving the number of cycles of stress in which a given value may be exceeded once. The next step is to interpret this in terms of the life of the ship. It can be done by estimating the number of stress cycles which the ship is likely to experience in its life based on information from the records of strain. For example, if from the records taken for a certain ship the length of a stress cycle is t seconds then the number of cycles per hour is

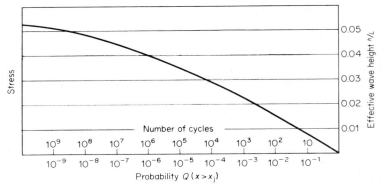

Figure 7.12

$3600/t$ and per day is $(3600 \times 24)/t$. If d is the days per annum that the ship is at sea and y is the lifetime of the ship in years then

$$\text{Total number of stress cycles} = \frac{3600 \times 24}{t} \times d \times y$$

As an example suppose $t = 6\,\text{s}$, $d = 300$ and $y = 25$. Then

$$\text{Total number of stress cycles} = \frac{3600 \times 24}{6} \times 300 \times 25$$

$$= 108\,000\,000 = 1.08 \times 10^8$$

The stress which might be exceeded once in the lifetime of the ship would then be obtained by setting of 1.08×10^8 cycles on the base and reading off the stress. Put the other way round, there is a probability of approximately 10^{-8} that this stress will be exceeded once in the life of the ship.

The probability of occurrence of 10^{-8} of a particular stress has been largely accepted as a reasonable value for ship structures and the problem in design on this basis would be to state what stress is acceptable and to design the structure so that the probability of its occurrence had this value.

Effective wave height

The method described above has shown how the overall strength of the structure of a ship can be examined on a probabalistic rather than a deterministic basis as would be the case in the standard static strength calculation which was described earlier. It is of interest, however, to examine how the two types of calculation can be related.

Suppose a series of static calculations are carried out for the ship poised on a wave of its own length and with varying heights. For each of these height/length ratios h/L a value of the peak-to-peak stress can be obtained which can be set off on the diagram in *Figure 7.12* and each labelled with its particular value of h/L. By drawing horizontal lines at these various levels it will be found that they intersect the probability curve at various values of the probability. It is thus possible to determine the probability of occurrence of a value of the stress determined from the simple static calculation for a particular value of h/L. h/L may be called the 'effective wave height ratio'. It will be found that at the value of $h/L = 0.05$ which was formally used in the static calculation the probability of encountering the corresponding stress is very low, and while it is not possible to generalise on this point it would seem that static calculations overestimate the stress which is likely to be obtained in service.

Use of model experiments for determining bending moments

The probability approach to structural strength described in the earlier section is very suitable for the purpose of analysing results obtained on ships at sea. It is not, however, so good from a design point of view. Another approach to the problem is by means of model experiments.

Over the last 20 years model experiments have been carried out in towing tanks to determine longitudinal bending moments.[2] The models are of two types. One is the hinged model in which the relative movement of the two parts is measured as the model moves through waves. The other is usually made of metal and the strain in the model is measured by strain gauges. In either case the models are calibrated by applying static moments.

Suppose that a model is run in a towing tank in regular waves of given height and length. The type of strain record obtained is cyclic in nature and under steady conditions would approximate to a sine curve. From the record and the static calibration it is possible to deduce the bending moment for any model speed. The dynamic behaviour of the model will depend upon the frequency of encounter of the waves with the model, which in turn will depend upon wave length and model speed. The bending moment will also depend upon the wave height and it is usually assumed to be directly proportional to the height h. It is thus possible to plot a coefficient

M/h (a response amplitude operator) to a base of frequency of encounter ω_e. In order to make this coefficient non-dimensional it needs to be divided by the square of the length, the breadth, the density, and the acceleration due to gravity, so that

$$\text{Bending moment coefficient} = \frac{M}{\rho g \, BL^2 h} \qquad (7.16)$$

The coefficient can be applied to full size ships, assuming that there is no scale effect. If experiments are carried out for different frequencies of encounter or different wave lengths a curve of the type shown in *Figure 7.13* can be plotted where the base can be the frequency ω_e or λ/L, i.e. the wave length/ship length ratio. Now

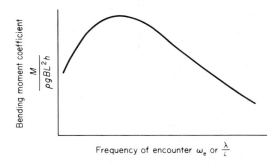

Figure 7.13

suppose that a sea spectrum is available. The ordinate of this spectrum is the square of the height of the wave component in the spectrum at some frequency ω. Hence, if the square of the ordinate of the bending moment coefficient curve at the same frequency is multiplied by the wave spectrum ordinate then the response of the ship to that frequency will be obtained, i.e. if the response spectrum ordinate is $E(\omega_e)$ and the wave spectrum ordinate is $S(\omega_e)$, the bending moment coefficient ordinate being $R(\omega_e)$, then

$$E(\omega_e) = S(\omega_e) \times R(\omega_e)^2$$

The total response of the ship is then found by summing these for all the frequencies in the wave spectrum. Thus

$$\text{Total response} = \int_0^\infty E(\omega_e)\,d\omega_e$$

The total response obtained in this way is the E in the Rayleigh distribution.

To find the long-term bending moment in the structure, the wave spectrum assumed would have to be appropriate to the various weather groups which were discussed earlier and allowance should strictly speaking be made for the fact that E will vary in a particular weather group. If information is not available concerning the variation of E in a particular weather group it would have to be assumed. The long-term probability that the stress deduced from the bending moment would exceed some value x_j would then be obtained as in the previous method by making use of data on the proportion of time the ship spends in the various weather groups.

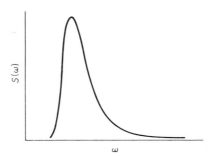

Figure 7.14

Wave spectra are usually given for a fixed point and the base to which they are plotted is ω, the frequencies of the wave components in the spectra. When a ship is moving through the sea a correction has to be made for the frequency of encounter ω_e. Consider the wave spectrum shown in *Figure 7.14*. The total energy is given by $\displaystyle\int_0^\infty S(\omega)\,d\omega$ where $S(\omega)$ is proportional to the wave height squared. If the ship is moving through the waves with velocity V the frequency of encounter becomes

$$\omega_e = \omega - \omega^2 V / g \qquad (7.17)$$

The spectrum for a given sea must have the same total energy whether the ship is stationary or moving. In terms of frequency of encounter:

$$\text{Energy} = \int_0^\infty S(\omega_e)\,d\omega_e$$

Now $d\omega_e = \{1 - (2\omega V / g)\}\,d\omega$ so that

$$\int_0^\infty \left(1 - \frac{2\omega V}{g}\right) S(\omega_e)\,d\omega = \int_0^\infty S(\omega)\,d\omega$$

Hence all ordinates in the modified spectrum must be multiplied by a factor

$$\frac{1}{1 - 2\omega V/g}$$

i.e.

$$S(\omega_e) = \frac{S(\omega)}{1 - (2\omega V/g)} = \frac{S(\omega)}{1 - (2V/g)[1 \pm \sqrt{(1 - \{4\omega_e V/g\})}]/(2V/g)}$$

$$= \frac{S(\omega)}{\sqrt{\{1 - (4\omega_e V/g)\}}} \tag{7.18}$$

Thus the spectrum has to be modified by altering the frequency from ω to ω_e according to equation 7.17 and by multiplying the ordinate at ω by $1/\sqrt{\{1 - (4\omega_e V/g)\}}$.

Horizontal bending and torsion

Attention has so far been centred on the longitudinal bending of a ship in the vertical plane only. Unless the ship is moving head on into long-crested seas two other effects can arise. The first is longitudinal bending in a horizontal plane and the second is twisting or torsion of the ship about its longitudinal centre line.

Horizontal bending will arise when a ship is moving obliquely across waves or when the waves are short-crested. Under these circumstances horizontal forces are generated which can result in horizontal acceleration of the masses making up the total mass of the ship. There will be no gravitational components of force in this case. There is no easy way of calculating the horizontal forces akin to the vertical force calculations, and they can only be evaluated by a detailed study of the hydrodynamic forces and the motions such as yawing and swaying which generate accelerations. In general the horizontal bending moments created are of much less magnitude than the vertical bending moments.

The question of the torsion of ships is one which has assumed some considerable importance in recent years because of the development of ships with large hatchways and the open deck type such as the container ship. Again the problem is a complex one. The torsional moment on the ship arises for the same reasons that horizontal forces exist, but even if the moment is accurately determined the response of the structure is not so readily obtained. This is because the simple torsion theory developed for circular shafts cannot be applied to the section of a ship which is really a hollow box girder. A great deal of research work has been carried out on this subject but a discussion of the results of such investigations is outside the scope of the present work.

Resistance of ship structures to buckling

So far the calculation of the longitudinal bending and shear stresses acting on the structure of a ship have been dealt with. It is necessary now to consider how the various parts of the structure can resist compressive stresses.

Whilst a part of the structure in tension may be capable of withstanding a very considerable stress, if the same members are subjected to compression it may collapse due to buckling at a very much lower value of stress. If a structural member is loaded axially in compression it is possible to distinguish what is known as a 'critical load' below which buckling will not take place but above which lateral deflection will develop and collapse will eventually result. For an ideally straight column with end loading this critical load is given by

$$P_{cr} = \frac{\pi^2 E I}{l^2}$$

where

l = length of the column
E = modulus of elasticity of the material
I = second moment of area of the cross section.

This is Euler's formula and assumes that the column is hinged-ended. Dividing by the cross-sectional area A it is possible to get this in terms of a critical stress p_{cr} rather than a critical load. Thus

$$p_{cr} = \frac{\pi^2 E I}{A l^2} = \frac{\pi^2 E}{(l/k)^2}$$

where k is the radius of gyration of the section.

The two important factors in this formula are the modulus E and the ratio l/k, the 'slenderous ratio' as it is sometimes called.

This formula can be applied to the plating of the structure of a ship. If for example a strip of plating between beams is considered in a transversely framed ship, then taking S as the spacing this will be the length of the strip, and if t is the thickness $k^2 = t^2/12$ so that

$$p_{cr} = \frac{\pi^2 E t^2}{12 S^2} \qquad (7.19)$$

which shows that the span thickness ratio is an important factor governing the buckling strength.

When a panel of plating is considered which is supported around all four edges as shown in *Figure 7.15* then support along the edges parallel to the application of the load has an important influence on the buckling stress.

It can be shown that for a long panel the buckling stress is approximately equal to

$$p_{cr} = \frac{\pi^2 E t^2}{3(1 - v^2) b^2} \qquad (7.20)$$

where b is the breadth of the panel and v is Poisson's ratio for the material. Also for a broad panel

$$p_{cr} = \frac{\pi^2 E t^2}{12(1 - v^2) S^2} \left[1 + \frac{S^2}{b^2} \right]^2 \qquad (7.21)$$

where in this case S is the length of the panel. The structures of the

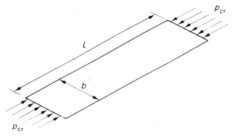

Figure 7.15

decks and the bottom of a ship are divided up into long or broad panels by means of longitudinal or transverse stiffening and it is of interest to compare the relative merits of the two methods of stiffening from a buckling point of view.

The ratio of the buckling stresses for the plating between stiffeners in the two cases for plates of the same thickness and with the same spacing of stiffeners is

$$\frac{4}{[1 + S^2/b^2]^2}$$

In a transversely stiffened panel S/b will be about 1/5 so that the ratio becomes $4/[1 + (\frac{1}{5})^2]^2 = 3.69$ so that it will be seen that in the longitudinal stiffened panel the buckling stress of the plating between stiffeners approaches four times the value for that for transverse stiffening. The advantage of longitudinal stiffening in ships' structures is clearly demonstrated and this form of stiffening has now been adopted almost universally.

The formula quoted for the buckling stress in columns and panels of plating are for absolutely straight members of homogeneous material axially loaded. In practice there is likely to be some initial curvature, particularly in plating. While initial curvature will not affect the elastic buckling stress it will increase the total stress since the member then becomes one which is subjected to a direct load and an eccentric load due to the initial curvature. The result is that the total stress on the concave side of the member may reach the yield stress for the material before instability takes place, and hence on unloading there will be permanent set and consequently damage to the structure. Practical formulae endeavour to take this effect into account. One formula of this type is the Rankine-Gordon formula which gives for the buckling load on a column:

$$P = \frac{f_c A}{1 + C(l/k)^2} \tag{7.22}$$

where

f_c and C are constants dependent upon the material
C is also dependent upon the fixing conditions
A = the cross-sectional area of the column.

A comparison between Euler's formula and the Rankine-Gordon formula is shown in *Figure 7.16*, where it will be seen that at high

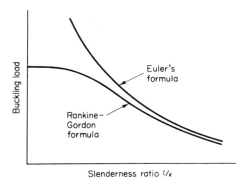

Figure 7.16

slenderness ratios the two give almost the same result while at low slenderness ratios the elastic stability stress is so high that failure due to yielding in direct compression takes place first.

In examining the buckling strength of the deck and bottom structure of the ship it is not sufficient to consider the plating only. The deck and bottom longitudinals must have sufficient strength to prevent their buckling, otherwise the stiffened plating would buckle over a wide area between bulkheads or transverses. This means that some minimum value for the second moment of area of the longitudinals is necessary in relation to their lengths to prevent this happening. Similarly in transversely stiffened decks the beams must have some minimum value of second moment of area to prevent them from bending and thus preventing the plating in between beams developing its full buckling strength.

Strength of transverse structure

Investigation of the strength of the transverse structure of a ship presents many more difficulties than the overall longitudinal strength problem. In both cases correct assessment of the loads on the structure is important but in the case of the strength of the transverse structure the response of the structure is not so easily determined. This is because the structure is highly redundant, which simply means that one or more members could be deleted without the structure necessarily collapsing. In consequence structural analysis becomes difficult.

A first approach to the transverse strength problem is to consider a 'slice' of the structure of one frame space in length in a transversely framed ship and to determine the forces acting upon the component parts of this piece of structure. The parts consist of a floor, a side frame and a deck beam or beams. Each of these parts could be considered separately. For instance, a floor plate with its associated strips of outer and inner bottom plating forms a beam of I section which is subjected to an upward force due to water pressure and a downward force due to cargo or whatever is carried in the space above the floor. The maximum bending moment on a beam depends on the end conditions. If, for instance, the beam is simply supported at its ends then the bending moment is $wl^2/8$, where w is the rate of loading and l is the length. If, however, the ends are fixed the greatest bending moment is $wl^2/12$ and occurs at the ends. When dealing with the floor as a beam it is necessary therefore to assume that it is either simply supported or full fixed at its ends. Neither condition really represents the truth and the degree of fixity will depend upon the rigidity and the loading of the adjacent structure, so that if either of these extreme conditions is assumed the bending moment will to some extent be fictitious. The required section modulus for the floor would be found by dividing the bending moment so derived by the working stress for the material. Because of the uncertainty of the end fixing conditions the results obtained in this method of approach are likely to be rather unreal, but nevertheless the calculation can be useful in comparing one structure with another.

Frames and beams can be treated separately in the same way, the loading on the former being due to water pressure on the outside and some cargo load on the inside if cargo is carried in the space. This load will be somewhat indeterminate for dry cargoes but can be easily calculated for liquid cargoes. The loading on beams is simply that due to cargo, passengers or a weather load for the exposed deck.

The procedure of treating each item of the transverse structure separately is unsatisfactory, as has been stated, because of the doubt about the end fixing conditions of the individual members. This led to the consideration of the complete transverse ring of structure as a whole and methods were developed for dealing with structure in this way.[3,4] The structure is a redundant one and the methods used in structural design for analysing such structures were employed. The method described in the two papers quoted was a strain energy method or the so-called 'theorem of least work method'. What was assumed was that a slice of the structure one frame space long was in equilibrium under the forces of buoyancy, cargo load and weight of structure. The forces and moments at any point in the structure can then be obtained in terms of these loads and certain unknown forces and moments at various points in the structure. These unknowns are required to have such values that the potential energy in the structure is a minimum. This potential energy is the strain energy of bending shear and direct loading. By calculating this for the complete ring of structure and differentiating with respect to the unknowns and equating to zero equations are obtained from which the unknown forces and moments may be determined and hence shearing force and bending moment at any point in the transverse ring of structure can be calculated.

This method of calculating the strength of the transverse structure was a great advance on the simpler method already discussed but required a great deal of calculation which was very tedious in pre-computer days. An easier method made use of what was called 'moment distribution' originally developed by Hardy Cross for land structures[5] and applied by others to ships.[6,7,8] It is not proposed to deal with this here but both this and the strain energy method should give the same results if the same basic assumptions are made.

The methods described here for dealing with transverse strength use the plane frame approach, i.e. the transverse structure being investigated is in two dimensions only and it is assumed that the adjacent structure in the fore and aft direction has no influence on it. In actual fact it has been seen that there are longitudinal girders in both the deck and the bottom of the structure and in order that a plane frame solution can be obtained these must be either ignored or considered to be completely rigid, in which latter case they form points of support for the transverse structure. Neither of these assumptions is correct because the longitudinal girders are themselves elastic and will deflect under the loads generated at the points of intersection between the longitudinal and transverse members. It will be seen, therefore, that the transverse strength problem becomes a three-dimensional one and a 3D approach to the problem becomes essential.

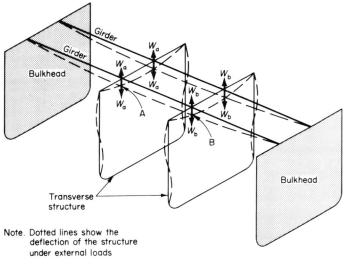

Figure 7.17

How this can be done will be apparent by considering *Figure 7.17*. A block of structure between two transverse bulkheads is considered. These bulkheads can be regarded in the first instance at any rate as being rigid since they are continuous sheets of plating extending right across the ship and stiffened in some way. Consider now a transverse ring of material which intersects a girder at some point A. When the structure is loaded the girder will deflect and so will the transverse material. At A the two deflections will be equal and a concentrated reaction, say W_a, will be generated. At some other position between the two bulkheads the transverse ring of material will intersect the girder at a point B and the condition for equal deflection will generate some other concentrated reaction W_b. There will thus be a whole series of concentrated reactions wherever a transverse ring of material intersects a longitudinal, and their values will be such that the deflection of the longitudinals and transverses must be equal at the intersections. It is thus possible to set up a system of linear simultaneous equations in these unknown reactions, there being as many equations as there are unknowns. The equations can be solved readily by means of computer. Only the briefest account has been given here of the 3D approach to transverse strength. The method has been developed by Yuille and Wilson[9] and the present author has demonstrated how longitudinals can affect the forces and moments on the transverse structure of a ship.[10]

The 3D approach to transverse strength is the direction in which study of the subject has gone in recent years. There is, however, still some difficulty in applying the method as outlined here. In modern

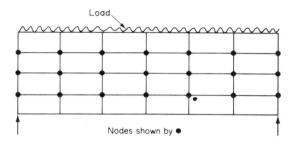

Figure 7.18

ships many of the parts of the transverse structure consist of plate girders which are very deep in relation to their lengths, and there are also large brackets attaching the parts of the structure. For these reasons it is inaccurate to regard the girders as simple beams and as the greatest bending moments often occur at the ends of these members in way of the brackets it is difficult to translate these into stresses in the structure. This has led to the introduction of a technique in which continuous structures are split up into a series of small but finite elements. The process is known as the 'finite element method' and has been adopted in ship structures from other disciplines. It can be described briefly by reference to *Figure 7.18*.

This figure shows a plate girder supported at its ends and loaded in some way. The simple beam approach would be to treat the girder as a whole and to work out the bending and shear stress from ordinary simple theory. The finite element method, however, would split the girder up into a network of small elements as shown by the horizontal and vertical lines. In this case the elements are simply two-dimensional and are of rectangular shape, but triangular elements could also be used. The method assumes that the elements are connected to one another at their corners only and these points are called 'nodes'. When the girder is under load it will be seen that the elements will distort in their own plane so that forces will be generated at the nodes. One of the conditions to be satisfied is that the nodal displacements of elements which are attached to each other must be the same and this condition together with the boundary conditions enable the nodal forces to be determined. It is necessary to relate nodal forces and displacements and to do this assumptions have to be made concerning the strain in the elements. It is thus

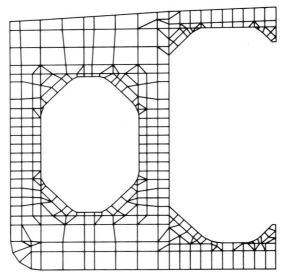

Figure 7.19

possible, when the nodal forces have been calculated, to determine the stress in the elements, and the state of stress so determined should approximate to the exact solution. Generally the finer the mesh the closer will be the result to the correct solution. *Figure 7.19* shows how the transverse structure of an oil tanker might be split up to use the finite element method. The method has proved very successful in the analysis of complex structures of this type and a more detailed discussion of it will be found in a paper by Kendrick.[11]

An attempt has been made to show how the study of the strength of the transverse structures of a ship has developed. The fact that the deflection of longitudinal members of the structure is taken into account in assessing the stresses in the transverse material would suggest that the loads in these members due to the overall longitudinal bending of the ship might influence the problem. This would seem to indicate that the two subjects of longitudinal bending and transverse bending of the structure are related and should not be treated as independent of one another. It is probable that future study of the strength of ships will move along these lines but it is still convenient to treat the two problems separately at the present time.

Structural discontinuities

A structural discontinuity arises where there is a sudden change in section of a structural member. At these positions simple theory will not predict the value of the stress, and while clear of the discontinuity the stress can be determined quite accurately by ordinary theory, at the change in section the local stress may be several times the value so calculated.

A simple example is the case of a tension member which carries a load W, the section of the member changing suddenly at some point from a to A. The stress on one side of this discontinuity is W/a and on the other side W/A but at the change in section the stress may be very much greater. Although the stresses W/a and W/A may be well within the safe limit for the material the local stress at the change in section may be several times either of these values and may therefore exceed the yield stress for the material. Local failure could then possibly take place and cracking of the material may occur. The only way to avoid these high local stresses where changes in section are unavoidable is to make the change as gradual as possible.

In ships' structures there are many places where changes in section take place and they need to be considered very carefully to avoid high local stresses.

Examples of severe discontinuities are the ends of large partial superstructures. Here there are sudden changes in the section resisting longitudinal bending and high local stresses arise. To avoid them the plating of the sheerstrake and deck stringer are thickened locally and the superstructure itself is graded down gradually into the rest of the structure.

Discontinuities occur at the corners of deck openings, particularly at the uppermost continuous deck which is most highly stressed. The practice is to put most of the material resisting bending into the plating abreast the openings and reduce the thickness of the plating within the line of openings. This has the effect of reducing the sudden change in going from a section clear of an opening to a section in way of an opening. It is usual to make the corners of all openings elliptical or parabolic or some other shape. In addition insert plates of greater thickness than the deck may be required at the corners of openings in the strength deck.

Other discontinuities occur where structural members, such as girders, end abruptly. They should be graded down gradually into the surrounding structure.

The importance of avoiding discontinuities cannot be over-stressed. Cracks may be initiated at these places in a structure which is otherwise quite moderately stressed and they could eventually lead to more general failure of the structure.

Materials used in ships' structures

Mild steel

When the use of wood was discontinued for large ships' structures in the nineteenth century, iron first of all took its place but this was quite rapidly succeeded by steel from about the latter quarter of the last century, which continued in use successfully up to the beginning of World War II. The steel used was mild steel and classification societies based their rules on this material. Mild steel is essentially iron with the addition of certain alloying elements in small percentages such as carbon, manganese and silicon. These have the effect of increasing the strength of the basic material, iron.

A stress/strain curve for mild steel shows the characteristics of the curve in *Figure 7.20*, which is for a tensile test on a structural specimen. There is a straight portion where stress is proportional

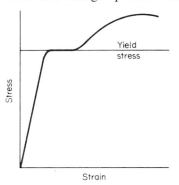

Figure 7.20

to strain, beyond which there is yielding of the material with a rapid increase in strain for no increase in stress. The curve then picks up again and failure eventually takes place at some value of the stress based on the original cross-sectional area of the specimen which is called the 'ultimate strength'. The stress at yield (the yield stress) is usually greater than one-half of the ultimate strength. The test specimen reduces in cross-sectional area when strained beyond the yield point and usually shows a considerable necking down locally at fracture. The ratio of the elongation at fracture measured over a given length (the gauge length) is taken as a measure of the ductility of the material. The ratio of stress to strain over the straight part of the curve is called the 'modulus of elasticity' or Young's modulus. Over the range where stress is proportional to strain the material is considered to be elastic, i.e. if the specimen is loaded and then the load is removed the strain will return to zero.

The properties of the mild steel used in shipbuilding are as follows:

Ultimate tensile strength	$400\text{–}495 \, \text{MN/m}^2$	$(26\text{–}32 \, \text{tonf/in}^2)$
Yield stress	$230\text{–}250 \, \text{MN/m}^2$	$(15\text{–}16 \, \text{tonf/in}^2)$
Elongation on a gauge length of $5.65 \sqrt{S_0}$ ($S_0 =$ cross-sectional area)	22%	
Modulus of elasticity	$201\,000\text{–}208\,000 \, \text{MN/m}^2$	$(13\,000\text{–}13\,500 \, \text{tonf/in}^2)$
Weight per unit volume	$76\,975 \, \text{N/m}^3$	$(490 \, \text{lbf/ft}^3)$

In the years before World War II very little more was required by classification societies than the simple tensile test and certain bend tests to determine the quality of steel to be used in ships, and in general this worked satisfactorily. During the war, however, there were a number of cases of cracking in welded ships, some of which led to complete structural failures especially at low temperatures. These fractures were brittle in character and often occurred at moderate stress levels which should not have caused failure. Thus, the brittle fracture problem made itself evident and resulted in extensive investigations on the quality of material and structural design both in the UK and the USA. Out of this emerged amongst other things the fact that the manganese/carbon ratio in steel had a marked influence on the brittle fracture problem and it was agreed that this ratio should not be less than 2:5. Also, the ability of steel to resist brittle fracture could to some extent be measured by the energy required to fracture a standard notched specimen. The Charpy V-notch specimen was chosen for this purpose and classification societies gave values for the energy for the higher grades of steel. A conference of classification societies led to the adoption of five grades of steel of varying abilities to resist brittle fracture, the highest grade being required for thick plate in highly stressed parts of the structure. The work which has been carried out on the brittle fracture problem has led to the development of steels which are not susceptible to this type of failure under ordinary climatic conditions down to a temperature of about $-40\,^\circ$C. Where steels have had to operate at extremely low temperatures (from -150 to $-200\,^\circ$C), as for example in the tanks of liquefied gas carriers, special steels have been developed which do not show a tendency to brittle fracture at these temperatures. These steels contain nickel as an alloying element and the 9% Ni steel is one which is often used for this purpose.

Higher tensile steel

One method of reducing the scantlings of ship structures and thus saving weight is to employ materials which have higher strengths and which can thus permit higher working stresses. There is quite a long history of the use of such materials and as far back as 1907 high silicon steels were used in the Atlantic liners *Lusitania* and *Mauritania*.[12] Some of the pre-war liners also employed these higher strength materials in their upper structural parts. With the increasing size of modern ships such as the oil tanker, the bulk carrier and the container ship, very thick mild steel plate has been required in the decks and bottoms, so that interest has been renewed in the use of high strength materials.

These steels have been called higher tensile steels by Lloyd's Register to distinguish them from really high tensile steels. The steels being considered have strengths in the range 400–620 MN/m^2 (29.5–40 tonf/in^2), with yield stresses ranging from 265–355 MN/m^2 (17–23 tonf/in^2). Over the years since World War II steels of these strengths which are weldable and free from the brittle fracture problem have been developed and are frequently employed in the types of vessels quoted. They are more expensive than mild steel but can justify their use on the basis of the weight saving which can be achieved.

Aluminium alloys

Aluminium alloys which have been used widely in the aircraft industry have found a more limited application in ship structures. Pure aluminium is a weak material which when alloyed with materials such as copper, manganese, silicon or magnesium produces alloys which are sufficiently strong to be of use structurally. Two types of alloys exist, referred to as 'heat treatable' and 'non-heat treatable'. The former owe their properties to a carefully controlled heat treatment whereas the latter require no such treatment. Both types have been used in ships, the heat treatable type at one time being favoured in North America while the non-heat treatable was used in the UK and on the Continent. Nowadays, with the widespread use of welding for joining aluminium, the non-heat treatable type is almost universally used because less reduction in strength is suffered due to the heat generated in welding than with the heat treatable type. The alloys used in shipbuilding are the aluminium magnesium type, in which magnesium is the principal alloying element. For small boat work alloys containing 3–3.5% Mg have been used but in larger ships the percentage is increased to 4–4.9. The former alloy is designated as N5 in the British Standards Classification and the latter as N8. The properties of these materials are as follows.

	N5	*N8*
Ultimate strength	$216 \, MN/m^2$	$278 \, MN/m^2$
	$(14 \, tonf/in^2)$	$(18 \, tonf/in^2)$
0.1% Proof stress	$93 \, MN/m^2$	$124 \, MN/m^2$
	$(6 \, tonf/in^2)$	$(8 \, tonf/in^2)$
Elongation, percentage on 2 in gauge length	18	16
Modulus of elasticity	$69 \, 500 \, MN/m^2$	$69 \, 500 \, MN/m^2$
	$(4500 \, tonf/in^2)$	$(4500 \, tonf/in^2)$
Weight	$26 \, 185 \, N/m^3$	$25 \, 980 \, N/m^3$
	$(166.7 \, lbf/ft^3)$	$(165.4 \, lbf/ft^3)$

Aluminium alloys have a weight little more than one third that of steel, which is one of the advantages of using these materials. They also have a high resistance to corrosion but require careful attention when used in conjunction with other materials. Direct contact with steel and copper and copper-bearing alloys should be prevented, otherwise galvanic corrosion is likely to take place in the presence of sea water. Where it is impossible to avoid the use of other metals with aluminium, effective electrical insulation should be provided between the materials. The aluminium magnesium alloys do not exhibit brittle fracture behaviour even at very low temperature and they are therefore very suitable for the tanks of liquefied gas carriers.

The main disadvantage of aluminium alloys as far as ship construction is concerned is their high cost, but in certain types of ships such as large passenger liners their use is justified because of the weight saving in superstructures, for example.[13] The use of aluminium alloys in the complete hulls of large cargo vessels would seem to be a doubtful economic proposition but the future may see developments in this direction.[14,15]

Glass reinforcement plastics

These materials consist of a relatively weak resin which is reinforced by means of glass fibres. The glass fibres are of very small diameter, being on the average about 0.01 mm. These fibres can be put loosely together to form a roving like an untwisted rope. Alternatively, the fibres can be twisted to form a yarn. The roving or yarn can be woven to form a woven roving which is like a coarse cloth, or a rather finer cloth can be obtained by weaving the yarn. The material is formed by impregnating layers of woven roving or cloth with resin, and laying one layer on top of another to build up the required thickness. The strength of the material depends very much on the ratio of glass to resin, increasing with increase in glass content.

It will be seen that glass reinforced plastic differs from other materials in that it is built or manufactured as required. Where shape is required, as in the construction of the hull of a ship, it is necessary to have a mould in which the material can be laid up. The construction of the mould can be expensive so that the material is more suitable for batch production where several items can be produced from the same mould.

The main advantages of glass reinforced plastics are their light weight, which is something of the order of $15\,720\,\text{N/m}^3$ ($100\,\text{lbf/ft}^3$) for the basic material, their resistance to corrosion, and the fact that they are non-magnetic. Against these must be set the extremely high cost and the low value of the elastic modulus, possibly about 1/15 that of steel.

Glass reinforced plastics have been used extensively in small boat construction and in ships' lifeboats, where they have been employed successfully in preference to other materials such as steel and aluminium alloys. It is unlikely at the present time that they would be used for complete hulls for merchant ships because of cost,[16] but they have obvious advantages in warship construction where weight saving and non-magnetic properties are important. Many of the problems involved in using these materials have been discussed in a Symposium in 1972.[17]

Structural failure

The whole or a part of the structure of a ship may fail in many different ways. Considering the longitudinal bending of the ship as a whole, failure may arise due to the yielding of the material itself, and on the basis of the plastic theory developed by Baker[18] when all the material in a structure in bending has reached the yield stress the structure is deemed to have failed. This concept has been applied to the structure of a ship by Caldwell[19] and on this basis the ultimate strength of the structure, i.e. the bending moment which will cause failure, can be determined. In the plastic theory it is assumed that when the material reaches the yield stress, further loading will produce strain but no additional stress. The situation is then eventually reached where all the material on the tension side of a beam is at the yield stress in tension and all the material on the compression side is at the yield stress in compression.

If it is assumed that these two yield stresses are equal then for equilibrium

$$A_c f_y = A_T f_y$$

where A_c and A_T are the areas in compression and tension respectively. It follows then that $A_c = A_T$ and the plastic neutral axis for the section is at such a position that the cross-sectional area above it is equal to the cross-sectional area below. For a beam of symmetrical cross section, say a simple rectangular section breadth b and depth d, the plastic neutral axis is at $d/2$ above the bottom. The total force in tension or compression in the fully plastic condition is $(bd/2) \times f_y$. The distance between the centroids of these two forces is $d/2$ so that the moment of resistance of the beam at failure is

$$\frac{bd}{2} \times f_y \times \frac{d}{2} = \frac{1}{4} bd^2 f_y$$

The moment which would just cause yielding at the outer surface is $\frac{1}{6} bd^2 f_y$ so that the ratio of the bending moment which would cause failure to that which would just initiate yielding is

$$\frac{1/4\, bd^2 f_y}{1/6\, bd^2 f_y} = \frac{3}{2}$$

The value of this ratio, often called the 'form factor', is dependent upon the shape of the section. In the limit if the beam was considered to have all the material concentrated in the flanges, the web thickness tending to zero, the factor would be unity. For the type of section which exists in a ship, i.e. a box girder, the form factor might be between 1.1 and 1.3 so that when yielding of the material in decks and bottom takes place there is not a great margin before complete plastic failure takes place. In the case of a ship, because the section is not symmetrical in the vertical direction the plastic neutral axis or the equal area axis is not at half-depth. If A is the half-area of the section, and $\bar{y}_c$ and $\bar{y}_T$ are the distances of the centroids of the areas above and below from the neutral axis, then the bending moment to cause failure would be

$$M = f_y A(\bar{y}_c + \bar{y}_T)$$

The assumptions made in this simple approach to ultimate failure of a structure in bending do not visualise buckling of the compression flange. However, in a plated structure it is almost certain that the material on the compression side will buckle before the yield stress is reached. The effect of this is that the bending moment which would cause failure is reduced. Failure due to buckling in a ship's structure is a mode which is likely to take place and there is experimental evidence to suggest that this is so from the few cases where ship structures have been tested to destruction. The bending moment which can cause failure of the capability of the structure is something about which it is not always possible to be precise. For instance, although the yield stress for the material of construction may be assumed to have a certain value, it can vary quite appreciably so that an element of probability that it will have a certain value enters into the problem. Also, although the various parts of the structure may have some nominal scantlings, they too can vary from point to point in the structure. It follows that there is some statistical distribution of the capability of the structure which might be assumed to be Gaussian and a curve such as that shown in *Figure 7.21* can be drawn which shows the probability that the capability will have a certain value. It has already been seen that the loads on the structure cannot be determined precisely so that a probability curve can also be drawn for the load which might be expected, and such a curve is also shown in *Figure 7.21*. If these two curves intersect at some point such as A then failure will occur and the position in the diagram will indicate the probability of this occurrence. This probability can then be related to the number of cycles of stress in the life time of the ship from which the life of the structure can be determined. It is clear that if both the load and the capability curves lie in a narrow band then the probability of failure is small, in which case it would be possible to move the load curve to the right in the diagram which is equivalent to saying that higher average loads can be accepted on the structure. The ratio of the average capability to the average load may be considered as a load factor and the determination of the value of this factor will be dependent on what is regarded as an acceptable probability of failure.

Another possible mode of failure is that of brittle fracture of the material. This was seen to be an important matter during World War II and the succeeding years. While it would be unwise to say

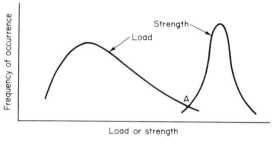

Figure 7.21

that this problem has been resolved completely the factors leading to its occurrence are now more widely understood. Progress in the development of materials which are notch tough at low temperature has been considerable and has been one of the major factors in preventing the occurrence of brittle fracture. Other factors are good design so as to avoid discontinuities and consequent high local stresses, and insistence on high standards of workmanship particularly in regard to welding.

One other possible cause of failure is fatigue. If a specimen of a material such as mild steel is subjected to a static load it will be found that it will fail at a quite high stress level. Should, however, the material be subjected to a cyclic load $\pm P$ say, it will be found that after a number of reversals failure takes place although the maximum stress might be quite moderate. If the value of P is reduced then the specimen would be able to withstand a much larger number of reversals before failure took place. This is the phenomenon known as fatigue and a curve of semi-range of stress plotted to a base of cycles to failure would be as shown in *Figure 7.22*. In some materials such as mild steel and certain aluminium alloys the curve shows a definite tendency to flatten out so that there is a fatigue limit. Others do not show this tendency so that structures designed in these materials and which are subjected to cyclic stresses have a definite life.

As far as the ship problem is concerned it has been seen that the structure is subjected to cycles of stress so that it would appear that fatigue could be a problem. The importance of avoidance of discontinuities in the structure is therefore again evident. The nominal stress in the structure might be well below the safe limit from a fatigue point of view but at a discontinuity there may be a large stress magnification factor so that local fatigue could occur with consequent cracking. Cracks initiated in this way may or may not

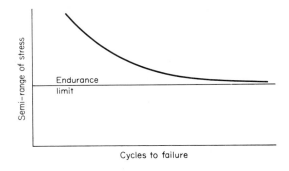

Figure 7.22

be dangerous, but there is always the possibility that such cracks may spread and lead to general failure of the structure. They may also increase the chances of brittle fracture.

Ideally, if there was sufficient knowledge covering the loading of ships' structures and their capability as well as knowledge of the fatigue and brittle fracture behaviour of structural details, it should be possible to use quite high stresses in relation to the yield stress for the material. However, the nominal stress for the structure of a ship must still be dependent to a great extent on what has proved satisfactory in practice. It is only by improvements in structural design to avoid stress concentrations and improvements in the properties of materials themselves that higher working stresses can be accepted. There must still therefore be a margin of safety inherent in the stresses which are permitted to allow for abnormal conditions. The modern view is that a structure cannot be designed to ensure a zero risk of failure. There must always be some probability of failure and the aim is to determine what is the acceptable probability.

REFERENCES

1. Isherwood, J. W., 'A new system of ship construction', *Trans. Instn. of Naval Architects*, 1908
2. Akita, Y., and Ochi, K., 'Model experiments on the strength of ships moving in waves', *Trans. Society of Naval Architects and Marine Engineers, U.S.A.*, 1955
3. Bruhn, J., 'On the transverse strength of ships', *Trans. Instn. of Naval Architects*, 1901
4. Bruhn, J., 'Some points in connection with the transverse strength of ships', *ibid.*, 1905
5. Cross, Hardy, 'Analysis of continuous frames by distributing fixed end moments', *Amer. Soc. of Civil Engrs.*, 1930
6. Hay, W. I., 'Some notes on ships' structural members', *Trans. Instn. of Naval Architects*, 1945
7. Adams, H. J., 'Notes on stresses in tanker members', *ibid.*, 1950
8. Adams, H. J., 'Some further applications of moment distribution to the framing of tankers', *Trans. North East Coast Instn. of Engrs. & Shpbdrs.*, 1952–1953
9. Yuille, I. M., and Wilson, L. B., 'Transverse strength of single hull ships', *Trans. Royal Instn. of Naval Architects*, 1960
10. Muckle, W., 'The influence of longitudinal girders on the transverse strength of ships', *Trans. North East Coast Instn. of Engrs. & Shpbdrs.*, 1960–1961
11. Kendrick, S., 'The structural designs of supertankers', *Trans. Royal Instn. of Naval Architects*, 1970
12. Luke, W. J., 'Some points of interest in connection with the design, building and launching of the *Lusitania*,' *Trans. Instn. of Naval Architects*, 1907
13. Muckle, W., 'Application of light alloys to superstructures of ships', *Trans. North East Coast Instn. of Engrs. & Shpbdrs.*, 1945–1946
14. MacIntyre, D., 'Aluminium ore carriers', *Trans. Society of Naval Architects and Marine Engineers, U.S.A.*, 1952
15. *Study of Aluminium Bulk Carrier*, Ship Structure Committee, USA, 1971
16. *Feasibility Study of a Glass Reinforced Plastic Cargo Ship*, Ship Structure Committee, USA, 1971
17. *Symposium on G.R.P. in Shipbuilding*, Royal Instn. of Naval Architects, 1972
18. Baker, J. F., and Roderick, J. W., 'Plastic theory—its application to design', *Trans. North East Coast Instn. of Engrs. & Shpbdrs.*, 1940–1941
19. Caldwell, J. B., 'Ultimate longitudinal strength', *Trans. Royal Instn. of Naval Architects*, 1965

8

Resistance

A ship when at rest in still water experiences hydrostatic pressures which act normally to the immersed surface. It has already been stated when dealing with buoyancy and stability problems that the forces generated by these pressures have a vertical resultant which is exactly equal to the gravitational force acting on the mass of the ship, i.e. is equal to the weight of the ship. If the forces due to the hydrostatic pressures are resolved in the fore and aft and transverse directions it will be found that their resultants in both of these directions are zero. Consider what happens when the ship moves forward through the water with some velocity V. The effect of this forward motion is to generate dynamic pressures on the hull which modify the original normal static pressure and if the forces arising from this modified pressure system are resolved in the fore and aft direction it will be found that there is now a resultant which opposes the motion of the ship through the water. If the forces are resolved in the transverse direction the resultant is zero because of the symmetry of the ship form.

Another set of forces has to be considered when the ship has ahead motion. All fluids possess to a greater or less extent the property known as viscosity and therefore when a surface such as the immersed surface of a ship moves through water, tangential forces are generated which when summed up produce a resultant opposing the motion of the ship. The two sets of forces both normal and tangential produce resultants which act in a direction opposite to the direction in which the ship is moving. This total force is the resistance of the ship or what is sometimes called the 'drag'. It is sometimes convenient to split up the total resistance into a number of components and assign various names to them. However, whatever names they are given the resistance components concerned must arise from one of the two types of force discussed, i.e. either forces normal to the hull surface or forces tangential to that surface.

The ship actually moves at the same time through two fluids of widely different densities. While the lower part of the hull is moving through water the upper part is moving through air. Air, like water, also possesses viscosity so that the above water portion of a ship's hull is subjected to the same two types of forces as the underwater portion. Because, however, the density of air is very much smaller than water the resistance arising from this cause is also very much less in still air conditions. However, should the ship be moving head on into a wind, for example, then the air resistance could be very much greater than for the still air condition. This type of resistance is, therefore, only to a limited extent dependent on the ship speed and will be very much dependent on the wind speed.

Types of resistance

It was stated above that it is sometimes convenient to split up the total resistance into a number of components; these will now be considered.

The redistribution of normal pressure around the hull of the ship caused by the ahead motion gives rise to elevations and depressions of the free surface since this must be a surface of constant pressure. The result is that waves are generated on the surface of the water and spread away from the ship. Waves possess energy so that the waves made by the ship represent a loss of energy from the system. Looked at in another way the ship must do work upon the water to maintain the waves. For this reason the resistance opposing the motion of the ship due to this cause is called 'wave-making resistance'. With deeply submerged bodies the changes in the normal pressure round the hull due to ahead motion have only a small effect on the free surface so that the wave resistance tends to be small or negligible in such cases.

The resistance arising due to the viscosity of the water is appropriately called 'viscous resistance' or often 'frictional resistance'. The thin layer of fluid actually in contact with the immersed surface is carried along with it but because of viscosity a shear force is generated which communicates some velocity to the adjacent layer. This layer in turn communicates velocity to the next layer further out from the hull and so on. It is clear then that there is a mass of fluid which is being dragged along with the ship due to viscosity and as this mass requires a force to set it in motion there is a drag on the ship which is the frictional resistance. The velocity of the forward moving water declines in going outwards from the hull and although theoretically there would still be velocity at infinite distance the velocity gradient is greatest near the hull and at a short distance outwards the forward velocity is practically negligible. Forward velocity is therefore confined to a relatively narrow layer adjacent to the hull. This layer is called the 'boundary layer'. The width of the layer is comparatively small at the bow of the ship but thickens in going aft, as will be seen from *Figure 8.1*, which shows the fall off in velocity at various positions in the length.

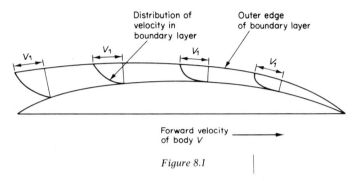

Figure 8.1

The actual thickness of the boundary layer is indeterminate but the point where the forward velocity has fallen to about 1% of what it would be if the water were frictionless is considered to be the outer extremity of the boundary layer. Thus, in *Figure 8.1* where the velocity V_1 of the water relative to the body is 0.99 of what it would be at the same point if the water was frictionless would be the outer edge of the boundary layer.

Theoretical investigations on flow around immersed bodies show that the flow follows the type of streamline pattern shown in *Figure 8.2*. However, where there are sharp changes of curvature on the surface of the body, and partly due to the viscosity of the fluid, the flow separates from the surface and eddies are formed. This separation means that the normal pressure of the fluid is not recovered as it would be according to theory and in consequence a resistance is generated which is often referred to as 'eddy-making resistance'.

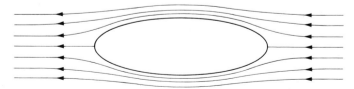

Figure 8.2   Streamline flow around elliptical body

This type of resistance, like wave-making resistance, arises from a redistribution of the normal pressures around the hull in contrast to the frictional resistance which arises because of tangential viscous forces.

The fourth type of resistance is that due to the motion of the above-water form through the air, as has already been mentioned, and could consist of a combination of frictional and eddy resistance.

Theoretical laws of motion of a body through a fluid

Before studying the various types of resistance in detail it is fitting to examine the basic laws upon which the resistance of a body moving through a fluid depends. This can be shown in general terms by the aid of dimensional analysis. This is a very powerful tool, often used for this purpose, although it is unable to state explicitly how the resistance depends upon various factors. It may be stated simply by saying that if some physical quantity depends upon a number of other physical quantities then the latter must be grouped together in such a way as to give the correct dimensions of the former. Thus, if some quantity F depends upon a, b, c, d, say, then F must be some function of these, i.e. $F = f(a, b, c, d)$, and a, b, c, d must be grouped so as to give the same dimensions as F. In carrying out this process it is important that all the quantities upon which F may depend are included in the analysis. In the ship resistance problem the resistance R can be expressed as a function of a number of quantities. They are the size of the ship as represented by some linear dimension l, a number of non-dimensional ratios r_1, r_2, r_3, etc., expressing the form of the ship, properties of the fluid, i.e. the density ρ and the viscosity μ, the static pressure of the fluid p, the speed of the ship V, and the acceleration due to gravity g. Hence

$$R = f(l, r_1, r_2, r_3, \text{etc.}, \rho, V, \mu, g, p) \qquad (8.1)$$

In what follows a comparison between geometrically similar ships will be considered, so that r_1, r_2, r_3, etc., can be left out of the analysis since they must be the same if the ships are geometrically similar. There are several ways of writing down the dimensional equation 8.1. One way is to imagine that the various quantities are raised to certain powers so that equation 8.1 can be written

$$R = f\{l^a \rho^b V^c \mu^d g^e p^f\} \qquad (8.2)$$

All the quantities shown in this equation can be expressed in terms of what are sometimes called the 'fundamental' dimensions of mass M, length L and time T. Thus R is a force and has dimensions

ML/T^2, ρ has dimensions M/L^3, V is L/T, μ is M/LT, g is L/T^2 and p is M/LT^2.

Hence

$$\frac{ML}{T^2} = f\left\{ L^a \left(\frac{M}{L^3}\right)^b \left(\frac{L}{T}\right)^c \left(\frac{M}{LT}\right)^d \left(\frac{L}{T^2}\right)^e \left(\frac{M}{LT^2}\right)^f \right\} \qquad (8.3)$$

Equating the indices of the dimensions on the two sides of the equation gives

$$
\begin{aligned}
1 &= b + d + f & \text{(i)} \\
1 &= a - 3b + c - d + e - f & \text{(ii)} \\
-2 &= -c - d - 2e - 2f & \text{(iii)}
\end{aligned}
\qquad (8.4)
$$

From　　(i)　$b = 1 - d - f$

and from　(ii)　$c = 2 - d - 2e - 2f$

From　　(iii)　when the values of b and c are substituted

$$a = 2 - d + e$$

The expression for resistance then becomes

$$R = f\{ l^{2-d+e} \rho^{1-d-f} V^{2-d-2e-2f} \mu^d g^e p^f \}$$

$$= \rho V^2 l^2 f\left\{ \left(\frac{\mu/\rho}{Vl}\right)^d, \left(\frac{gl}{V^2}\right)^e, \left(\frac{p}{\rho V^2}\right)^f \right\}$$

The quantity μ/ρ is denoted by v and is called the 'kinematic coefficient of viscosity'.

The expression for resistance may be written more generally as

$$R = \rho V^2 l^2 \left\{ f_1\left(\frac{v}{Vl}\right), f_2\left(\frac{gl}{V^2}\right), f_3\left(\frac{p}{\rho V^2}\right) \right\} \qquad (8.5)$$

It will be seen that $\rho V^2 l^2$ has the dimensions of a force while the three quantities v/Vl, gl/V^2 and $p/\rho V^2$ are all non-dimensional and are simply numbers. It will be seen also that since v/Vl contains the coefficient of viscosity this term is bound up with frictional resistance, whilst the term gl/V^2 containing the gravitational constant is concerned with motion under gravity and is connected with wave-making resistance. The term $p/\rho V^2$ relates the static pressure of the fluid to the dynamic pressure. It may, under certain circumstances, be of importance but in most cases of ship resistance it can be ignored. It is concerned with the problem of cavitation which will be seen later to be of importance in propellers. This leaves the expression for resistance as

$$R = \rho V^2 l^2 \left\{ f_1\left(\frac{v}{Vl}\right), f_2\left(\frac{gl}{V^2}\right) \right\} \qquad (8.6)$$

Dimensional analysis can take the problem no further than equation 8.6. It cannot predict if the frictional term and the wave-making term are dependent upon one another or not. A major assumption made at least in the first instance when dealing with ship resistance is that these two terms are independent, so that the equation for resistance can be rewritten

$$R = \rho V^2 l^2 \left\{ f_1 \left(\frac{v}{Vl} \right) + f_2 \left(\frac{gl}{V^2} \right) \right\} \tag{8.7}$$

Consider first the term $f_2(gl/V^2)$, which is concerned with the wave-making resistance of a ship, and suppose that there are two geometrically similar ships or a ship and a geometrically similar model.

Let ρ_1, V_1, l_1, etc., refer to one of these and ρ_2, V_2, l_2 refer to the other. Then

$$R_1 = \rho_1 V_1^2 l_1^2 f_2 \left(\frac{gl_1}{V_1^2} \right)$$

and

$$R_2 = \rho_2 V_2^2 l_2^2 f_2 \left(\frac{gl_2}{V_2^2} \right)$$

assuming that g is the same for both. It follows that

$$\frac{R_2}{R_1} = \frac{\rho_2}{\rho_1} \times \frac{V_2^2}{V_1^2} \times \frac{l_2^2}{l_1^2} \frac{f_2(gl_2/V_2^2)}{f_2(gl_1/V_1^2)}$$

The function f_2 is unknown and would, in general, depend in a complicated way upon the ship form, the speed, etc. If, however, it is arranged that $gl_1/V_1^2 = gl_2/V_2^2$ then no matter what the nature of f_2 is, this function will cancel out and

$$\frac{R_2}{R_1} = \frac{\rho_2}{\rho_1} \times \frac{V_2^2}{V_1^2} \times \frac{l_2^2}{l_1^2}$$

but since $gl_1/V_1^2 = gl_2/V_2^2$ then $V_2^2/V_1^2 = l_2/l_1$ and therefore

$$\frac{R_2}{R_1} = \frac{\rho_2}{\rho_1} \times \frac{l_2}{l_1} \times \frac{l_2^2}{l_1^2} = \frac{\rho_2 l_2^3}{\rho_1 l_1^3}$$

Now a property of the ship form which depends upon the density of the fluid and the cube of the linear dimensions is the displacement. Hence

$$\frac{R_2}{R_1} = \frac{\Delta_2}{\Delta_1} \tag{8.8}$$

The relation between the speeds and the length for this result to hold may be written $V_1/\sqrt{gl_1} = V_2/\sqrt{gl_2}$

or
$$\frac{V_1}{V_2} = \sqrt{\frac{l_1}{l_2}}$$
(8.9)

The result represented by equations 8.8 and 8.9 may be expressed as follows: the wave resistances of geometrically similar ships are in the ratio of their displacements when their speeds are in the ratio of the square roots of their lengths.

This is a particularisation of a wide general law which goes back probably to Newton. Its application to the wave resistance problem was stated in the form given above by William Froude,[1] who was the pioneer of the model experiment method for the determination of ship resistance. It has come to be known as Froude's law of comparison. The quantity given above, $V/\sqrt{gl}$, is called the 'Froude number' and if expressed in a consistent system of units is independent of the system used as it is a non-dimensional quantity. Often g is left out since it is, for all practical purposes, constant for any part of the world, and speed is given in knots with length in feet. In this form it is referred to as the 'speed/length ratio'. It is the value of this ratio or the Froude number which defines whether a ship is fast or slow, and not the absolute speed. Thus, two ships may have the same absolute speed and one of them could be a fast ship and the other a slow one, since the former may be short and the latter much longer.

When two ships are running at speeds which conform to the Froude law then they are said to be running at corresponding speeds.

The Froude law suggests an experimental method whereby the wave resistance of a ship could be inferred from the wave resistance determined from a geometrically similar model. Froude was the first to do this in a specially built tank or channel filled with water which he had constructed at Torquay. If a scale model of the ship is run at the corresponding speed in the tank and the wave resistance measured or deduced then the wave resistance for the ship is simply this value multiplied by the ratio of the displacements of ship and model.

The application of this method of finding the wave resistance of a ship will be considered in more detail later, but first of all the first term in the dimensional equation must be discussed, i.e. that dealing with viscosity.

It will be seen that frictional resistance can be written

$$R_f = \rho V^2 l^2 f_1 \left(\frac{v}{Vl}\right)$$

and if once again two geometrically similar ships are considered then

$$R_{f1} = \rho_1 V_1^2 l_1^2 f_1 \left(\frac{v_1}{V_1 l_1} \right)$$

$$R_{f2} = \rho_2 V_2^2 l_2^2 f_1 \left(\frac{v_2}{V_2 l_2} \right)$$

Following the same procedure as in the case of wave-making resistance

$$\frac{R_{f2}}{R_{f1}} = \frac{\rho_2}{\rho_1} \times \frac{V_2^2}{V_1^2} \times \frac{l_2^2}{l_1^2} \times \frac{f_1(v_2/V_2 l_2)}{f_1(v_1/V_1 l_1)}$$

The condition to be satisfied for dynamical similarity in this case is that $v_1/V_1 l_1 = v_2/V_2 l_2$ which gives

$$\frac{R_{f2}}{R_{f1}} = \frac{\rho_2}{\rho_1} \times \frac{V_2^2}{V_1^2} \times \frac{l_2^2}{l_1^2}$$

If, for argument's sake, v and ρ are considered to be the same in both cases, then $V_1 l_1 = V_2 l_2$ and

$$\frac{R_{f2}}{R_{f1}} = \frac{l_1^2}{l_2^2} \times \frac{l_2^2}{l_1^2} = 1 \tag{8.10}$$

In other words the frictional resistances of both ships are the same when $V_1 l_1 = V_2 l_2$. More generally it can be stated that

$$\frac{R_{f1}}{\rho_1 V_1^2 l_1^2} = \frac{R_{f2}}{\rho_2 V_2^2 l_2^2} \quad \text{when} \quad \frac{v_1}{V_1 l_1} = \frac{v_2}{V_2 l_2}$$

This is what is known as Rayleigh's law. The quantity Vl/v is called Reynolds' number after Osborne Reynolds who carried out experiments on the nature of frictional resistance in pipes. The Rayleigh law suggests a method of plotting frictional resistance data, i.e. as $R_f/\rho V^2 l^2$ (or the wetted area A may be substituted for l^2) against Reynolds' number.

The resistance of a ship has been seen to be compounded of both wave-making and frictional resistance so that two laws have to be satisfied to achieve complete dynamical similarity between the two motions when dealing with geometrically similar bodies or ships. These are:

$$\frac{V_1}{\sqrt{l_1}} = \frac{V_2}{\sqrt{l_2}} \quad \text{Froude's law}$$

$V_1 l_1 = V_2 l_2$ Rayleigh's law for two bodies working in the same fluid

The only way in which these two conditions can be satisfied is if

$V_1 = V_2$ and $l_1 = l_2$ so that if a model and a ship are being considered then the model must be the same size as the ship! It is thus not possible to satisfy both conditions at one and the same time for a ship and a model and one is left to choose between satisfying the Froude law or the Rayleigh law. A little consideration will show which is the more reasonable procedure. Suppose the model scale was 1/25 of the ship, then the speed of the model to satisfy the Froude law would be $1/\sqrt{25}$ ship speed, which is 1/5 of the speed. If on the other hand an attempt was made to satisfy the Rayleigh law, the model speed would be 25 times the ship speed. Even supposing that such a speed was obtainable on the model scale, the disturbance created in the water would, in all probability, destroy all dynamical similarity between the two motions. Reason suggests therefore that in attempting to estimate ship resistance from experiments on models the Froude law should be satisfied.

The problem of determining the resistance of a ship from that of a model consists in towing the model in a tank or channel of water at a speed corresponding to Froude's law. The force necessary to tow the model is measured and this is its total resistance. The frictional resistance is then calculated and deducted from this total, leaving a resistance (which is sometimes called the 'residuary resistance') which can be considered to be almost entirely wave-making. This resistance is scaled up in the ratio of ship and model displacements to determine the wave-making resistance of the ship. To this is added the calculated frictional resistance of the ship to obtain its total resistance. It will be seen that the accuracy of this procedure depends to a large extent upon the accuracy with which frictional resistance for both model and ship can be calculated. It is consequently necessary to have reliable data on which to base calculations and this involves a very detailed study of the frictional resistance problem.

Frictional resistance

In the previous section the dependence of frictional resistance on Vl/v has been demonstrated. This quantity known as Reynolds' number is associated with Reynolds' name because of the investigations which he carried out on the flow of water in pipes.[2] These investigations demonstrated that there are two distinct types of flow, one of which is called 'laminar flow' and the other 'turbulent flow'. In the case of laminar flow in a straight tube the particles of fluid move in straight lines parallel to the axis of the tube, whereas when the flow is turbulent the motions of the particles are in the form of spiral eddies.

Reynolds demonstrated this by observing the flow in glass tubes. He not only showed the existence of these two types of flow but also showed that different laws of resistance applied to the two types. In a particular tube Reynolds showed that if care was taken to ensure that the water entered the mouth of the tube smoothly the flow started off as laminar but at some distance from the mouth of the tube the flow broke down and became turbulent. This occurred at some critical velocity which was dependent upon the diameter of the tube and the temperature (i.e. upon the viscosity of the water). He found that for tubes of different diameters the critical velocity was inversely proportional to the diameter and the coefficient of viscosity v. In other words Vd/v was constant. Below the critical velocity the resistance to flow was proportional to velocity. At, and for some small range above the critical velocity there was an unstable region where it was not possible to identify any simple law of resistance. At higher velocities, however, where the motion was fully turbulent, the resistance followed a relatively simple law in which it was found to be proportional to a power of the velocity between one and two. The figure which Reynolds deduced from his experiments was 1.723. A further conclusion from his experiments was that $R/\rho v^2 d^2$ was the same for the same values of vd/v and this applied to both the laminar and turbulent regimes.

Reynolds' experiments were carried out on flow in tubes or pipes and although the results have little quantitative application to the resistance of ships, qualitatively the conclusions are most important. They are that the two regimes of laminar and turbulent flow exist, that the change-over from one to the other is dependent on Reynolds' number and that the law of resistance is different for the two types of flow. Calculations have been made for laminar flow of fluid past a flat surface, and a well known formula developed for this case is that by Blassius which gives

$$C_f = \frac{R_f}{\frac{1}{2}\rho S V^2} = 1.327 \left(\frac{VL}{v} \right)^{-\frac{1}{2}} \qquad (8.11)$$

where

$C_f =$ what may be called the specific resistance coefficient
$S =$ wetted surface area of the surface
$V =$ velocity
$\rho =$ density of the fluid
$L =$ length of the surface.

If the values of C_f from Blassius' formula are plotted to a base of Reynolds' number and on the same diagram results for turbulent flow past flat surfaces are plotted, the various conclusions arrived at

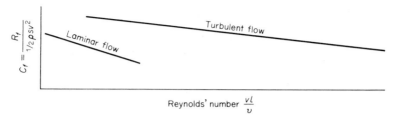

Figure 8.3

from Reynolds' experiments can be seen clearly. This has been done in *Figure 8.3*. It will be noted that at low Reynolds' numbers the Blassius formula could apply to resistance but at higher Reynolds' numbers the turbulent line is above the laminar line, indicating that for a given Reynolds' number the resistance would be greater in turbulent flow than in laminar flow. Experimental results show that there is a regime of mixed flow corresponding more or less to Reynolds' region of instability where the flow is of uncertain type and this may extend to Reynolds' numbers as high as 10^6. This has an important bearing on model experiments which are some of the main sources of knowledge of ship resistance. If the model used is too small it is possible that it will be running in the region of mixed or unstable flow and the calculation of its frictional resistance is uncertain. This places a lower limit on the size of model which might be used so as to ensure that the model is in the turbulent regime.

If one knew for certain that the flow around the model was completely laminar there would be no problem, since the resistance could be calculated for this type of flow. However, it is probable that with a small model there would be laminar flow over the forward part which would develop into turbulent flow further aft.

To overcome this, in addition to putting a lower limit on the size of the model it is customary nowadays to use some turbulence stimulating device at the fore end to prevent the occurrence of laminar flow. These devices can take several forms. Often what is called a 'trip wire' is used, which consists of a small diameter wire fitted round a forward section of the model, at about $9\frac{1}{2}$ station. Alternatively, the forward end of the model can be roughened or sometimes studs fitted which project from the surface. These devices all tend to stimulate turbulence and thus prevent the occurrence of laminar flow.

Should the presence of laminar flow in the model be ignored, then it might happen that the frictional resistance would be calculated assuming the flow to be turbulent and when this was deducted from the measured total resistance an erroneously low value for the wave-making resistance would be obtained, which in turn would result in the wave resistance for the ship being too low. Thus, the total ship resistance would also be low and hence an underestimate of the power necessary to drive the ship would be obtained.

Experiments on frictional resistance

The first important experiments on the frictional resistance of ships were made by W. Froude and were reported by him in the British Association Proceedings for 1872 and 1874. Froude carried out the experiments in the tank which he had constructed at Torquay and they consisted in towing vertically planes or planks and measuring the resistance by a specially designed dynamometer. The planks varied in length from 0.6 to 15.0 m (2 to 50 ft), they were 0.48 m (19 in) deep and 4.75 mm (3/16 in) thick. They were towed fully immersed with the 0.48 m (19 in) dimension vertical so that wave-making resistance was almost completely eliminated. The planks were tested with different surfaces to investigate the influence of roughness on resistance. The surfaces tested were calico, varnish, tinfoil, paraffin wax and three grades of sand. Froude attempted to express the resistance in terms of an empirical formula, as follows:

$$R_f = f S V^n \tag{8.12}$$

where

S = wetted surface area
V = speed
f and n are empirical coefficients.

(Froude was not the first to suggest this type of formula, the same form being used by Beaufoy.) Typical results for some of the planks tested by Froude are shown in *Table 8.1*.

The results in *Table 8.1* are for a speed of 600 ft per minute but have been converted so that the resistance in equation 8.12 is in pounds with the speed in knots and the area S in square feet. Column A represents the f values for the whole plank while Column B gives the f values for the last square foot.

Some points to be noted from these results are that the values of both n and f depend upon the nature of the surface and for very rough surfaces n seems to tend to a value of 2. There is some tendency for n to decline with length, at least for the smoother surfaces. The value of f shows a definite tendency to diminish as length increases.

Froude attributed the decline in the f values with length to the

Table 8.1 TYPICAL FRICTIONAL RESISTANCE RESULTS OF FROUDE

Nature of surface	Length of surface or distance from cutwater															
	2 ft			8 ft			8 ft			20 ft		
		f			f			f			f	
	n	A	B	n	A	B	n	A	B	n	A	B
Varnish	2.00	0.0117	0.0111	1.85	0.0121	0.0098	1.85	0.0104	0.0089	1.83	0.0097	0.0087
Paraffin	1.95	0.0119	0.0115	1.94	0.0100	0.0083	1.93	0.0088	0.0077			
Tinfoil	2.16	0.0064	0.0063	1.99	0.0081	0.0076	1.90	0.0089	0.0083	1.83	0.0095	0.0090
Calico	1.93	0.0281	0.0234	1.92	0.0106	0.0166	1.89	0.0184	0.0155	1.87	0.0170	0.0152
Fine sand	2.00	0.0231	0.0147	2.00	0.0166	0.0128	2.00	0.0137	0.0110	2.06	0.0104	0.0086
Medium sand	2.00	0.0257	0.0208	2.00	0.1078	0.0139	2.00	0.0152	0.0133	2.00	0.0139	0.0130
Coarse sand	2.00	0.0314	0.0251	2.00	0.0204	0.0148	2.00	0.0168	0.0140			

fact that the after ends of these planes were moving in water which already had an ahead velocity so that the rubbing velocity was reduced.

In order to make use of data such as these it is necessary to be able to extrapolate the results of the resistance experiments up to ship lengths. At the end of his second report to the British Association, Froude left open the question of extrapolation of the results to ship lengths but suggested two alternative methods. These were (1) to continue the law of decline of the f values to lengths greater than 50 ft; or (2) to assume that the reduction in f values ceased above 50 ft and to calculate f for greater lengths, assuming that the first 50 ft had the value for the 50 ft plank and that the remainder of the length had an f value equal to that for the last square foot of the 50 ft plank. The latter procedure would yield the formula

$$f_L = \frac{f_{50} \times 50 + (L-50)f_{49-50}}{L} \qquad (8.13)$$

It is not very clear what method was eventually adopted but for a fuller discussion on this subject the reader is referred to a paper by Payne.[3] At a later stage, however, Froude's son, R. E. Froude, produced a set of coefficients based on his father's work in which he standardised the power of speed with which the resistance is supposed to vary as 1.825, so that the formula for resistance became

$$R_f = fSV^{1.825} \qquad (8.14)$$

Table 8.2 shows values of f based on R. E. Froude's work. They have been extended up to a length of 1200 ft.

Table 8.2 FRICTIONAL RESISTANCE RESULTS BASED ON WORK OF R. E. FROUDE

Length, ft	f	Length, ft	f	Length, ft	f	Length, ft	f
8	0.011 96	50	0.009 61	400	0.008 83	1100	0.008 55
9	0.011 74	75	0.009 30	500	0.008 78	1200	0.008 52
10	0.011 57	100	0.009 21	600	0.008 73		
15	0.010 93	150	0.009 06	700	0.008 68		
20	0.010 52	200	0.008 99	800	0.008 64		
30	0.010 07	300	0.008 90	900	0.008 61		
40	0.009 79			1000	0.008 57		

The coefficients shown in this table are intended to apply to a wax surface for the model and a clean painted surface for a full size ship. They are for speed V in knots and wetted surface area S in square feet. If the area is in square metres then the coefficients would be found by multiplying the figures in the table by $1/0.304\,82^2 = 10.76$.

Alternative formulae for frictional resistance

The dependence of frictional resistance on Reynolds' number as demonstrated by dimensional analysis suggests a type of formula based on this quantity. Thus, the resistance can be written in the form

$$C_f = \frac{R_f}{\frac{1}{2}\rho S v^2} = F\left(\frac{vl}{v}\right) \qquad (8.15)$$

In more modern approaches to the frictional resistance of ships formulae of this type have tended to replace the Froude empirical coefficients. The function of Reynolds' number is of course empirical in the sense that the value to be used is based on experimental results.

One of the formulae which was developed some 40 or more years ago was that due to Schoenherr in America.[4] Schoenherr developed a formula based on theoretical work by Prandtl and von Kármán. The formula can be written in the form

$$A/\sqrt{C_f} = \log_{10}(\text{R.N.} \times C_f) + M \qquad (8.16)$$

where A and M are constants and R.N. is Reynolds' number.

Schoenherr examined the available experimental information which had been produced since Froude's time. Many experiments on planks, pontoons, etc., had been carried out in that period and Schoenherr found that good agreement could be obtained between the theoretical formula and these data if M was put equal to zero and a value of 0.242 was assigned to A. The Schoenherr formula then became

$$0.242/\sqrt{C_f} = \log_{10}(\text{R.N.} \times C_f) \qquad (8.17)$$

The value of C_f cannot be obtained explicitly from this formula but the results for a range of Reynolds' numbers can be plotted as a curve as shown in *Figure 8.4* or C_f can be tabulated against Reynolds' number.

The Schoenherr formula applies to a smooth surface, and to take into account the roughness of a ship surface additions to the values of C_f obtained from the formula have to be made. This will be considered later.

The Schoenherr formula, in common with all formulae for frictional resistance, extends the estimation of resistance to well beyond the experimental limit and to this extent the values of C_f at the higher Reynolds' numbers must be conjectural. The correctness of the co-efficients at higher Reynolds' numbers can only really be checked by full scale experiments.

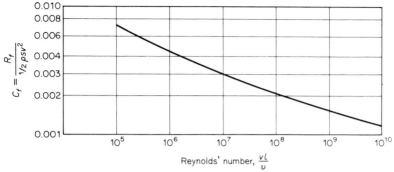

Figure 8.4 Schoenherr line

A more recent formula which has been developed by the International Towing Tank Conference was produced in 1957 and the curve of C_f drawn to a base of Reynolds' numbers is known as the I.T.T.C. 1957 'line'. It is given by

$$C_f = \frac{R_f}{\frac{1}{2}\rho S v^2} = \frac{0.075}{(\log_{10} \text{R.N.} - 2)^2} \qquad (8.18)$$

This formula was developed in order to obtain a better correlation between estimates of ship resistance and those of models. As shown in *Figure 8.5* it gives essentially the same resistance coefficients as Schoenherr's curve at high Reynolds' numbers but is somewhat higher than Schoenherr at low Reynolds' numbers. Both formulae require additions for roughness on the full scale. A comparison of the coefficients from the Schoenherr and I.T.T.C. formulae is shown in *Table 8.3*.

Table 8.3 COMPARISON OF COEFFICIENTS FROM SCHOENHERR AND I.T.T.C. FORMULAE

Reynolds' number	Schoenherr	I.T.T.C. 1957
10^6	0.004 41	0.004 687
10^7	0.002 934	0.003 000
10^8	0.002 072	0.002 083
10^9	0.001 531	0.001 531
10^{10}	0.001 172	0.001 172

It will be seen that there is about 6% difference in the coefficients at a Reynolds' number of 10^6, while at 10^9 the two coefficients are identical.

Comments on frictional resistance formulae Two different approaches to the frictional resistance problem have been considered here: firstly

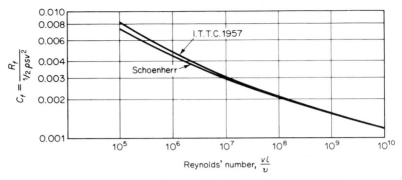

Figure 8.5   Comparison between Schoenherr and I.T.T.C. 1957 lines

what may be called the Froude method and secondly the Reynolds' number method. The latter approach is on a more sound theoretical basis and it is of interest to investigate if the Froude data might be plotted in this way. Using the Froude formula, the frictional resistance coefficient C_f would become

$$C_f = \frac{R_f}{\frac{1}{2}\rho S V^2} = \frac{fSV^{1.825}}{\frac{1}{2}\rho S V^2} = \frac{2fV^{-0.175}}{\rho}$$

$$= \frac{2fL^{0.175}}{\rho v^{0.175}} \times \left(\frac{VL}{v}\right)^{-0.175} \tag{8.19}$$

In order that for a given viscosity and density of fluid the term in front of the function of Reynolds' number should be constant, f should vary as $L^{-0.175}$. This is not in fact the case and it has been shown that the f values can be written

$$f = 0.008\,71 + \frac{0.053}{8.8 + L} \qquad \text{for salt water and}$$

$$f = 0.008\,49 + \frac{0.0516}{8.8 + L} \qquad \text{for fresh water}$$

The result is that if the Froude data are plotted on a base of Reynolds' number, a series of lines will be obtained for different lengths instead of one line as would be expected from the theoretical formula. One factor which may in part account for this is that the Froude planks were not geometrically similar, in that they all had the same depth while the length varied. Another point in connection with the Froude formula is that the coefficients as given in *Table 8.2* would require correction for density of the water and they do not allow for variation in viscosity with changes of temperature. Allowance of course can be made for changes in these properties of the fluid but with the Schoenherr and I.T.T.C. formulae they are automatically taken into account if the appropriate values of density and coefficient of kinematic viscosity are used for any given temperature.

Another important point has to be considered when applying the results of data obtained from plane surfaces to the frictional resistance of a curved surface such as that of a ship. In the first place, if the flow around a solid body is considered it will be found that there is high pressure near the bow, the pressure is reduced over the middle portion and then increases again at the stern. These pressure changes result in reduced velocity at the bow and stern with increased velocity over the middle portion. The local rubbing velocity, therefore, varies over the surface with the result that because the frictional resistance varies at a power greater than the first, the total frictional resistance of the ship surface will be greater than that of a flat plane of the same total area moving at the same average speed. This augment of resistance which may be called 'form drag' will vary with the fullness of the form, being greater in the bluffer forms.

There is another cause which would account for the difference in the frictional resistance of a ship and the equivalent plane. If a body is imagined to be moving totally submerged in an inviscid fluid the pressure changes over the forward part would be completely balanced by the pressure recovery over the after part and the total resistance would be zero. Because, however, the fluid possesses viscosity complete recovery of the normal pressure does not take place over the after part, so that if the normal pressures were integrated over the whole surface there would be a resultant drag. Although this arises due to loss of normal pressure over the after part of the ship, it occurs because of viscosity and is called 'viscous pressure drag'.

For the two reasons given above it can be expected that the actual resistance of a model or ship attributable to viscosity will be greater than that of the equivalent flat surface of the same area. It is of importance, therefore, to consider how this will affect the determination of the resistance of a ship from that of a model. The whole of the resistance can be expressed in non-dimensional form in the same way as the frictional resistance has been dealt with. Thus, the total resistance can be written in terms of a coefficient

$$C_t = \frac{R_t}{\frac{1}{2}\rho S V^2}$$

It has been seen that this resistance can be split up into two major components, R_f and R_r (the residuary resistance, i.e. the resistance left after the frictional resistance has been deducted from the total).

The latter resistance is largely wave making. It follows that

$$C_t = \frac{R_t}{\frac{1}{2}\rho S V^2} = \frac{R_f}{\frac{1}{2}\rho S V^2} + \frac{R_r}{\frac{1}{2}\rho S V^2} = C_f + C_r$$

Now if a model and a ship are considered, then

$$C_{tm} = C_{fm} + C_{rm}$$

and

$$C_{ts} = C_{fs} + C_{rs}$$

If the Froude procedure is adopted for finding the ship resistance from that of a model, then in coefficient form this is given by

$$C_{ts} = C_{tm} - (C_{fm} - C_{fs}) \qquad (8.20)$$

This assumes that $C_{rm} = C_{rs}$ which will be seen to be correct when the model and ship are running at the same speed/length ratio. The quantity $C_{fm} - C_{fs}$, which may be called the skin friction correction to the model C_t, represents the difference in the skin friction for the model and the ship. The values of C_t for model and ship are shown plotted in *Figure 8.6* to a base of Reynolds' number and the process of extrapolation described can be clearly seen. It will also be observed from this figure that if the friction lines used were displaced up or down by some constant amount there would be no difference in the value of C_t estimated for the ship. This means that the actual position of the line or the actual frictional resistance is not so important when estimating the resistance of the ship, provided always that the error in C_f for ship and model are the same and also that any residuary resistance included in C_r which cannot be attributed to wave-making resistance can be considered to obey the Froude law of comparison. It is clear from this that it is the slope of the friction line which is

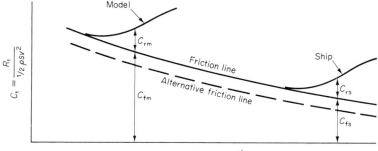

Figure 8.6

important rather than its absolute position in the diagram. It should not be imagined, however, that it is unimportant to assess as nearly correctly as possible the frictional resistance of model and ship. The determination of wave-making resistance by deducting the frictional resistance from the measured total resistance is one of the main (but not the only) means for finding the value of the former, so that it is necessary to know the frictional resistance accurately, particularly when comparing experimental results with calculations for wave resistance.

Experimental determination of resistance due to normal pressure

It is possible to determine experimentally the resistance of a model arising from the normal pressure on the hull. This resistance when deducted from the measured total resistance would then leave that due to the tangential drag on the surface, which would be the viscous resistance. In order to explore the pressure distribution on the hull small holes are made in the surface at various positions in the length and at different levels. The holes are connected by tubes to a manometer. The level of the water in a particular tube when the model is run in the tank will measure the dynamic pressure at the point on the surface to which the tube is attached. If the levels in all the tubes are measured simultaneously then it is possible to obtain a complete record of the pressure over the hull surface.

Now if p is the normal pressure at any point and dA is the element of area on which it acts, then the normal force is pdA. Also if θ is the angle which the small element of area makes with the centre line, then the fore and aft force is $pdA \sin \theta$. But, $dA \sin \theta$ is the projection of the area on a transverse plane, so that if the pressures are plotted in their appropriate positions on a body plan of the model they represent, when summed over the total area of the transverse section of the model, the fore and after force, i.e. the force due to the normal pressure on the hull.

It should be noted that the force acting on the fore body and that on the after body are opposite in sign so that the total force, i.e. the pressure resistance, is the difference of these two forces. Therefore, to obtain accurate values for the resistance, pressure should be measured very accurately since the net force is the difference of two large quantities.

Having obtained the pressure resistance in this way the frictional resistance will then be obtained by deducting this from the measured total resistance. This resistance should be very much closer to the true frictional resistance than that deduced from plank results.

If this procedure is carried out on a series of geometrically similar models (geosims) all run at the same Froude number then the frictional resistance could be obtained at a range of Reynolds' numbers, from which the general character of the frictional resistance line could be inferred.

Roughness and fouling

Froude's experiments on frictional resistance have shown that resistance increases with the roughness of the surface and since this type of resistance represents the major part of the total in many ships it is important to keep the immersed surface as smooth as possible. Roughness may be considered to be of three types: (a) structural roughness, (b) corrosion and (c) fouling.

Structural roughness arises because of the method of constructing the shell of the ship. In a ship with riveted shell plating the overlaps of the edges and ends of individual plates and the rivet points constitute a form of roughness. Welded joints would make a much smaller contribution to this form of roughness so that one would expect the modern all-welded hull to be less rough than its riveted counterpart. On the other hand, waviness of shell plating in between frames also contributes to structural roughness and this is more likely to occur in welded than in riveted structures. Other features which are likely to contribute to structural roughness are openings in the shell such as circulating water openings in steam ships, although they may be considered to increase eddy-making rather than increase frictional resistance.

A material such as steel is subject to corrosion in the marine environment in spite of the fact that it is painted. Corrosion produces a roughening of the surface and will increase the frictional resistance of the ship. It is important, therefore, from time to time to clean the outside surface and repaint to protect the steel and prevent wastage, not only from a structural point of view but also to ensure that as smooth a surface as possible is maintained.

Fouling is caused by the attachment of marine organisms such as grass, shells, barnacles, etc., to the outside surface. This could represent a very severe form of roughening if measures were not taken to prevent it. The usual procedure is to paint the hull with an anti-fouling composition. Such a composition contains toxic materials usually based on compounds of mercury or copper. These leach out into the water surrounding the ship and prevent the marine growths from attaching themselves to the hull and spreading. After a time the anti-fouling paints on the hull have to be renewed so as to keep the surface of the ship clean. Fouling is very much dependent on the amount of time a ship spends in port relative to its time at sea and also varies with the zone where the ship trades.

From the shipowner's point of view, deterioration of the surface of a ship from whatever cause either results in loss of speed for a given power or increased power for a given speed. This means that the running costs of the ship will increase with time and it becomes a problem for the owner to decide when to dock his ship so as to avoid undue wastage of power and fuel. The problem is therefore an economic one since the cost of docking has to be weighed against the increased running costs incurred by roughening. It is not appropriate to pursue this further here but it should be realised that the hull surface will generally deteriorate with time and will hence give more resistance to motion through the water.

Frictional resistance formulae such as those of Schoenherr and the I.T.T.C. are intended to give the resistance of an absolutely smooth surface. Such a degree of smoothness is not attainable in practice and for this reason percentages have to be added to the frictional coefficient C_f to allow for the amount of roughening which can be expected on an actual ship. The addition varies a great deal with the painting system employed. A value which has often been used in America is an addition of 0.0004 to C_f as determined from the Schoenherr formula. As the value of C_f at a Reynolds' number of 10^9 might be 0.001 53, it will be seen that this addition for roughness represents an increase of resistance of the order of 25%.

Wetted surface area

In whatever method is used for the calculation of frictional resistance, the area of the outside surface of the model or ship in contact with the water (called the 'wetted surface area') is required.

The problem is to find the area of the surface of a solid which is generally of non-mathematical form and which is curved in two directions. An exact value for the area is not generally obtainable so that various approximate formulae have been developed, based on what are considered to be the principal parameters likely to affect it.

Some idea can be obtained of these parameters by considering the wetted surface area of an extremely simplified form with vertically sided sections. If V is the volume of displacement of the ship and d the draught at which it floats, then the waterplane area, which will also be the area of the flat bottom, will be V/d. Now for a ship of this form, length L, the approximate area of the vertical sides is given by $2 \times L \times d$. This neglects the increased length on the sides due to the curvature at the ends. It follows that

$$\text{Wetted surface area } S = \frac{V}{d} + 2Ld$$

The volume V can be written as $C_B \times L \times B \times d$, where C_B and B are the block coefficient and the breadth respectively. Hence

$$S = \frac{C_B\,LBd}{d} + 2Ld = L(C_B\,B + 2d) \qquad (8.21)$$

For an actual ship form having the same principal dimensions and block coefficient, this would overestimate the wetted surface area. A modification to the formula due to Denny gives

$$S = L(C_B\,B + 1.7d) \qquad (8.22)$$

An alternative approximate formula for wetted surface area due to Froude is

$$S = L(V^{\frac{1}{4}} + V^{\frac{3}{4}}/L) \qquad (8.23)$$

Taylor, in America, expressed wetted surface area in terms of the displacement and length in the formula

$$S = C\sqrt{\Delta L} \qquad (8.24)$$

where

Δ = displacement in tons
L = length in feet
C = a coefficient which depends upon the breadth/draught ratio and the midship section coefficient. It varies from about 15.2 to 16.5.

Gertler,[5] in a re-analysis of Taylor's work, expressed the wetted surface in terms of volume instead of displacement, as follows:

$$S = C_S\sqrt{VL} \qquad (8.25)$$

where V is the volume of displacement.

A more accurate means of calculating the wetted surface area of a ship is to lift girths of the immersed sections from a body plan. If these girths are then integrated using say Simpson's rule, an approximation to the wetted surface area will be obtained. The reason why the area so obtained is not exact is because no allowance is made for the slope of the sections at the ends. In order to allow for this, a mean length on the surface should be taken instead of the length between perpendiculars. This mean should be the mean of the lengths measured on the various waterlines. The length on a waterline can be approximated as follows.

Suppose that the length is divided into *n* equal divisions, the ordinate at the perpendicular being designated y_1. Then, as shown in *Figure 8.7*, the distance round the waterline between the kth and the

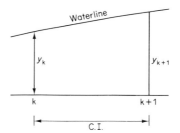

Figure 8.7

$(k+1)^{th}$ stations is given approximately by $\sqrt{\{C.I.^2 + (y_{k+1} - y_k)^2\}}$ where C.I. is the spacing of the ordinates. It follows that the total length is

$$\sum_{k=0}^{k=n} \sqrt{\{C.I.^2 + (y_{k+1} - y_k)^2\}}$$

If this is done for each waterline in turn then the mean length for all the waterlines can be obtained and this length can then be used in the calculation for wetted surface area. It will be found that the error involved by neglecting this obliquity correction, as it is sometimes called, is not great, especially in ships of fine form, but in very full ships it may be necessary to make such a correction.

Wave-making resistance

It was stated in a previous section that when a ship moves through water, changes are created in the normal pressure on the immersed portion of the hull, and if the vessel were completely immersed at considerable depth below the free surface there would be no surface disturbance. In an inviscid fluid the resultant force on the vessel due to these changes in pressure would be zero. With surface vessels, however, the situation is different. Because the free surface must be one of constant pressure, i.e. the pressure of the atmosphere, then elevations of the surface take place to satisfy this condition. These elevations cause waves on the surface which modify further the normal pressure on the hull and the resultant force in the fore and aft direction caused by these pressures is no longer zero. This force is the wave-making resistance of the ship. The magnitude of this resistance is very much dependent on the speed and form of the ship and as might be expected would be greater in a fast ship than in a slow ship. Also, the resistance would be expected to be greater in a ship of full form than in one of fine form.

Although much of the knowledge of ship wave resistance has been obtained by experiment, particularly on models, a great deal of theoretical investigation has been carried out based on pure hydro-dynamic theory and it is possible to make fair estimates of ship wave resistance by this means, at least for certain types of craft. Such investigations are outside the scope of the present work but nevertheless it is instructive to look at the results of some of the early theoretical work.

The nature of the wave pattern accompanying a surface ship was demonstrated by Kelvin. He considered a pressure point moving with constant velocity through a stationary fluid and showed that the resultant disturbance was as shown in *Figure 8.8*. The pattern consists of two types of waves: diverging waves on each side of the pressure point with their crests inclined at an angle to the direction of motion, and a system of transverse waves with curved crests intersecting the centre line at right angles. This type of wave pattern can be clearly observed when a ship is moving through the water, the diverging waves in particular being noticeable, indicating that the region of high pressure near the bow is responsible for this type of disturbance.

The distance between the crests of the transverse waves depends upon the velocity of the travelling pressure point. In fact this distance is equal to the wave length of a free running wave on the surface of the fluid of the same velocity as the travelling pressure point. If this velocity is v then the wave-length according to wave theory is

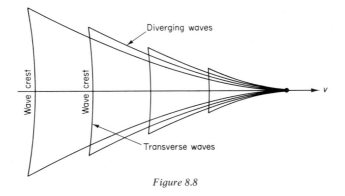

Figure 8.8

$l = 2\pi v^2/g$, so that if speed increases the waves left behind by the travelling disturbance will increase in length.

It is a matter of observation that not only are waves generated from the bow of a ship but similar types of waves can be seen originating from the stern. These waves may be considered to be due to a negative disturbance in the neighbourhood of the stern. The complete system of waves accompanying a ship would follow the pattern shown in *Figure 8.9*, consisting of two sets of transverse

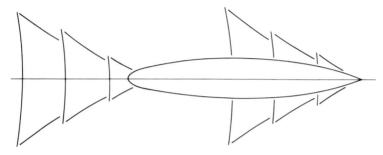

Figure 8.9   Bow and stern wave systems

waves originating from the bow and stern respectively and two sets of diverging waves spreading outwards from the ship originating from these positions. Observations indicate that the diverging waves do not extend for the distance along their crests as suggested by the Kelvin wave pattern, but tend to fade out into the undisturbed fluid.

Theoretical work by Wigley[6] has shown that other wave systems can arise from a ship form. His work, on a simple wedge shaped form, has demonstrated that wave systems can emanate from the fore and after shoulders of the form so that the total wave pattern might become more complex than indicated in *Figure 8.9*.

Interference effects

If it is considered that the wave-making resistance of a ship is due mainly to the disturbance created at the bow and a similar disturbance at the stern, it is possible to explain qualitatively interference phenomena which can arise due to the interaction between these two systems.

The transverse waves generated at the bow move aft their length along the crests, diminishing in the process and their heights also reducing. Since the distance between crests is given by $l = 2\pi v^2/g$

it follows that as speed increases the length increases so that it is possible for a crest from the bow system to coincide with a crest from the stern system. If the speed is increased further it will be seen that a position can be reached where a trough from the bow system is coincident with a crest from the stern system. In the former case pressure on the after body of the ship will be reduced, or looked at in another way the wave disturbance left behind the ship will be increased. From whichever point of view this is considered, the resistance will be increased. In the second case the trough from the bow system will tend to wipe out the crest from the stern system with the result that the pressure on the after body is increased, resulting in a reduced resistance.

This cycle can be repeated again and again until finally the first trough from the bow system is coincident with the first trough from the stern system. It would be expected therefore that the curve of wave resistance to a base of speed, while showing a general increase, would oscillate about a mean curve depending upon whether the interference effect arising from the two systems yields a maximum or minimum resistance. The curve would have the character of that shown in *Figure 8.10*, where it will be seen that there is a series of 'humps' and 'hollows' corresponding to maximum and minimum values of the resistance. These interference effects have been demonstrated theoretically by Havelock,[7] Wigley[6] and others, but in addition have also been shown to occur experimentally. The first to demonstrate the existence of humps and hollows was W. Froude[8] and his work on this subject will now be discussed.

Froude's work on humps and hollows   Froude's work in this field was carried out on models tested in the Torquay tank. His approach to the problem was to use a given bow and stern and to insert various

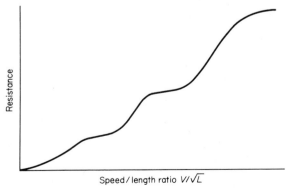

Figure 8.10   Humps and hollows in resistance curve

lengths of parallel middle body in between. Thus, if a series of such models is run at constant speed then the length of the waves generated will be constant. As successive lengths of parallel middle body are inserted between the bow and the stern the distance of separation between the origins of the bow and stern wave systems is increased so that a curve of wave resistance plotted to a base of length of ship or length of parallel middle should show a succession of humps and hollows. If now the speed is increased then a similar curve should be obtained, the average resistance being higher and the humps and hollows displaced in the length direction.

The model used by Froude represented in the full size a bow and

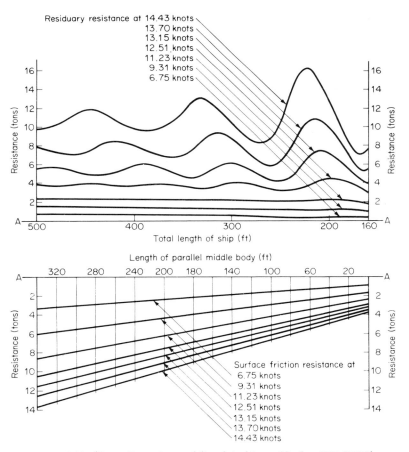

Figure 8.11   (From *Trans. Instn. of Naval Architects*, **18**, plate VIII (1877))

a stern each of length 24.4 m (80 ft) with varying lengths of parallel middle from 0 to 103.6 m (0 to 340 ft). The breadth was 11.6 m (38.4 ft) and the draught 4.3 m (14.4 ft). The smallest ship was therefore 103.6 m (340 ft) long and the largest 152.4 m (500 ft). Froude measured the total resistance of the models and deducted the calculated frictional resistance, the remainder being considered to be almost entirely wave-making resistance. His results are shown plotted in *Figure 8.11* and indicate quite clearly the humps and hollows which are attributed to interference phenomena. Each curve is for a constant speed and by calculating the wave length appropriate to that speed it can easily be shown that this approximates closely to the distance between the humps in the curve, thus justifying the assumption that they are due to interference effects arising from the bow and stern wave systems.

Scale effect on wave resistance

When considering the basic laws for the resistance of a body moving through a fluid it was shown that when two geometrically similar bodies or ships, or a ship and a model, are run at the same Froude number, their wave resistances are simply proportional to their displacements. It was also assumed that wave resistance was independent of viscosity. This assumption has been the basis of the determination of the wave resistance of a ship from that of a model for over a hundred years since Froude's time. Water, however, possesses viscosity and this must have some influence on the waves formed. Comparisons between calculated and measured wave resistance show, for example, that the humps and hollows in the curve of resistance to a base of speed are less pronounced in the latter and this could be attributed to the effect of viscosity. Another factor which could influence the wave-making resistance is the presence of the boundary layer around the hull. It has been seen that viscosity is confined mainly to the boundary layer and outside of this the fluid behaves very much as though it were non-viscous. The influence of the boundary layer, therefore, is to alter the dimensions of the hull by lengthening it and making the form wider locally. As such viscous effects are likely to be more pronounced on the model scale than on the full scale, it follows that this will produce a scale effect on wave-making resistance. Apparently, differences between calculations and measurements of wave resistance attributable to scale effect are a function of speed, tending to rise to a maximum at moderate Froude numbers and then diminishing at higher Froude numbers.

Eddy-making resistance and separation

The theoretical flow of fluid past a long slender body follows a smooth pattern and for a deeply submerged body in an inviscid fluid the resistance can be shown to be zero. It is possible to develop theoretically the flow perpendicular to a flat plate and this would show the same type of pattern on the rear surface as on the front surface and again the resistance would be zero. Actual observations of the flow in such a case, however, in real fluids indicate that the fluid leaves the edges of the plate and the region in behind the plate is filled with fluid in the form of eddies. The result is that the recovery of pressure on the rear surface which would balance the pressure on the front surface is not fully achieved and there is in consequence a resistance created. This resistance is often referred to as 'eddy-making resistance', and will exist wherever there are abrupt obstructions to the flow.

In the case of a ship there may be many such obstructions and in order to reduce their resistance the items in question have to be 'streamlined'. This simply means that abrupt changes in shape have to be avoided. A typical example of an obstruction causing eddy-making is the sternframe. Formerly, in ordinary single-screw ships the sternframe consisted of a propeller post and a rudder post, both of which were of rectangular cross section. It is clear that these would behave in very much the same way as flat plates and eddies would be generated behind them, thus creating resistance. A great improvement was achieved by rounding off these sections and thus converting what were bluff bodies into more slender streamlined bodies.

Another feature of the ship which can create additional resistance is the rudder. In the past the rudder consisted simply of a single plate which was fitted between arms attached to the rudder stock. The arms presented obstructions to the flow with the result that eddy-making resistance was created. A great improvement was again obtained by replacing the single plate by double plates, the arms providing the stiffening being placed in the space between the two plates. Thus, the double-plate streamlined rudder emerged, which presented a smooth surface to the fluid, and consequently the resistance became little more than that due to the additional frictional resistance created by the wetted surface area provided by the rudder.

Most ships are fitted with what are called 'bilge keels', which are plates extending outwards perpendicular to the shell of the ship for a distance in some cases up to 1 m and extending in the fore and aft direction for about one-half of the length of the ship. Their purpose is to increase roll damping and hence reduce the angles of roll when the ship is at sea amongst waves. Whilst the efficiency of bilge keels depends upon the fact that they have a large resistance to motion in the transverse direction, if not properly aligned they can also create resistance to ahead motion of the ship. They will, of course, always cause additional resistance due to their wetted surface area but they may also cause eddy-making resistance. In order to minimise this they are usually carefully aligned so as to be in the lines of flow around the hull. The flow pattern is determined experimentally on the model in the towing tank or perhaps in a circulating water channel. The pattern can be obtained by means of small flags projecting from the surface of the model and capable of aligning themselves in the streamline or by fitting capillary tubes at the bow from which dye can be drawn which will mark the model and thus show the directions of the streamlines. By this means it is possible to find the correct position on the hull where the bilge keels should be attached. One difficulty which arises is that strictly speaking the keels can only be correctly placed for one condition of loading of the ship, since obviously the flow pattern will be affected by draught and trim.

In multiple-screw ships the wing shafts must emerge from the hull at some point in the length. They are either what are called 'open' shafts supported near the propellers by brackets attached to the hull, or are enclosed in specially designed bossings. In either case the open shafting and brackets or the bossings can create additional resistance to the flow. Little can be done about open shafting since the shafts must follow the appropriate line from the machinery to the propeller, but the brackets (usually called A-brackets) should be streamlined to avoid undue resistance. As far as bossings are concerned, they should lie parallel to the lines of flow around the hull, so that flow patterns in the neighbourhood of these appendages should be determined in a similar manner to that described for bilge keels.

Separation of flow can take place on the hull itself and can generate additional resistance. The explanation of this phenomenon can be found by considering the velocity and pressure distribution in the boundary layer. Outside the boundary layer the fluid behaves as though it were inviscid and the pressure distribution in the direction of motion is essentially that which would be obtained in a frictionless fluid. The pressure has been seen to increase towards the after end of a ship so that the velocity is reduced. The pressure is transmitted through the boundary layer without any significant change so that this pressure gradient in the fore and aft directions is the same inside the boundary layer as it is outside. Consequently, the velocity in the boundary layer is reduced due to this cause. Because of the presence of the surface, the layers of fluid close to the boundary are already slowed up and hence the energy and momentum of the fluid is small. The further reduction in velocity due to the adverse pressure gradient may be sufficient to bring the fluid to rest at some point along the surface and further along it may even have a small velocity in the direction of motion of the ship. At this position the fluid would separate from the surface. The effect is that the pressure in the after body of the ship is not recovered and hence resistance is created. Since the phenomenon is dependent very much on the pressure gradient it is likely to be more pronounced where there is a steep gradient, i.e. in bluff bodies or in ships of full form.

Appendage resistance

It has been seen that appendages such as bilge keels, bossings, shaft brackets and rudders will give additional resistance to motion through the water over and above the resistance of the 'naked' hull. The resistances of these features will be partly frictional but may be also partly due to eddy-making. It has sometimes been considered that the resistances of the appendages are likely to be subject to scale effect, that is they may represent a different fraction of the total resistance on the full scale than they do on the model scale. For this reason, appendages are sometimes omitted in the model and percentages are added to the ship resistance estimated from the naked model. Some light has been thrown on the scale effect of appendages by model experiments carried out on series of geometric-ally similar models, usually nowadays called 'geosims'. If these geosims are run at the same Froude numbers then they will represent the same condition at different Reynolds' numbers. A series of experiments was carried out by the British Ship Research Association on a set of geosims of the former Clyde paddle steamer *Lucy Ashton* and the results compared with full scale experiments on the ship.[9] The models were 2.74, 3.66, 4.88, 6.09, 7.32 and 9.14 m (9, 12, 16, 20, 24 and 30 ft) in length, the ship being 57.9 m (190 ft) long. Both models and ship were tested with and without appendages and the increments in resistance determined due to these. The ratio of ship increment/model increment are shown in *Table 8.4* for these models run at a range of speeds.

From the results in *Table 8.4* it will be seen that at very low speeds there appears to be no scale effect on the resistance of appendages, but at high speeds the effect is very clearly marked. It would seem, therefore, that for this ship at any rate scale effect reduces the resistance of the bossing by about one-half at normal speeds with

Table 8.4   RATIO OF SHIP RESISTANCE TO MODEL RESISTANCE INCREMENTS

Ship speed, knots	Model length, m(ft)					
	2.74(9)	3.66(12)	4.88(16)	6.09(20)	7.32(24)	9.14(30)
5	1.00	1.00	1.00	1.00	1.00	1.00
8	0.44	0.48	0.52	0.56	0.58	0.61
10	0.49	0.54	0.57	0.60	0.62	0.65
12	0.52	0.57	0.60	0.62	0.65	0.68
$13\frac{1}{2}$	0.22	0.25	0.28	0.30	0.32	0.35
$14\frac{1}{2}$	0.10	0.12	0.14	0.16	0.17	0.20

models of normal length, whereas at the higher speeds the effect is very much greater. The conclusion is that model experiments with bossing exaggerate very much the resistance of the appendages.

If resistance estimates for ships are made from experiments on naked models, additions will have to be made for appendage resistance. This addition has often been considered to range from 10 to 12%, but results quoted by Mandel in America[10] suggest the figures quoted in *Table 8.5*. Some of these figures seem to be very high and Mandel points out that the upper limits for $V/\sqrt{L} = 0.7$ may be biased by scale effect.

Table 8.5   ESTIMATED PERCENTAGE ADDITIONS FOR APPENDAGE RESISTANCE

Ship type	Speed/length ratio		
	0.70	1.0	1.6
Large fast quadruple-screw ships	10–16	10–16	
Small fast twin-screw ships	20–30	17–25	10–15
Small medium speed twin-screw ships	12–30	10–23	
Large medium speed twin-screw ships	8–14	8–14	
All single-screw ships	2–5	2–5	

Air resistance

It was stated above that the motion of the above water form of the ship through air created resistance, but because of the very much smaller density of air as compared with water, the resistance in still air is likely to be small.

However, because in most cases there will be a wind blowing the resistance could be very much greater. The wind direction is independent of the direction of motion of the ship so that the relative wind may be in any direction from 180° ahead to 180° astern. When the wind is astern and the wind speed is in excess of the ship speed, the resistance will be negative or in other words the wind force will help to propel the ship. Should the wind be directly ahead, then the relative wind speed will be the sum of its speed and the speed of the ship. If the wind force on the ship is assumed to be proportional to the square of the speed, with the wind directly ahead, it will be proportional to $(V_s + V_w)^2$, where V_s and V_w are the ship and wind speeds respectively. The still air resistance will vary as V_s^2 so that the ratio of these two forces is $(V_s + V_w)^2/V_s^2$.

Suppose $V_s = 15$ knots and $V_w = 50$ knots, then

$$\left(\frac{V_s + V_w}{V_s}\right)^2 = \left(\frac{15 + 50}{15}\right)^2 = 18.8$$

This shows the order of increase in the still air resistance due to a head wind.

The wind resistance of the hull might be expressed by a formula of the type

$$R = kAV^2$$

where k is a coefficient and A is the projected area of the above water portion of the ship on a plane perpendicular to the centre line. This type of formula has been used for the wind force on flat plates. The value of k will be dependent on wind direction. Experimental work by Stanton on flat plates has given a value of 0.0032 for k with the speed in miles per hour and the area A in square feet and the force R in pounds.

A comprehensive series of experiments on wind resistance was carried out by Hughes.[11] The series consisted of tests on models which were towed upside down in water in the towing tank. Although the tests were carried out in water they can be applied to air by using the appropriate density. Hughes expressed the force on the above water portion by the following formula:

$$R = 0.0068 \, KV^2 \times \frac{A\sin^2\theta + B\cos^2\theta}{\cos(\alpha - \theta)} \tag{8.26}$$

where

R = force in pounds
V = speed in ft/s
A = longitudinal projected area of the above water form
B = equivalent transverse projected area of the above water form
θ = angle of the wind relative to the centre line of the ship measured from the bow
α = direction of the resultant force relative to the centre line.

The area B depends on the total projected area of the above water portion B_2, and the projected area of the ship from the waterline to the upper deck B_1.

Hughes gives B as $(B_2 - B_1) + 0.3 B_1$. To evaluate the Hughes formula for any angle of wind θ other than $0°$ or $180°$ it is necessary to find the angle α of the direction of the force relative to the ship. This angle is given in a series of curves in terms of θ. However, for $\theta = 0$, $\alpha = 0$ and hence the formula for the head on resistance would become

$$R = 0.0068\, KV^2 B$$

The value of K could be taken reasonably as 0.6, so that

$$R = 0.004\,08\, V^2 B \tag{8.27}$$

Presentation of model resistance data

Some reference has already been made to the presentation of model data in the form of dimensionless coefficients and it has been seen that the total resistance of a ship or model can be expressed as

$$C_t = \frac{R_t}{\frac{1}{2}\rho S V^2}$$

and this can be separated out into its components of wave and frictional resistance

$$C_w = \frac{R_w}{\frac{1}{2}\rho S V^2} \quad \text{and} \quad C_f = \frac{R_f}{\frac{1}{2}\rho S V^2}$$

This presentation can be very useful and has often been used as, for example, in Gertler's presentation of D. W. Taylor's model experiment data.[5]

A much earlier method of presenting data, however, which has found acceptance in many countries and is still in use today is that by R. E. Froude.[12] It is sometimes referred to as the 'constant' system and in fact this was the term used by Froude himself in his original paper. It is fitting to consider here the basis of the system.

Froude constant system

Froude started off by referring length dimensions to a unit which he called U and which is the cube root of the volume of displacement of the ship. With displacement Δ in tons of salt water at $35\,\text{ft}^3/\text{ton}$, this gives $U = (35\Delta)^{\frac{1}{3}}$.

What was called the 'length constant' related the length of the ship to U. This length constant was called $\text{\textcircled{M}}$ (circular M), so that

$$\text{\textcircled{M}} = \frac{L}{U} = \frac{L}{(35\Delta)^{\frac{1}{3}}} = 0.3057 \frac{L}{\Delta^{\frac{1}{3}}}$$

$\text{\textcircled{M}}$ measures in non-dimensional form the relation between the length of the ship and the amount of displacement in that length. It will be seen that it gives an indication of the 'slimness' or 'stumpiness' of the form. Thus, if a large displacement is packed into a short length the form will be stumpy and $\text{\textcircled{M}}$ will be small. On the other hand, if the displacement is small for a given length a slim form would be produced which would be shown by $\text{\textcircled{M}}$ being large.

Froude measured the wetted surface area of a ship or model in non-dimensional form and since the wetted surface S is proportional to the square of the dimensions then the skin constant $\text{\textcircled{S}}$ should be given by

$$\text{\textcircled{S}} = \frac{S}{U^2} = \frac{S}{(35\Delta)^{\frac{2}{3}}} = 0.0935 \frac{S}{\Delta^{\frac{2}{3}}}$$

Speed was measured in two ways: firstly in terms of the unit U and secondly in terms of L, the length of the ship. Using the first method the speed constant is defined as $\text{\textcircled{K}}$ which is equal to the ratio of the ship speed in knots to the speed of a wave of length $U/2$. Therefore

$$\text{\textcircled{K}} = \frac{V \times (6080/3600)}{\sqrt{\{(gU/2)/2\pi\}}} = \frac{V \times (6080/3600)}{\sqrt{\{32.2(35\Delta)^{\frac{1}{3}}/4\pi\}}}$$

$$= 0.5834 \frac{V^{\frac{1}{6}}}{\Delta^{\frac{1}{6}}}$$

In the second method the speed constant called $\text{\textcircled{L}}$ is defined as the ratio of the speed of the ship to that of a wave, length $L/2$. Hence,

$$\text{\textcircled{L}} = \frac{V}{\sqrt{(gL/4\pi)}} = 1.0552 \frac{V}{\sqrt{L}}$$

The most important constant in this system is that concerning resistance. It has been seen from the law established by W. Froude that wave resistance is proportional to the displacement for geometrically similar ships at corresponding speeds, i.e. at constant values of $\text{\textcircled{K}}$ or $\text{\textcircled{L}}$. This suggests that resistance might be expressed simply as the ratio of resistance ÷ displacement, i.e. R/Δ. Froude started from this position and called his resistance constant $\text{\textcircled{C}}$ which was made proportional to R/Δ.

The intention was that resistance as expressed by $\text{\textcircled{C}}$ should be plotted against $\text{\textcircled{K}}$ or $\text{\textcircled{L}}$. In the types of ships with which Froude was dealing, i.e. warships, there was a considerable speed range, with the result that $\textcircled{C}$ varied from a low value at low values of $\textcircled{K}$ to a very much larger value at high $\textcircled{K}$. For this reason and for convenience in plotting the results, Froude introduced $\textcircled{K}^2$ in the denominator of $\textcircled{C}$ which had the effect of making it more nearly constant over a range of speeds. The modified form of $\textcircled{C}$ then became

$$\textcircled{C} \, \alpha \, \frac{R}{\Delta \, \textcircled{K}^2}$$

and this will still be seen to be non-dimensional.

Introducing a scale factor of 1000 in the numerator gives the following:

$$\textcircled{C} = \frac{1000R}{\Delta \, \textcircled{K}^2}$$

The next step was to relate $\textcircled{C}$ to horsepower for ship purposes rather than resistance and this can be achieved by multiplying top and bottom by V so that

$$\textcircled{C} \, \propto \, \frac{RV}{\Delta \times V \times \textcircled{K}^2}$$

Since $\textcircled{K} = 0.5834(V/\Delta^{\frac{1}{3}})$ and since R and Δ are in tons and V in knots, to get the numerator in terms of horsepower, i.e. units of 33 000 ft lbf/min, the product RV will require to be multiplied by $\{2240 \times (6080/60)/33\,000\}$ and to preserve the scale of $\textcircled{C}$ the denominator would require to be multiplied by the same constant, so that

$$\textcircled{C} \, \propto \, \frac{RV(6080/60) \times (2240/33\,000)}{\Delta \times V \times (0.5834^2/33\,000) \times (V^2/\Delta^{\frac{2}{3}}) \times (6080/60) \times 2240}$$

$$= 0.4271 \frac{\text{E.H.P.}}{\Delta^{\frac{2}{3}} V^3}$$

As stated above, a scale factor of 1000 was introduced so that finally the resistance constant became

$$\textcircled{C} = 427.1 \frac{\text{E.H.P.}}{\Delta^{\frac{2}{3}} V^3} \tag{8.28}$$

where

E.H.P. = effective horsepower
Δ = displacement in tons
V = speed in knots.

The term 'effective horsepower' requires definition. It is the power necessary to overcome the resistance of a ship at a speed V. It is sometimes referred to as the 'towrope horsepower', i.e. the power which would have to be exerted if the ship were towed through the water.

The value of Ⓒ should be the same for geometrically similar ships or a ship and a geometrically similar model at the corresponding speed if all the resistance obeyed Froude's law of comparison. It has, however, been seen that the frictional component of the resistance does not obey the Froude law and is in fact a greater fraction of the total in the model than in the ship.

In other words Ⓒ is subject to scale effect, so that its value determined for a model requires correction before it can be applied to the ship. This correction was made by Froude as follows.

Ⓒ may be regarded as being made up of one component dependent on wave resistance and one dependent on frictional resistance, so that

$$\text{Ⓒ} = \text{Ⓒ}_w + \text{Ⓒ}_f$$

and for model and ship

$$\text{Ⓒ}_m = \text{Ⓒ}_{wm} + \text{Ⓒ}_{fm}$$

$$\text{Ⓒ}_s = \text{Ⓒ}_{ws} + \text{Ⓒ}_{fs}$$

hence

$$\text{Ⓒ}_s = \text{Ⓒ}_m - (\text{Ⓒ}_{fm} - \text{Ⓒ}_{fs})$$

on the assumption that Ⓒ$_w$ is the same for model and ship. The quantity Ⓒ$_{fm}$ − Ⓒ$_{fs}$ is called the 'skin friction correction'.

Now

$$\text{Ⓒ}_f = 1000 \times \frac{R_f}{\Delta \text{Ⓚ}^2}$$

and if R_f is in pounds

$$\text{Ⓒ}_f = \frac{1000}{2240} \times \frac{R_f}{\Delta \text{Ⓚ}^2}$$

if the coefficient is to be independent of units. Since $R_f = f S V^{1.825}$ then

$$\text{Ⓒ}_f = \frac{1000}{2240} \times \frac{f S V^{1.825}}{\Delta \text{Ⓚ}^2}$$

and because $S = \text{Ⓢ}\Delta^{\frac{2}{3}}/0.0935$

and

$$V = \frac{\text{Ⓛ}\sqrt{L}}{1.0552}$$

and it can be shown that $(K)^2 = (L)^2 (M)$ then

$$(C)_f = \frac{1000}{2240} \times \frac{f(S)\Delta^{\frac{2}{3}} \times (L)^{1.825}(\sqrt{L})^{1.825}}{0.0935 \times \Delta \times 1.0552^{1.825}(K)^2}$$

$$= \frac{1000 \times f(S)\Delta^{\frac{2}{3}} \times (L)^{1.825}(\sqrt{L})^{1.825}}{2240 \times 0.0935 \times \Delta \times 1.0552^{1.825}(L)^2 \times 0.3057(L/\Delta^{\frac{1}{3}})}$$

$$= 14.165 f (S) (L)^{-0.175} L^{-0.0875} \qquad (8.29)$$

Froude denoted by 'O' the quantity $14.165 f/L^{0.0875}$ so that the value of $(C)_f$ is 'O' $(S) (L)^{-0.175}$. From this it follows that the skin friction correction is

$$(C)_{fm} - (C)_{fs} = ('O'_m - 'O'_s) (S) (L)^{-0.175} \qquad (8.30)$$

The 'O' values were calculated by Froude from the values of f, the two being related by the equation

$$'O' = \frac{14.165 f}{L^{0.0875}} \qquad (8.31)$$

Values of 'O' determined from equation 8.31 are shown in *Table 8.6* for different lengths of ship. The values are those given in R. E. Froude's paper presented to the Institution of Naval Architects in 1888.

Table 8.6   VALUES OF 'O' FOR DIFFERENT LENGTHS OF SHIP

L, ft	'O'	L, ft	'O'
8	0.140 90	25	0.109 76
10	0.134 09	50	0.096 64
12	0.128 59	100	0.087 16
14	0.124 06	200	0.080 12
16	0.120 35	300	0.076 55
18	0.117 27	400	0.074 12
20	0.093 80	500	0.072 19
		600	0.070 51

As it was much more convenient on the model scale to use displacement and resistance (δ and r) in pounds in fresh water and speed (v) in feet per minute, Froude gave the values of the constants in these units and they are as follows:

$$\text{(M)} = 3.966 \frac{l}{\delta^{\frac{4}{3}}}$$

$$\text{(S)} = 15.73 \frac{S}{\delta^{\frac{2}{3}}}$$

$$\text{(K)} = 2.074 \frac{v}{\delta^{\frac{1}{6}}}$$

$$\text{(L)} = 1.041 \frac{v}{\sqrt{l}}$$

$$\text{and} \quad \text{(C)} = 232.5 \frac{r}{\delta^{\frac{2}{3}} v^2}$$

In these expressions the speed v is actually in units of 100 ft/min.

From these latter values of the constants it is possible to derive the value of (C) by measuring the resistance in pounds and the speed in hundreds of feet per minute. If the skin friction correction is made as described, then the value of (C) for the ship can be obtained. This is a rapid way of finding the ship horsepower from the model results and it is, of course, equivalent to the procedure described at the beginning of this chapter, i.e. the total resistance of the model is measured, from which the calculated frictional resistance is deducted. The residuary resistance which is considered to be wave-making, is then put up according to the law of comparison and the ship frictional resistance is added on to it to obtain the total ship resistance, from which the effective horsepower for the ship can be calculated.

Since (C) is subject to scale effect, i.e. it varies with the length, it has been customary to standardise its value for a particular length for comparative purposes. The practice at the Admiralty Experiment Works in Froude's time was to give (C) for a length of 300 ft. The practice adopted in the Ship Division of the National Physical Laboratory was to standardise on a length of 400 ft.

The (C) method of presenting resistance data enables the merits of different forms to be compared very easily. The value of (C) for two different forms at the same speed/length ratio will show which is superior, the one with the lower value of (C) being the better.

For a given form the curve of (C) to a base of speed or speed/length ratio will indicate when the form is being driven at too high a speed. Over quite a wide range of speed, the value of (C) for mercantile forms is often fairly constant, but as the speed continues to increase the (C) curve rises and can eventually become very steep,

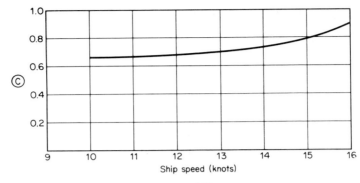

Figure 8.12

showing that it is uneconomical to drive the ship faster. A typical curve of Ⓒ for a merchant ship is shown in *Figure 8.12*.

Taylor's method of plotting model experiment data

Admiral D. W. Taylor, who was in charge of the Washington Tank in America towards the latter part of R. E. Froude's career in England, developed a method for plotting model data and although this is now mainly of historical interest it is of interest to examine the basis of his method. In the first place, Taylor decided to separate frictional resistance from wave-making resistance and express both in pounds per ton of displacement. Fundamentally this is the same style of presentation as Froude's.

For frictional resistance Taylor used a fairing of Froude's data due to Tideman in which the index of speed was taken to be 1.83 instead of 1.825. It follows that

$$\frac{R_f}{\Delta} = \frac{fSV^{1.83}}{\Delta}$$

If $S = C\sqrt{\Delta L}$, $x = V/\sqrt{L}$, $y = \Delta/(L/100)^3$ then

$$\frac{R_f}{\Delta} = \frac{fC\sqrt{\Delta L}\,x^{1.83}(\sqrt{L})^{1.83}}{y^3 L/10^6} = \frac{fC \times 1000 \times L^{-0.085}\,x^{1.83}}{y^{0.5}}$$

For similar ships the value of C is constant, f depends upon length and $f/L^{0.085}$ varies slowly with length. It will thus be seen that x and y are the two principal variables and for a given length they will be the only variables. Taylor standardised his results on a length of

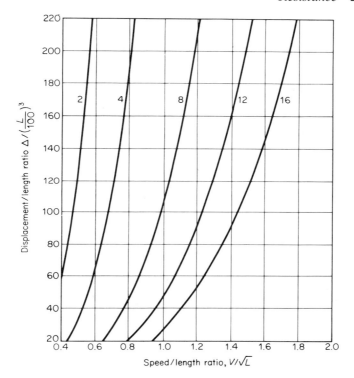

Figure 8.13 Contours of frictional resistance in pounds per ton of displacement for ship 500 ft long

500 ft and plotted contours of R_f/Δ with $V/\sqrt{L}(x)$ and $\{\Delta/(L/100)^3\}(y)$ as the co-ordinates as shown in *Figure 8.13*.

Taylor gave a correction factor for length which at 500 ft would be 1.0, for lengths below 500 ft would be greater than 1.0 and for lengths above 500 ft would be less than 1.0. The diagram was also drawn for a value of the wetted surface coefficient C of 15.4, so that for other values C_A, say, there would be a wetted surface correction factor $C_A/15.4$. The frictional resistance of the ship is then given by the following equation:

$$\text{Frictional resistance} = \left(\frac{R_f}{\Delta}\text{from contours}\right) \times \Delta$$

$$\times \text{(Wetted surface correction factor)}$$

$$\times \text{(Length correction factor)}$$

Taylor plotted wave-making resistance (which he called 'residuary resistance') as R_w/Δ, i.e. pounds per ton of displacement as contours against the parameters which he considered to affect this resistance, such as prismatic coefficient, displacement/length ratio $(\Delta/(L/100)^3)$ and breadth/draught ratio for various values of the speed/length ratio. This will be discussed later but it is sufficient to say here that when R_w/Δ has been obtained from the contours and this is added to R_f/Δ from the frictional resistance chart it is then possible to calculate the total resistance for a particular ship.

Example of calculation of ship resistance from results of model experiments

To illustrate the use of the model experiment in calculating ship resistance a worked example is given below to show how the Froude method can be applied and also to illustrate the application of the method using the resistance coefficient C_t. As a great many of the data used are given in imperial units these will be employed in the calculations.

The particulars of the ship are as follows:

Length	$= 450\,\text{ft}\,(137.18\,\text{m})$
Breadth	$= 62\,\text{ft}\,(18.89\,\text{m})$
Draught	$= 28\,\text{ft}\,(8.53\,\text{m})$
Block coefficient	$= 0.65$
Midship area coefficient	$= 0.98$
Wetted surface area	$= 35\,450\,\text{ft}^2\,(3293.4\,\text{m}^2)$
Speed	$= 15\,\text{knots}$
Density of sea water	$= 64\,\text{lb/ft}^3\,(1025\,\text{kg/m}^3)$
Froude's coefficient f	$= 0.0088\,(0.421\,34)$

Tests on a geometrically similar model 16 ft (4.877 m) long run at the corresponding speed gave a total resistance of 4.25 lbf (18.904N) in fresh water whose density was 62.4 lb/ft³ (1000 kg/m³).

$$\text{Speed of model} = 15 \times \sqrt{\frac{16}{450}} = 2.83 \text{ knots}$$

$$= 286.6\,\text{ft/min}\,(87.35\,\text{m/min})$$
$$= 4.78\,\text{ft/s}\,(1.46\,\text{m/s})$$

$$\text{Wetted surface area of model} = 35\,450 \times \left(\frac{16}{450}\right)^2$$

$$= 44.8\,\text{ft}^2\,(4.16\,\text{m}^2)$$

Froude's coefficient f for model (fresh water) $= 0.010\,84\,(0.506\,36)$

Total resistance of model R_{tm} $= 4.250\,\text{lbf}\,(18.904\,\text{N})$

Frictional resistance of model R_{fm}
$= 0.010\,84 \times 44.8 \times 2.83^{1.825}$
$(0.506\,36 \times 4.16 \times 2.83^{1.825})$ $= 3.161\,\text{lbf}\,(14.062\,\text{N})$

Wave resistance of model R_{wm} $= 1.089\,\text{lbf}\,(4.842\,\text{N})$

$$\text{Wave resistance of ship } R_{ws} = 1.089 \times \frac{64}{62.4} \times \left(\frac{450}{16}\right)^3 \text{lbf}$$

$$\left(4.842 \times \frac{1025}{1000} \times \left(\frac{137.18}{4.877}\right)^3 \text{N}\right)$$

$$= 24\,848\,\text{lbf}\,(110\,449\,\text{N})$$

Frictional resistance of ship R_{fs}
$= 0.0088 \times 35\,450 \times 15^{1.825}$
$(0.421\,34 \times 3293.4 \times 15^{1.825})$ $= 43\,699\,\text{lbf}\,(194\,352\,\text{N})$

Total resistance of ship R_{ts} $= 68\,547\,\text{lbf}\,(304\,801\,\text{N})$

$$\text{Effective power } P_E = \frac{68\,547 \times (6080/60) \times 15}{33\,000} = 3157\,\text{hp}$$

$$\text{or} \quad \frac{304\,801 \times 15 \times (6080/3600) \times 0.3048}{1000} = 2353\,\text{kW}$$

$$\textcircled{C} = \frac{427.1\,P_E}{\Delta^{\frac{2}{3}}\,V^3} = \frac{427.1 \times 3157}{14\,508^{\frac{2}{3}} \times 15^3} = 0.672$$

$$\text{Alternatively,} \quad \textcircled{C} = \frac{579.7 \times P_E}{\Sigma^{\frac{2}{3}} \times V^3} \text{ (kW)}$$

where $\Sigma =$ ship mass in tonnes.

In this case $\Sigma = 14\,735$ tonnes

$$\text{Hence} \quad \textcircled{C} = \frac{579.7 \times 2353}{14\,735^{\frac{2}{3}} \times 15^3} = 0.672$$

Instead of calculating the resistance as above the R. E. Froude method can be employed by first finding $\textcircled{C}$ for the model and then making the skin friction correction

$$\textcircled{C}_m = \frac{232.5 \times r_{tm}}{\delta^{\frac{2}{3}} v^2}$$

Now $\delta = 14\,500 \times \left(\dfrac{16}{450}\right)^3 \times \dfrac{62.4}{64} \times 2240 = 1423\,\text{lb}$

$\delta^{\frac{2}{3}} = 1423^{\frac{2}{3}} = 126.5$

$\therefore$ $\textcircled{C} = \dfrac{232.5 \times 4.25}{126.5 \times 2.87^2} = 0.948$

The skin friction correction is $(O_m - O_s)\,\textcircled{S}\,\textcircled{L}^{-0.175}$.

For the model and the ship $O_m = 0.120\,35$ and $O_s = 0.073\,15$

$$\textcircled{S} = 0.0935 \times \dfrac{35\,450}{14\,500^{\frac{2}{3}}} = 5.57$$

$$\textcircled{L} = 1.0552\dfrac{V}{\sqrt{L}} = 1.0552\dfrac{15}{\sqrt{450}} = 0.746$$

Skin friction correction $= (0.120\,35 - 0.073\,15) \times 5.57 \times 0.746^{-0.175}$
$= 0.277$

$\textcircled{C}_{\text{ship}} = 0.948 - 0.277 = 0.671$

It will be seen that this value of $\textcircled{C}$ is practically the same as that obtained by the first method, as of course it should be. Having obtained $\textcircled{C}$ by the Froude skin friction correction, the horsepower can then be obtained.

It is of interest to re-calculate the resistance using the Schoenherr and the I.T.T.C. formulae. To apply either of these, additional information is required, namely, the kinematic coefficient of viscosity and the value of C_f at the Reynolds' number for both model and ship. For fresh water $\nu = 1.2285 \times 10^{-5}\,\text{ft}^2/\text{s}$ and for sea water $\nu = 1.2817 \times 10^{-5}\,\text{ft}^2/\text{s}$.

$\therefore$ Reynolds' number for model $= \dfrac{16 \times 4.78}{1.2285 \times 10^{-5}} = 6.225 \times 10^6$

Reynolds' number for ship $= \dfrac{450 \times 15 \times 6080/3600}{1.2817 \times 10^{-5}} = 8.89 \times 10^8$

Consider first Schoenherr. The values of C_f for model and ship are 3.172×10^{-3} and 1.553×10^{-3} respectively.

Now $C_{\text{tm}} = \dfrac{R_{\text{tm}}}{\frac{1}{2}\rho S v^2} = \dfrac{4.25}{\frac{1}{2}(62.4/32.2) \times 44.8 \times 4.78^2} = 0.004\,284$

C_{fm} $\qquad\qquad\qquad\qquad\qquad\qquad = 0.003\,172$

$C_{\text{wm}} = C_{\text{ws}}$ $\qquad\qquad\qquad\qquad\qquad = 0.001\,112$

C_{fs} $\qquad\qquad\qquad\qquad\qquad\qquad\; = 0.001\,553$

C_{ts} $\qquad\qquad\qquad\qquad\qquad\qquad\; = 0.002\,665$

$$v_{\text{ship}} = \frac{15 \times 6080}{3600} = 25.33\,\text{ft/s}$$

$$\therefore \quad R_{\text{ts}} = \tfrac{1}{2}\rho S V^2 \times C_{\text{ts}} = \tfrac{1}{2}(64/32.2) \times 35\,450 \times 25.33^2 \times 0.002\,665$$
$$= 60\,240\,\text{lbf}$$

The Schoenherr formula makes no allowance for roughness. If an addition of 0.0004 on the value of C_f were made as was common practice in the USA then C_{ts} would be 0.003 065 and the resistance would be $0.003\,065/0.002\,665 \times 60\,239 = 69\,280\,\text{lbf}$. This will be seen to be not very different from the result obtained from the Froude method.

Consider now the I.T.T.C. 1957 formula for frictional resistance. It gives

$$C_f = \frac{0.075}{(\log_{10} RN - 2)^2}$$

For the model $C_{\text{fm}} = \dfrac{0.075}{(\log_{10} 6.225 \times 10^6 - 2)^2} = 0.003\,262$

Hence $C_{\text{wm}} = C_{\text{ws}} = 0.004\,284 - 0.003\,262 = 0.001\,022$

$$C_{\text{fs}} = \frac{0.075}{(\log_{10} 8.89 \times 10^8 - 2)^2} = 0.001\,553$$

$$C_{\text{ts}} = 0.001\,022 + 0.001\,553 = 0.002\,575$$

$$R_{\text{ts}} = \tfrac{1}{2} \times (64/32.2) \times 35\,450 \times 25.33^2 \times 0.002\,575$$

$$= 58\,204\,\text{lbf}$$

The effective horsepower based on this result will be

$$\frac{58\,204 \times 15 \times 6080/60}{33\,000} = 2680$$

As in the case of the Schoenherr formula some addition would require to be made to this power to allow for roughness.

Model experiments

Because of the great cost involved in carrying out full scale experiments on the resistance of ships, and for that matter because of the limited field which could be covered by such experiments, a great deal of attention has been focused in the last hundred years on experiments on models. Much of the knowledge of resistance of ships has been derived from this type of experiment and it has been demonstrated throughout this chapter, and particularly in the last section, how the resistance of a ship and hence the power necessary to drive it can be estimated from a model.

W. Froude was the pioneer of the model experimental method and the tank which he constructed at Torquay, which was opened in 1872, was the first of its kind. The tank was a channel of water 84.7 m (278 ft) long by 11 m (36 ft) broad at the top and 3 m (10 ft) deep. Over this channel ran a truck or carriage on a light railway, the truck carrying the dynamometer gear for measuring resistance. The truck was towed by an endless rope driven by a double cylinder steam engine with a governing device to ensure a uniform speed. The maximum speed available was 305 m/min (1000 ft/min). The models to be tested were attached to the truck through the dynamometer and the resistance was measured by means of a spring, the extension of which was recorded on a revolving drum driven from one of the axles of the truck. Time was recorded on the drum and the distance travelled by the truck was also recorded, so that from the time/distance records speed could be derived. The models tested by Froude were made of paraffin wax, a material which is easily shaped and which can easily be altered when necessary.

Since Froude's time great developments have taken place both in the size of tanks and the type and extent of the equipment used for making measurements, but basically the principles involved have remained unchanged. Today, every maritime nation throughout the world has its own towing tank and many are in excess of 305 m (1000 ft) in length and some considerably so. As would be expected the cost of running these establishments goes up very rapidly with increase in size.

In the modern large tank the carriage spanning the tank is of course driven by electric motors on the carriage itself, this being sufficiently large to enable experimental staff to ride on the carriage when carrying out experiments. Recording devices have become largely electrical and their scope has extended greatly since Froude's time. Most of the early experimental work on ship models was carried out in smooth water but as ships rarely, if ever, meet such conditions in service the idea of running models in artificially created waves developed and tanks today are usually fitted with wave makers. The wave maker is a device fitted at one end of the tank which enables waves of any given height and length within prescribed limits to be created. These waves run down the tank and if the model is towed from the other end then it will run head on into the waves. For experiments of this sort the model must be free to heave and pitch and the instrumentation on the carriage enables these motions to be recorded. The effect of heaving and pitching is in general to increase the resistance as compared with its value in smooth water so that a comparison of the results will indicate by how much the resistance has been increased due to this cause.

Modern wave-making equipment not only permits regular waves to be produced but also enables irregular waves to be generated by analysing an irregular sea into a series of harmonic components and programming the wave maker accordingly. Even with this equipment, however, in the normal long tank it is only possible to test the model in long-crested waves approaching it head on or, if possible, to run the model in reverse in following seas. This has led to the development of what are called 'sea keeping tanks' which are more nearly square in shape, the breadth being almost as great as the length. These tanks sometimes have wave makers on two sides at right-angles so that if they are both working simultaneously it is possible to generate confused seas. Free running radio controlled models are used which can move in any direction and the various motions such as heaving, pitching and rolling can be recorded as well as the resistance, propeller revolutions, thrust and torque, etc. A complete record of the behaviour of the model can thus be obtained.

In the development of experimental tank methods many problems have arisen with which it is not possible to deal here. One, however, has already been discussed, i.e. the question of laminar flow on models and the methods adopted to overcome this. Another concerns the question of model size in relation to tank size. Should the cross section of a model be too great relative to the cross section of the tank, boundary effects could be created which would distort the results of experiments. This puts an upper limit on the size of model which can be used in a tank of any particular size.

Form parameters affecting resistance

Before discussing some of the standard series of models which have been tested, it is important to consider in broad general terms the features of form which affect resistance.

Generally, the effect of the features which will be discussed is on wave resistance rather than on frictional resistance. To some extent, however, there will be an influence on frictional resistance if the flow round the hull is altered and also if the wetted surface area of the hull is changed by alteration of the feature in question.

Fullness of form

One of the basic features of the form of a ship which affects resistance is the general fullness of the lines. This may be represented by either the block coefficient or the prismatic coefficient. For most merchant ships the resistance at a given speed will increase as the block or prismatic coefficient is increased. This is quite understandable since the full ship will have steeper lines at the fore and after ends or larger angles of entrance and run and may therefore be expected to create a greater disturbance than ships of finer form. There is evidence of an optimum value of the block or prismatic coefficient on either side of which the resistance might be expected to increase. This optimum could be in the working range in high speed forms, but in slow speed ships is usually well below practical values. Generally, however, block or prismatic coefficient should reduce as speed of ship increases so that in designing a ship there is a limit of fullness to be observed for a given speed. A formula of the type

$$C_B = A - B(V/\sqrt{L})$$

for what is called the 'economical' block coefficient has often been used, where A and B are constants, V is the speed in knots and L the length in feet. Typical values for the constants are $A = 1.08$ and $B = \frac{1}{2}$ so that

$$C_B = 1.08 - V/2\sqrt{L} \qquad (8.32)$$

Some explanation of the term 'economical block coefficient' is required. In moderate speed ships power can always be reduced by reducing the block coefficient so that the machinery and fuel weights would be reduced. However, for given principal dimensions reduction in block coefficient means reduction in total displacement, which means a reduction in the displacement available for the carriage of cargo and hence in the payload. It is therefore advantageous to increase the block coefficient so that the payload is increased until the situation is reached where the increase in fuel and machinery weights because of the increased block coefficient more than counterbalances the increased cargo carrying capacity. The value of the block coefficient given in equation 8.32 is only a very rough guide and the actual economical value could only be determined by a full economic investigation to obtain the values of the constants A and B in the basic formula. The formula does, however, show the dependence of the value of the block coefficient on speed or rather speed/length ratio. It is only valid for speed/length ratios up to about unity and should not be used for higher values.

Slimness

It was shown above that slimness can be defined by the ratio of the length to the cube root of the volume of displacement. Froude used the coefficient $\widehat{M} = 0.3057(L/\Delta^{\frac{1}{3}})$ to define this. Taylor used what was virtually the inverse of this, i.e. $\Delta/(L/100)^3$ and Ayre[13] used simply $L/\Delta^{\frac{1}{3}}$. Gertler's presentation uses volumetric coefficient which is $C_V = V/L^3$, where V is the volume of displacement. It will be seen that C_V is equal to $LBdC_B/L^3 = BdC_B/L^2$ and for a given block coefficient and midship area coefficient it measures the ratio of the midship area to the square of the length. It follows that a ship with a large midship area on a given length will be stumpy, which will result in steep angles of entrance and run for the waterlines at the ends. Increase in the volumetric coefficient or reduction in the value of $\widehat{M}$ might therefore be expected to have the effect of increasing the resistance of the form.

Another way of looking at the question of the slimness of the form is by considering the relation of length to breadth. The volumetric coefficient can be obtained in terms of breadth/draught ratio and length/breadth ratio, since

$$C_V = \frac{BdC_B}{L^2} = \frac{B^2}{L^2} \times \frac{d}{B} \times C_B$$

For a given block coefficient and breadth/draught ratio the coefficient varies as $1/(L/B)^2$.

The influence of slimness or stumpiness of the form as represented by any of these parameters is affected by the fullness of the form and the speed/length ratio. Generally speaking in high speed forms, where the block or prismatic coefficient is low, the displacement/length ratio must be kept low in order to avoid undue resistance. For slow speed ships with fairly full block coefficients this factor is not nearly so important. In terms of the dimensions of the ship it follows then that fast ships require larger length/breadth ratios than slow ships.

Breadth/draught ratio

Generally, resistance increases with increase in breadth/draught ratio in the normal working range of this variable. This again can be explained by the fact that as breadth/draught ratio increases the angles of the waterlines at the ends will be increased and thus a greater disturbance created. It is probable, however, that with extremely high values of breadth/draught ratio the flow around the hull would tend to be more in the vertical than in the horizontal direction and this may lead to a reduction in resistance.

Longitudinal distribution of displacement

For a ship of a given length and displacement, when block coefficient, displacement/length ratio and breadth/draught ratio are fixed, all the dimensions are fixed. However, there are still many other features of the form which can be varied and these can have an influence on resistance.

One of the most important is the longitudinal distribution of displacement as expressed by the longitudinal position of the centre of buoyancy of the under water form. The influence of this parameter on resistance arises because for a given block coefficient the position of the centre of buoyancy which is the centroid of the under water volume governs the relative fullness of the ends. Thus, a ship with the centre of buoyancy well forward of amidships will have a bluff fore end and a fine after end and conversely if the centre of buoyancy is aft of amidships the fore end will be finer and the after end fuller. Now fining the fore end will reduce its resistance but as this is done at the expense of filling the after end the resistance of the latter will be increased. It can be seen, therefore, that a position can be reached where the gain at the fore end is more than compensated by the loss at the after end. The same result will apply if the after end is fined by filling the fore end. It would appear then that there is some optimum position of the centre of buoyancy where the resistance will be minimal and on either side of which the resistance will increase. The designer should aim to put the centre of buoyancy in this ideal position, but it should be pointed out that he is not always free to do so. In dealing with the trim of a ship it has been seen that for a ship to float on an even keel the centre of buoyancy must be under the centre of gravity, so that to avoid excessive trim the position of the centre of buoyancy is dictated by the loading of the ship.

Length of parallel middle body

For a ship of given dimensions, block coefficient and position of centre of buoyancy, the length of the parallel middle can be varied without altering any of these parameters, so that this must be considered as another feature of the form which can affect resistance. In high speed ships where the block coefficient is low there will usually be no parallel middle, but in ships of moderate and high block coefficients parallel middle body is necessary to obtain satisfactory form at the ends. In a ship where the length of the parallel middle body is l_1 say, if this is increased to a length l_2 then the ends must be fined to give the same block coefficient. On the other hand if the length of the parallel middle is reduced to l_3 then the ends must be filled to give the same block coefficient. It can be seen that as in the case of the position of the longitudinal centre of buoyancy there will be some best length of parallel middle for a given block coefficient to ensure minimum resistance.

Shape of sections

The features of the form so far discussed will fix to a very great extent the shape of the immersed cross-sectional area curve, but none of these parameters will fix the shape of the body sections. A rough indication of the shape of the sections can be obtained by considering the shape of the load waterline in relation to the sectional area curve.

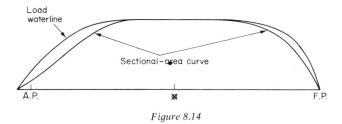

Figure 8.14

If these two are plotted non-dimensionally as in *Figure 8.14*, then when they both lie close together the sections will be of U-form whilst if they lie well apart V-shaped sections will be obtained. Another factor which might be used to give an indication of section shape is the vertical prismatic coefficient. A high vertical prismatic will tend to give U-sections whilst a low value will give V-type sections.

It is not possible to generalise on the shape of sections adopted but it will be found that slow and moderate speed merchant ship forms often tend to have U-shaped sections in the fore body and V-shaped sections aft.

Bulbous bows

The lines of a ship normally run into a stem bar or a plate stern at the bow. In modern ships it is quite common for the form to be swelled out below the waterline near the stem to form a bulb.

Historically, this probably dates back to the times when warships were fitted with rams at the fore end. The influence of the bulb on ship resistance has been studied both experimentally and theoretically and it has been shown that this device can under certain circumstances have a favourable effect in reducing resistance. The bulb is usually well below the surface and the theory is that it creates a pressure system which tends to suppress the normal bow pressure system and thus reduce the wave-making resistance.

The bulb was usually thought to be favourable only for moderate to high speed ships but it has been found that it is also beneficial in relatively low speed ships such as tankers and bulk carriers, and ships of these types are often fitted with bulbous bows today. The reduction in resistance in these lower speed ships where the wave resistance is only a small percentage of the total has suggested that the effect of the bulb is to reduce the frictional resistance as well and this would seem to arise because the redistribution of velocity created by the bulb reduces the average rubbing velocity of the water with the hull. Some of the theoretical work has shown that the bulb has a greater effect when it is situated some distance forward of the stem and this has led to the development of the ram bulb which might extend some appreciable distance beyond the forward perpendicular.

Standard series of resistance experiments

The use of a model experiment for an individual ship has already been demonstrated. It is now usual practice for a shipowner to have a model of a new ship tested in a towing tank. The purpose is twofold: firstly, a series of tests can be made to ascertain the best form of the ship to give minimum resistance and this would involve tests run with various alterations to some basic form; secondly, having decided upon some particular form the model can be run to obtain an estimate of power for the ship. The example worked previously has shown how this can be done. The extent of the testing to be done on an individual model will be dependent on the requirements of the owner. In general, nowadays it would include tests with a self-propelled model and more will be said about this later. It may or may not include tests in waves.

Apart from this type of testing a great deal of work has been carried out in towing tanks to ascertain the influence of various features of the ship form on resistance. The tests are done by starting with some basic or parent form and then varying systematically a number of parameters which are considered to influence the resistance. A series of tests of this sort is referred to as a 'standard series' or a 'methodical series'. The results obtained from such tests can be of value in showing what features of the form are important as far as resistance is concerned and are also useful in estimating power for new designs before the stage has been reached where a model for the design in question can be tested.

The last section dealing with features of the ship form which influence resistance has shown that there are many variables concerned. In that section six features were considered apart from bulbous bows. To cover a range of these variables it might be necessary to examine four or five values of each. If only four of each were tested then this would involve $4^6 = 4096$ models which would all have to be run at a range of speeds. It will be seen that the amount of work is enormous and would take a very long time to carry out. For this reason when planning a standard series of model experiments it is customary to settle on a few primary variables which are considered to have the main influence on ship resistance and to restrict examination of other variables to limited ranges. The primary variables which might be considered to be of most importance are the fullness as represented by block or prismatic coefficient, the displacement/length ratio and the breadth/draught ratio. A further variable which might be considered is the longitudinal position of the centre of buoyancy.

The procedure adopted is to choose a parent form which has good general characteristics as far as resistance is concerned and then to derive a series of forms from this by varying the parameters which it has been decided to examine.

Many tests on standard series have been carried out in the past and brief reference will be made to a few of these only. Of these, one which has become very well known is that due to Admiral D. W. Taylor.[14] It was produced in the early part of this century.

Taylor chose as his basic form one which had a cruiser stern and a slightly bulbous bow. The form had no parallel middle body and the centre of buoyancy was at amidships. The variables chosen were prismatic coefficient (called by Taylor 'longitudinal coefficient'), displacement/length ratio $\Delta/(L/100)^3$ and breadth/draught ratio. Five values of displacement/length ratio, eight values of prismatic coefficient and two values of breadth/draught ratio were chosen, making in all a series of eighty models. Taylor's method of plotting the frictional resistance data has already been discussed. The method of plotting the wave resistance (or residuary resistance) was to draw contours of R_r/Δ with prismatic coefficient as absisca and $\Delta/(L/100)^3$ as ordinate. For each of the two breadth/draught ratios chosen, i.e. 2.25 and 3.75, there was a series of diagrams of this type, each one of the series being for a particular value of the speed/length ratio $V/\sqrt{L}$. The range of $V/\sqrt{L}$ was from 0.60 to 2.0 originally but at a later stage the results were extended downwards to a value of $V/\sqrt{L} = 0.30$. Typical charts of residuary resistance are shown in *Figure 8.15*. The procedure in using these is to lift off the values of R_r/Δ for the required prismatic coefficient and displacement/length ratio for a series of values of $V/\sqrt{L}$ and for each of the values of breadth/draught ratio. For the actual breadth/draught ratio R_r/Δ is obtained by linear interpolation. It is thus possible when the values of R_f/Δ are obtained from the Taylor chart discussed earlier to obtain R_t/Δ, and hence R_t and the horsepower can be calculated.

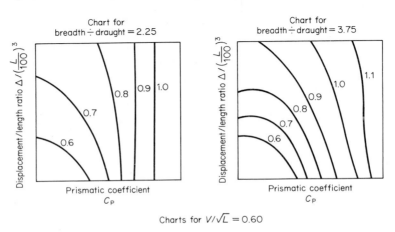

Figure 8.15   Taylor's contours of residuary resistance in pounds per ton of displacement

In 1954 a re-analysis of the Taylor data was published by M. Gertler. In this presentation the cofficients C_f and C_r were used in preference to Taylor's original method of using resistance in pounds per ton of displacement. Frictional resistance was calculated from the Schoenherr formula and residuary resistance was determined from a series of charts, each chart being for a particular value of prismatic coefficient C_p and breadth/draught ratio.

The residuary resistance coefficient $C_r = R_r/\frac{1}{2}\rho S V^2$ was plotted as a series of contours for constant volumetric coefficient $C_v = V/L^3$ to a base of $V/\sqrt{L}$. A typical chart is shown in *Figure 8.16*. In Gertler's presentation there are charts for three values of breadth/draught ratio, namely 2.25, 3.0 and 3.75, and for intermediate values the results can be obtained by interpolation.

An early series of model experiments carried out in England was

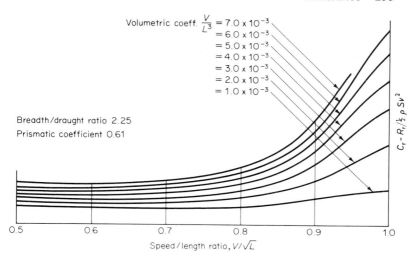

Figure 8.16   Typical chart from Gertler's analysis of Taylor's standard series data

due to R. E. Froude.[15] These were on forms with cruiser sterns and with rams or bulbous bows and were intended to represent warship types rather than merchant ships. Six different types were considered which represented on the full scale the ships shown in *Table 8.7*.

Table 8.7   SIX SHIP TYPES EXPERIMENTED ON BY R. E. FROUDE

Type	Dimensions and draught, ft (m)	Block coefficient	Displacement, tons-force (kN)
1	$350 \times 57 \times 22$ $(106.68 \times 17.37 \times 6.71)$	0.4865	6100 (60 780)
2	$340 \times 57 \times 22$ $(103.63 \times 17.37 \times 6.71)$	0.5000	6083 (60 610)
3	$330 \times 57 \times 22$ $(100.58 \times 17.37 \times 6.71)$	0.5140	6056 (60 310)
4	$325 \times 57 \times 22$ $(99.06 \times 17.37 \times 6.71)$	0.521	6048 (60 260)
5	$320 \times 57 \times 22$ $(97.53 \times 17.37 \times 6.71)$	0.528	6037 (60 150)
6	$310 \times 57 \times 22$ $(94.49 \times 17.37 \times 6.71)$	0.541	6008 (59 860)

For each of these types two ratios of breadth to draught were considered: the first, called A, had the breadth/draught ratio shown in the table, i.e. 57/22, and in the second, B, the breadth/draught ratio was 66/19, both of these ratios referring to a ship of length

107 m (350 ft). For the two variations A and B, a range of $\widehat{M}$ was examined: in A from $\widehat{M}$ = 7.884 to 4.886, and in B from $\widehat{M}$ = 9.933 to 5.407. It will be seen that Froude covered the same variables as did Taylor in his experiments—fullness, displacement/length ratio and breadth/draught ratio. It is not intended to discuss in detail here how these results were plotted, and for more detailed information the reader is referred to Froude's original paper. It is sufficient to say that the data were plotted as $\widehat{C}$ for a 91.4 m (300 ft) ship to a base of $\widehat{M}$ for each type covered by the experiments and this resulted in a set of diagrams, each one of the set being for a constant value of $\widehat{K}$.

Another set of experiments which Froude carried out was on a series of fine forms representing ships 146 m (480 ft) long with breadths from 22.4 to 23.0 m (73.5 to 75.5 ft) and draughts from 7.6 to 8.0 m (25.0 to 26.25 ft).[16] The purpose of these experiments was to consider the relative merits of hollow versus straight waterlines. A special feature was that the models were not only tested in still water but were also run in waves and they are therefore probably the first models to be tested in such conditions.

More recently, methodical series of resistance experiments have been carried out on behalf of the British Ship Research Association and the results of these experiments have been recorded in a number of papers, in two of which the data have been summarised.[17,18] The variables used in these series were block coefficient, which ranged from 0.65 to 0.80, length/displacement ratio $L/V^{\frac{1}{3}}$ ($\widehat{M}$), breadth/ draught ratio and longitudinal position of the centre of buoyancy. The basic parent form represents a ship 122 × 16.8 × 7.9 m (400 × 55 × 26 ft) load draught, and for this form curves of $\widehat{C}$ were plotted to a base of block coefficient for various constant speeds. Curves giving correction factors for variation of length/displacement ratio, breadth/draught ratio and position of centre of buoyancy from those of the standard form were included. Similar sets of diagrams were given for reduced draughts and for a trimmed ballast draught. The forms represent typical single-screw merchant ship forms with cruiser sterns of the period after World War II and from the data given in these papers it is possible to estimate the effective horse-power for a wide range of the principal variables affecting resistance.

Another set of results for merchant ships has been given by Lackenby and Milton.[19] These were a re-analysis of results obtained from the David Taylor Model Basin Series 60 forms tested in the USA. The results cover the same variables as in the B.S.R.A. Series, the range of block coefficient being from 0.60 to 0.80 with variations of the position of the centre of buoyancy from 2.48% aft to 3.51% forward of amidships. Length/breadth ratio varied from 6 to 8 and breadth/draught ratio from 2.5 to 3.5. The values of Ⓒ for a 122 m (400 ft) ship were plotted to a base of block coefficient for a range of constant speeds with correction factors for variation in the other three variables from the values for the parent form. In this case Ⓒ was given using the Froude skin friction correction and also using the I.T.T.C. 1957 line.

In this section only very brief reference has been made to standard or methodical series results and it is impossible to reproduce all the results of even one of these series here. An attempt has, however, been made to show how the results of these experiments have been analysed and plotted. For detailed information the reader should consult the original papers.

Estimation of power from standard series results

The standard or methodical series forms a very suitable basis for making estimates of power, particularly in the early stages of a design. It has been seen that in such series only a limited number of variables can be considered and they are usually block or prismatic coefficient, length/displacement ratio, breadth/draught ratio and longitudinal position of the centre of buoyancy. It has also been demonstrated that even when these parameters are fixed there are still many form variables which could be different in two forms. In estimating power from a standard series, therefore, it is essential that these form variables in the ship for which the estimate is being made should have values as near as possible to those in the standard series, otherwise errors in the estimate are likely to occur. As an example it would be unwise to estimate the power of a ship with a cruiser stern or a bulbous bow from a series which did not have these features. Similarly, errors would occur if there were a marked difference in section shape or length of parallel middle body. To avoid these sources of error it is important when making estimates of power to choose a standard series which has the same general characteristics as the form which is being designed.

Full scale tests on resistance

The model experiment is one of the main sources of knowledge of resistance of ships but ultimately the final test of the accuracy of extrapolation is what happens in the actual ship. Full scale tests have from time to time been made but because of the expense involved in carrying out such tests they have been made much more infrequently than might otherwise be considered desirable.

The earliest tests were those carried out by W. Froude.[1] They were made on a screw sloop, the *Greyhound*, which was 52.6 m (172 ft 6 in) long and 10.1 m (33 ft 2 in) beam. This ship was towed by the *Active* in three different conditions of loading as shown below.

Condition	Displacement, tons-force (kN)	ft	Draught, in	(m)
Normal	1161 (11 570)	13	9	(4.19)
Medium	1050 (10 500)	12	11.5	(3.95)
Light	938 (9 345)	12	1	(3.68)

The towrope was attached to a dynamometer on the *Greyhound* and experiments were carried out over a range of speeds from 3 to 12 knots. The speed was measured by means of what Froude called a 'log ship' towed by the *Greyhound*. The full scale results were compared with those inferred from a model of the *Greyhound*, and showed in general that the actual curve of the resistance to a base of speed was of the same character as that derived from the model but was somewhat higher. This was attributed by Froude to the greater roughness of the *Greyhound* surface than that assumed in the estimates from the model. There was general confirmation of the validity of the law of comparison which had been put forward by Froude and these experiments helped to establish the method of determining the resistance of a ship from that of a model which has now been accepted for a century.

A much more extensive programme of full scale tests was carried out after World War II by the British Ship Research Association and the results have been recorded in a number of papers.[9,20,21,22] These experiments were made on a former Clyde paddle steamer, the *Lucy Ashton*. This ship was 58.1 m (190 ft 6 in) long by 6.4 m (21 ft) breadth moulded and 2.2 m (7 ft 2 in) depth moulded. Most of the experiments were made at a draught of 1.6 m (5 ft 4 in) corresponding to a displacement of 390 tons in salt water. The block coefficient was 0.685 and the prismatic coefficient 0.705. The hull was stripped of its paddle wheels and jet engines were mounted on deck, the purpose of which was to propel the ship without any disturbance of the surrounding water. The speed range tested was from 5 to 15 knots. The influence of different conditions of the surface of the ship was investigated, including different types of paint. The results were compared with tests on a series of geosims which had lengths of 2.7, 3.7, 4.9, 6.1, 7.3 and 9.1 m (9, 12, 16, 20, 24 and 30 ft) respectively.

Estimates of the ship resistance were made from each of these models using various skin friction formulae, including Froude and

259

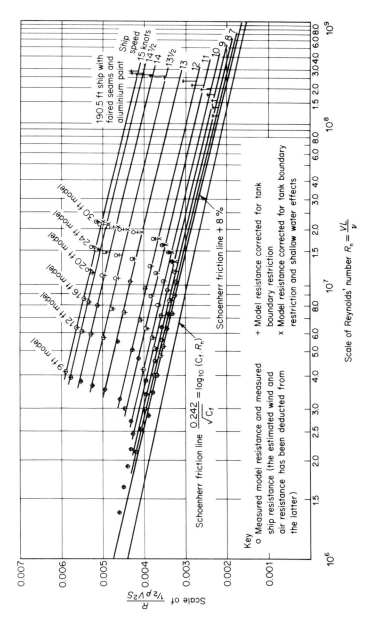

Figure 8.17

Schoenherr, and the results were compared with the full scale measurements. Generally the most satisfactory correlation between model and full scale was obtained using the Schoenherr formula. The results are shown in *Figure 8.17*. Total resistance coefficient $C_t = R_t/\frac{1}{2}\rho S v^2$ is plotted against Reynolds' number and the actual observed results for the various lengths of models and for the ship have been plotted to a base of Reynolds' number, and Schoenherr's line has been plotted on the same diagram. If then lines are drawn through the experimental results at the same Froude number for the different models corresponding to various ship speeds, the observations for the models for a particular ship speed should lie on one of these lines, and furthermore if the Schoenherr line is correct they should be parallel to this line. This apparently was so for the smaller models and when the results for the larger models were corrected for tank boundary and shallow water effects they were brought into line as well. By extrapolating the lines for the various constant speeds to the ship Reynolds' numbers the ship resistance should be obtained. When this was done it was found that there was a discrepancy between the two, the ship resistance being higher. This was attributed to increased roughness in the actual ship, the roughness allowances to be made in C_f to bridge the gaps being as follows for various ship surfaces:

Sharp seams with red oxide paint	+0.000 3
Fair seams with red oxide paint	+0.000 2
Sharp seams with aluminium paint	+0.000 19
Fair seams with aluminium paint	+0.000 13

Only the briefest reference has been made here to the correlation of model results with full scale measurements and in general it can be concluded that there is broad agreement. The subject is, however, a very extensive one and analysis of the *Lucy Ashton* experiments has continued up to the present day. The subject cannot really be discussed fully without reference to propulsion problems, which will be considered in the next chapter.

REFERENCES

1. Froude, W., 'On experiments with H.M.S. *Greyhound*', *Trans. Instn. of Naval Architects*, 1874
2. Reynolds, O., 'An experimental investigation of the circumstances which determine whether the motion of water shall be direct or sinuous and the law of resistance in parallel channels', *Philosophical Trans. of the Royal Society*, 1883
3. Payne, M. P., 'Historical note on the derivation of Froude's skin friction constants', *Trans. Instn. of Naval Architects*, 1936
4. Schoenherr, K. E., 'Resistance of flat surfaces moving through a fluid', *Trans. Society of Naval Architects and Marine Engineers, U.S.A.*, 1932

5. Gertler, M., *A Re-analysis of the Original Test Data for the Taylor Standard Series*, Navy Department, Washington D.C., 1954
6. Wigley, W. C. S., 'Ship wave resistance. An examination and comparison of the speeds of maximum and minimum resistance in practice and theory', *Trans. North East Coast Instn. of Engrs. and Shpbdrs*, 1930–1931
7. Havelock, T. H., 'Some aspects of the theory of ship waves and resistance', *ibid.*, 1925–1926
8. Froude, W., 'Experiments upon the effect produced on the wave-making resistance of ships by length of parallel middle body', *Trans. Instn. of Naval Architects*, 1877
9. Lackenby, H., 'B.S.R.A. resistance experiments on the *Lucy Ashton*', Part III, 'The ship model correlation for the shaft appendage conditions', *ibid.*, 1955
10. Mandel, P., 'Some hydrodynamic aspects of appendage design', *Trans. Society of Naval Architects and Marine Engineers*, 1953
11. Hughes, G., 'Model experiments on the wind resistance of ships', *Trans. Instn. of Naval Architects*, 1930
12. Froude, R. E., 'On the constant system of notation of results of experiments on models used at the Admiralty Experiment Works', *ibid.*, 1888
13. Ayre, A. L., 'Essential aspects of form and proportions as affecting merchant ship resistance and a new method of estimating E.H.P.', *Trans. North East Coast Instn. of Engrs and Shpbdrs*, 1926–1927
14. Taylor, D. W., *Speed and Power of Ships*, United States Shipping Board, revised 1933, now out of print
15. Froude, R. E., 'Some results of model experiments', *Trans. Instn. of Naval Architects*, 1904
16. Froude, R. E., 'Model experiments on hollow versus straight waterlines in still water and among artificial waves', *ibid.*, 1905
17. Moor, D. I., Parker, M. N., and Pattullo, R. N. M., 'The B.S.R.A. Methodical Series. An overall presentation. Geometry of forms and variation of resistance with block coefficient and longitudinal centre of buoyancy', *Trans. Royal Instn. of Naval Architects*, 1961
18. Lackenby, H., 'The B.S.R.A. Methodical Series. An overall presentation. Variation of resistance with breadth/draught ratio and length/displacement ratio', *ibid.*, 1966
19. Lackenby, H., and Milton, P., 'D.T.M.B. Standard Series 60. A new presentation of the resistance data for block coefficient, L.C.B., breadth/draught ratio and length/breadth ratio variations', *ibid.*, 1972
20. Denny, Sir Maurice, 'B.S.R.A. Resistance experiments on the *Lucy Ashton*', Part I, 'Full scale measurements', *Trans. Instn. of Naval Architects*, 1951
21. Conn, J. F. C., Lackenby, H., and Walker, W. P., 'B.S.R.A. Resistance experiments on the *Lucy Ashton*', Part II, 'The ship model correlation of the naked hull', *ibid.*, 1953
22. Livingston Smith, S., 'B.S.R.A. Resistance experiments on the *Lucy Ashton*', Part IV, 'Miscellaneous investigations and general appraisal', *ibid.*, 1955

9

Propulsion

Propulsive devices may take many forms but in this chapter only mechanical means of propulsion will be discussed. Of the mechanical means the screw propeller has found most use and the chapter will be devoted almost exclusively to this.

Before actually discussing propeller theory and experiments it is important to understand the terms used, and also to have some knowledge of certain principles of hydrodynamics, so that in the first part of the chapter these will be considered.

Definitions

The helicoidal surface

Consider a line AB perpendicular to a line AA' and suppose that AB rotates with uniform velocity about AA' and at the same time moves along AA' with uniform velocity (see *Figure 9.1*). The surface swept out by AB is a helicoidal surface and is in fact the surface of a screw. The blade of a propeller is part of such a surface, or, more correct, the face of what is called a flat faced blade is part of such a surface.

Now suppose that the line AB makes one complete revolution and arrives at A'B' having travelled an axial distance AA'. The distance AA' represents the pitch of the surface and since the propeller blade is part of that surface it is the pitch of the blade, usually denoted by P.

Consider next some point distance r from the line AA'; then the circumferential distance travelled by the point in one revolution will by $2\pi r$ and the axial distance is the same as before, P. Suppose the line AB rotated at N revolutions in unit time, then the circumferential velocity $v_c = 2\pi Nr$ and the axial velocity is $v_a = NP$. The

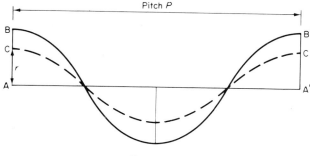

Figure 9.1

point travels in a direction inclined at an angle θ to the centre line which is given by $\tan \theta = v_\mathrm{c}/v_\mathrm{a} = 2\pi N r/NP = 2\pi r/P$. It will be seen that for any radius r, $\tan \theta$ is constant, and this suggests a simple method for describing the path of a point on the blade. If the path is unwrapped and laid out flat the point at radius r will move along a straight line as shown in *Figure 9.2*. The face of a flat faced propeller blade will be on the line OA, and this really defines what is meant by a flat faced blade; when a radial section through the blade at some radius r is laid out flat the face is a straight line.

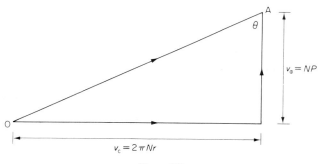

Figure 9.2

Propellers can have two or more blades, and three, four or five are common in marine propellers and even six blades are sometimes used. Each blade is part of a different helicoidal surface spaced at angular distances $360/n$ degrees, where n is the number of blades. Often in modern propellers the pitch of a blade is not constant over the radius and this means that sections at different radii are not on the same helicoidal surface.

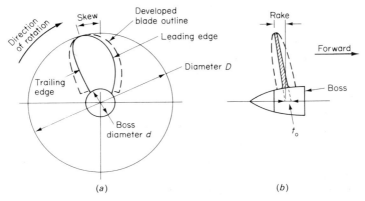

Figure 9.3   (a) *View perpendicular to shaft axis* (b) *Side elevation*

Propeller diameter

If in *Figure 9.3* (*a*) a circle with its centre on the shaft axis is drawn which is tangential to the tips of the blades then the diameter of this circle is the propeller diameter *D*. The diameter is one of the most important features of a propeller.

Boss diameter

The blades at their lower ends or roots are attached to a boss which in turn is attached to the propeller shaft. The maximum diameter of this boss is called the boss diameter *d*. The boss diameter is usually made as small as possible and should be no larger than will be sufficient to accommodate the blades. It is usually expressed as a fraction of the propeller diameter, i.e. as d/D. At one time propeller blades were manufactured separately from the boss, but modern fixed pitch propellers have the boss and blades cast together. However, in controllable pitch propellers it is of course necessary for blades and boss to be manufactured separately.

Blade outline

The inclination of any blade section to the centre line of the shaft varies with the radius since $\tan \theta = 2\pi r/P$. It follows therefore that the blade is twisted about the radius and it thus cannot be truly developed into a flat surface. This, however, can be done approximately and the resulting outline is called the 'developed outline' of the blade. The shape of the outline varies a great deal nowadays, although at one time it was part of an ellipse whose major axis was $D/2$. Many different shapes are at present in use and the determination of the best outline is one of the aims in propeller design.

In most propeller designs it is customary to show a projected outline of the blade, this being the view seen looking along the axis of the propeller. It gives the projection on a transverse plane; a similar projection can be made on the longitudinal plane and is useful in showing the clearance which exists between the propeller and the hull. It enables a satisfactory stern frame aperture to be decided upon, especially in single-screw ships.

Blade sections

When blade sections are referred to, radial sections through the blade are to be understood and the shape of these sections is their shape when laid out flat. Blade sections were formerly of the flat face circular back type, the face being the side of the blade looking aft and the back being the side looking forward. This type of section required only one parameter to define its shape completely, the thickness of the blade (measured perpendicular to the face at the thickest point) expressed as a fraction of the blade width or chord as shown in *Figure 9.4(a)*. If the blade width is also known then the blade section

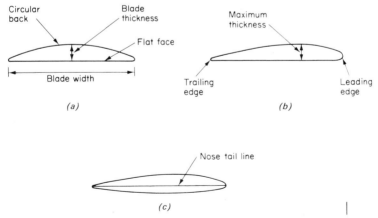

Figure 9.4   (a) *Flat face circular back section* (b) *Aerofoil section* (c) *Section with cambered face*

is completely defined. In this type of section the face is a straight line and the back is a circular arc. Where the face and back meet at the edges of the section, sharp edges would result and it is customary to round these off for manufacturing purposes.

The pitch of a flat face circular back section is easily defined, being simply the pitch of the flat face. It should be noted, however, that the back of the section could be considered as having a pitch which varies across the breadth since if tangents are taken to different points in the breadth they will make different angles with the centre line, depending upon which point in the breadth is considered.

The blade sections of modern propellers are often aerofoil sections, i.e. they are like the section of an aeroplane wing. The characteristics of such sections are shown in *Figure 9.4(b)*. The maximum thickness of the section is forward of the middle of the chord and the leading edge (the edge which meets the on-coming fluid) is bluffer than the trailing edge. The face of an aerofoil section may be flat but the back cannot be defined by any single dimension as in the circular back type. It is necessary therefore to specify the exact shape by a series of ordinates at different positions in the breadth. The face of a section may also be shaped as well as the back, as shown in *Figure 9.4(c)*. It is not uncommon for certain sections of the blade to have the leading and trailing edges 'lifted' as shown. In such cases it is not easy to define pitch but it can be done by considering the pitch of some line such as the nose tail line of the section.

The characteristics of a blade section can be defined by two ratios, the thickness ratio which is the maximum thickness divided by the chord or blade breadth, t/b, and the centre line camber ratio which is the ratio of the camber of the line drawn midway between the face and the back to the blade breadth, c/b. In the case of a flat face circular back section the camber ratio is half the thickness ratio and for a symmetrical section the centre line camber ratio would be zero.

The maximum thickness of blade sections decreases in going outwards from the boss to the tip and this decrease is often linear. Lines showing the blade thickness at any radius are usually drawn on the propeller plan as shown in *Figure 9.3(b)*. If these lines are produced down to the shaft axis an artificial thickness t_0 is obtained. The ratio of this thickness to the propeller diameter t_0/D is called the blade thickness fraction.

Rake and skew

In considering the generation of the helicoidal surface it was assumed that the line AB in *Figure 9.1* was perpendicular to the shaft axis. Generally, however, this line is tilted aft at some angle and under these circumstances the propeller blades are said to be raked. Rake can be measured either as the angle which the generator line makes with the vertical or alternatively as the longitudinal distance of the line at the tip from the point of its intersection with the shaft axis. The purpose of rake is to increase clearance of the tip from the hull, the rake always being aft.

Referring to the transverse view of the propeller in *Figure 9.3 (a)* it will be seen that the blade tip is some distance from the vertical in the opposite direction to the direction of rotation. This distance is called skew.

Pitch ratio

The ratio of the pitch of a propeller to its diameter is an important feature in considering propeller performance, as will be seen later. It is called the pitch ratio P/D and is denoted by p. Where the pitch is not constant then the pitch ratio has to be defined by the pitch at some chosen position in the length of the blade, usually the tip. In such cases it is necessary in comparing propellers to define how the pitch varies over the radius.

Blade area

Blade area is usually considered as the area enclosed by the developed blade outline and is related to the area of the screw disc, which is $(\pi/4)D^2$ where D is the propeller diameter. R. E. Froude used the term 'disc area ratio', which is defined as the blade area, including the part which is taken up by the boss, divided by $\pi D^2/4$. Thus if A_0 is the blade area down to the shaft axis then

$$\text{Disc area ratio} = \frac{A_0}{\pi D^2/4}$$

This was quite a convenient way of expressing the fraction which the developed area is of the propeller disc area for the elliptical type of blade and its wide tipped modification used by Froude. It has the advantage that for a given blade outline for different boss diameters the disc area ratio would be the same. In modern practice, however, it is usual to use a blade area ratio which can be written

$$\text{Blade area ratio} = \frac{\text{Total blade area clear of boss}}{\pi D^2/4}$$

Sometimes the term 'projected area' will be met with and this is simply the area of the blades projected on to a plane perpendicular to the axis of rotation.

Forces on a blade section

Some of the concepts used in the aeronautical field are useful in considering the forces acting on propeller blades. The aerofoil is a body which is totally immersed in a fluid and this can be considered to be true for a propeller blade section under working conditions. Referring back to the dimensional equation for the force on a body developed in the previous chapter it will be seen that the gravitational term can be omitted in this case and the force becomes

$$F = \rho v^2 \, l^2 f\left(\frac{v}{vl}\right)$$

If b is the breadth of the aerofoil, i.e. the dimension in the direction of motion, and S is the span or the dimension perpendicular to the direction of motion then l^2 in the above equation could be replaced by $S \times b$ or the area A of one surface of the aerofoil. It is then possible to write the force in non-dimensional form as

$$\frac{F}{\rho v^2 A} = f\left(\frac{v}{vl}\right) \tag{9.1}$$

Another factor which comes into the problem when considering an aerofoil is its attitude to the flow. This is called the 'angle of incidence' and is taken as the angle between some reference line on the aerofoil (say the lower surface) and the flow as shown in *Figure 9.5*. The expression for the force in non-dimensional terms then becomes

$$\frac{F}{\rho v^2 A} = f\left\{\left(\frac{v}{vl}\right)\alpha\right\} \tag{9.2}$$

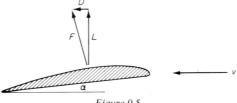

Figure 9.5

The resultant force F can be resolved into two components, one L perpendicular to the direction of the flow called the 'lift', and another D tangential to the direction of the flow called the 'drag'. They are expressed in terms of non-dimensional lift and drag coefficients C_L and C_D given by

$$C_L = \frac{L}{\frac{1}{2}\rho v^2 A} \text{ and } C_D = \frac{D}{\frac{1}{2}\rho v^2 A}$$

Both of these coefficients are functions of the angle of incidence and Reynolds' number and for a given aerofoil at a given Reynolds' number will be functions of angle of incidence only. The lift and drag coefficients when plotted to a base of angle of incidence give curves as shown in *Figure 9.6*. The lift coefficient curve is practically

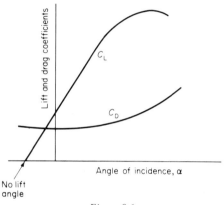

Figure 9.6

a straight line starting from a small negative angle of incidence called the 'no lift angle', but eventually curving downwards, and at some positive angle there is a steeper drop in the curve where stalling due to breakdown of flow occurs. The drag coefficient curve has its minimum value near zero angle of incidence and then rises slowly at first and then more rapidly at the higher angles.

Generation of lift on an aerofoil

Figure 9.7 shows an aerofoil section inclined at an angle of incidence α to the direction of flow. The paths of the streamlines are shown. If the velocity of the aerofoil relative to the main body of the fluid is v it will be found that the streamlines are slowed up on the lower

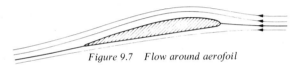

Figure 9.7   Flow around aerofoil

surface or the face and that the velocity on the back is increased. The effect of this is to increase the pressure on the lower face and to reduce it on the upper face, with the net result that a force is generated on the aerofoil.

Theoretical investigations of the flow around a circular cylinder in two dimensions (or around an infinitely long cylinder) show that the type of flow in *Figure 9.8* is obtained with stagnation points

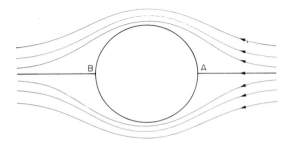

Figure 9.8

or points of zero velocity at A and B. The total force on the cylinder would be zero. This flow around a cylinder can be transformed to the flow around an aerofoil shape as shown in *Figure 9.9* where the stagnation points move to positions A′ and B′ and the total force

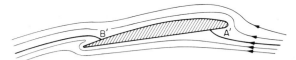

Figure 9.9   Flow around aerofoil without circulation

on the aerofoil is again zero. By imagining that there is super-imposed on the flow a circulatory flow in the anti-clockwise direction the stagnation points will be displaced because the velocity will be increased on the top of the aerofoil and reduced on the face. If the circulation is adjusted properly the rear stagnation point can be brought to the trailing edge so that the fluid leaves the aerofoil at that point with finite velocity and the flow is as shown in *Figure 9.10*. The circulation around any curve in the fluid is taken as the line

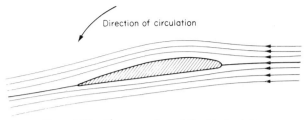

Figure 9.10   Flow around aerofoil with circulation

integral of the velocity around the curve as in *Figure 9.11* where
it will be seen that

$$\text{Circulation } \Gamma = \int_C u \cos \alpha \, ds$$

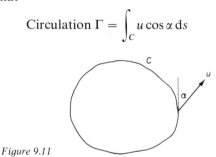

Figure 9.11

The simplest type of circulation is where there is no radial velocity
and the velocity around a circle of radius r is constant and is
inversely proportional to the radius. Thus, if the tangential velocity
at radius r is u then

$$\Gamma = 2\pi r u \text{ or } u = \frac{\Gamma}{2\pi r}$$

If this circulation is superimposed on the flow around any body
in two dimensions then it will be found that the force on the body
perpendicular to the direction of motion will be

$$L = \rho v \Gamma \tag{9.3}$$

where ρ is the density of the fluid and v is the velocity of the
body relative to the undisturbed fluid. In the case of the aerofoil
the value of Γ has to be adjusted to make the trailing edge a
stagnation point so that there will be a lift $L = \rho v \Gamma$.

For a certain type of thin aerofoil Γ can be shown to be
$\pi A v \times \sin(\alpha + \beta)$ where A is the surface area of one side of the
aerofoil and α and β are the angle of incidence and the no lift angle

respectively. Hence

$$L = \rho v \times \pi A v \sin(\alpha + \beta)$$
$$= \pi \rho v^2 A \sin(\alpha + \beta)$$

Consequently the lift coefficient is

$$C_L = \frac{L}{\frac{1}{2}v^2 A} = 2\pi \times \sin(\alpha + \beta)$$

and the slope of the lift coefficient curve is $dC_L/d\alpha = 2\pi \cos(\alpha + \beta)$ which for small angles of incidence gives a slope of 2π.

Three-dimensional flow

The theory outlined in the previous section is for flow in two dimensions. Because the aerofoil is of finite span perpendicular to the flow, there tends to be span-wise flow. This arises because there is increased pressure on the lower face and reduced pressure on the upper face. At the ends of the aerofoil these pressures tend to equalise with the result that there is flow span-wise from the high pressure region to the low pressure, and vortices are formed at the ends. The effect is to reduce the lift near the ends of the aerofoil so that for a given angle of incidence the lift of the aerofoil as a whole is reduced. The effect is greater the less is the span in relation to the chord. The ratio of these dimensions S/C is called the 'aspect ratio' of the aerofoil and as aspect ratio increases the lift characteristics approach more nearly those for the aerofoil in two dimensions.

Pressure distribution around aerofoils

When the flow around an aerofoil has been determined for any given angle of incidence it is possible to calculate the pressure distribution across the chord. The type of distribution obtained is as shown in *Figure 9.12*, where it will be seen that most of the lift force is generated by reduced pressure on the back. The maximum reduction in pressure occurs at some point between mid chord and the leading edge. If the reduction in pressure is too great in relation to the static pressure in a fluid like water it is possible for bubbles to form filled with air and water vapour. These bubbles are swept along the chord until they reach regions of higher pressure where they collapse, generating high localised forces which can damage the material of the blade in the case of a marine propeller. This is the phenomenon known as 'cavitation' and will be discussed more fully later.

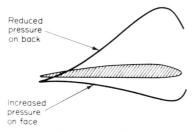

Figure 9.12   Pressure distribution on aerofoil

Theories of action of the marine propeller

Since the introduction of the screw propeller many theories of its action have been put forward, but none of the earlier theories are of any value quantitatively. One nineteenth century theory, however, is of considerable value in explaining qualitatively how a propeller works and this is the momentum theory. The theory was first put forward by Rankine and was later developed by R. E. Froude.[1] Froude's development will be discussed here.

Momentum theory

In the momentum theory the propeller is regarded simply as a means for accelerating water. In the first instance it is assumed that only axial acceleration is imparted to the water. The change in velocity of the water in passing through the screw implies a change in pressure as shown in *Figure 9.13*. At infinity ahead of the propeller the pressure is p_0, at the screw disc it receives an increment of pressure dp, after which the pressure ultimately returns to p_0 again at infinity behind the screw. From velocity considerations the water is originally at rest, it achieves a velocity of aV_a at the screw, goes on accelerating and eventually has a velocity bV at infinity behind, where V_a is the velocity of the propeller relative to still water. The increase in velocity results in contraction of the column of water entering the screw as shown in *Figure 9.13*.

In developing the momentum theory it will be assumed that the increment in velocity is constant across the screw disc, that motion is communicated only to the column of water of area A at the screw disc, and that friction losses can be neglected. It will be seen that

Velocity of water relative to the screw $= V_a(1 + a)$

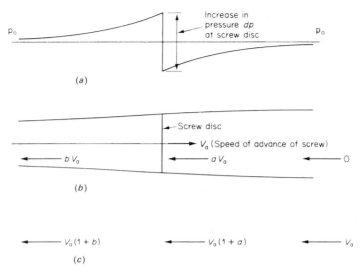

Figure 9.13   (a) *Pressure of water* (b) *Absolute velocity of water* (c) *Velocity of water relative to screw*

so that the mass of water being acted upon in unit time is $\rho A V_{\text{a}}(1 + a)$.

Since this mass eventually achieves a velocity bV_{a} then

Change of momentum in unit time $= \rho A V_{\text{a}}(1 + a)bV_{\text{a}}$

The time rate of change of momentum is a measure of the force imparted to the water and is therefore equal to the thrust generated on the propeller. Hence

$$T = \rho A V_{\text{a}}(1 + a)bV_{\text{a}} \qquad (9.4)$$

The work done by the thrust on the water is

$$TaV_{\text{a}} = \rho A V_{\text{a}}^3(1 + a)ba \qquad (9.5)$$

This must be equal to the kinetic energy of the water in the propeller race, so that

$$\text{Kinetic energy} = \tfrac{1}{2}\rho A V_{\text{a}}(1 + a)(bV_{\text{a}})^2 \qquad (9.6)$$

Equating equations 9.5 and 9.6:

$$\rho A V_{\text{a}}^3 (1 + a) ba = \tfrac{1}{2}\rho A V_{\text{a}}^3 (1 + a) b^2$$

from which it will be seen that

$$a = b/2 \qquad (9.7)$$

or in other words half the velocity ultimately achieved by the water is acquired by the time it reaches the propeller.

The useful work done by the propeller is equal to the thrust T generated on it, multiplied by its forward velocity V_a through the water, and the total work done is this plus the work done in accelerating the water. Therefore

$$\text{Total work} = \rho A V_a^3 (1 + a) ba + \rho A V_a^3 (1 + a) b$$

The efficiency of propulsion is the ratio of the useful work to the total work and is given by

$$\eta = \frac{\rho A V_a^3 (1 + a) b}{\rho A V_a^3 (1 + a) ba + \rho A V_a^3 (1 + a) b}$$
$$= 1/(1 + a) \tag{9.8}$$

This is what may be called the ideal efficiency and it shows that for high efficiency a must be as small as possible. Referring back to the expression for thrust (equation 9.4), since $b = 2a$ it follows that

$$T = 2 \rho A V_a^2 (1 + a) a$$

from which it will be seen that as a becomes small for given A and V_a the thrust will also be small, and if a were zero then thrust would also be zero. From this it will be seen that 100% efficiency is not possible unless the thrust is zero.

The simple axial momentum theory does, however, show how high efficiency can be achieved. The object of any system of propulsion is to produce a thrust T which will overcome the resistance of the ship at some speed V_a. It follows then that if these two quantities are fixed in any particular case the quantity $A (1 + a) a$ must be constant. If a is to be small for high efficiency then A must be large, so that the general conclusion can be drawn that for high efficiency the propeller diameter must be large. This conclusion, although drawn from a theory which omits many factors, does in fact agree with the results of experiments.

Extension of momentum theory

The theory considered in the previous section assumed that the water in passing through the propeller disc received an axial increase in velocity only. It is evident, however, that because of the rotation of the blades in a real propeller the water will also have rotational motion imparted to it. The momentum theory can be extended to take this effect into account.

The rotation of the water involves a further loss of energy and if the increase in rotational velocity at any radius is given by $a' 2 \pi N r$, where N is the revolutions in unit time and a' is a small fraction similar to a, it can be shown that the propulsive efficiency now becomes

$$\eta = (1 - a')/(1 + a) \tag{9.9}$$

so that it will be seen that the 'ideal' efficiency is reduced further.

Blade element theory

The momentum theory considers the propeller merely as a means for accelerating water with no regard as to how forces are generated on the propeller blades. The blade element theory attempts to take into account the mechanism of the creation of the forces on the blades. An early theory of this type was put forward by W. Froude,[2] but this is now only of historical interest. The theory which will be described here considers the forces on a radial section of the propeller blade and takes into account the increase of axial and rotational velocity occurring at the blade as considered in the momentum theory.

It has been seen that the path of a blade element can be un-wrapped and laid out flat, and this is a very convenient way of examining the forces on the element. Suppose a radial section distance r from the axis of rotation is considered, then as shown in *Figure 9.14* the rotational velocity is $2\pi Nr$, where N is the revolutions in unit time. If the blade was a screw working a nut it would at the same time move in an axial direction with a velocity NP, i.e. along the line OC where P is the pitch of the blade. However, because the water in which the blade is working is not a solid medium, it moves forward at some less speed V_a and therefore the direction of motion of the blade relative to the stationary fluid is along OB, the direction of flow making an angle of incidence ϕ_2 with the face of the blade. An old concept was that the blade had slipped back a distance $NP - V_a$ in unit time and the ratio $(NP - V_A)/NP = 1 - V_A/NP$ was called the slip S, the angle ϕ_2 in the diagram being called the 'slip angle'. The term 'slip' has to a large extent fallen into disuse and a coefficient $J = V_a/ND$, known as the 'advance coefficient', is now more often used. Now $P = pD$, so that

$$S = 1 - (V_a/p\,ND) = 1 - J/p$$

so that these two quantities are related.

The line OB in *Figure 9.14* represents the direction of motion of the blade relative to still water, but it has been seen that the water at the screw disc has received an axial increase in velocity aV_a and a rotational increase in velocity $a'2\pi Nr$. The effect of this is that

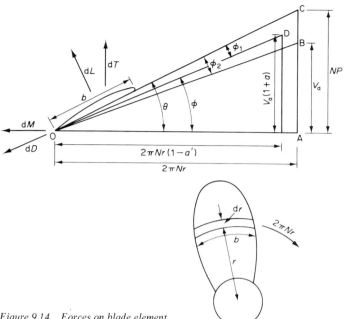

Figure 9.14 Forces on blade element

the flow relative to the blade is along the line OD, making an angle of incidence ϕ_1 with the face. On the blade element breadth b and radial depth dr there will thus be lift and drag forces dL and dD generated which will be respectively perpendicular and tangential to OD. It follows that

$$dL = \tfrac{1}{2}\rho\,dAv^2 = \tfrac{1}{2}\rho C_L\{V_a^2(1+a)^2 + 4\pi^2 r^2(1-a')^2\} \times b\,dr$$

and

$$dD = \tfrac{1}{2}\rho\,dAv^2 C_D = \tfrac{1}{2}\rho C_D\{V_a^2(1+a)^2 = 4\pi^2 r^2(1-a')^2\}b\,dr$$

The element of thrust on the blade will be obtained by resolving the lift and drag forces in the axial direction so that

$$dT = dL\cos\phi - dD\sin\phi = dL\{\cos\phi - (dD/dL)\sin\phi\}$$

$$= \tfrac{1}{2}\rho v^2 C_L\{\cos\phi - (dD/dL)\sin\phi\}b\,dr$$

$$= \tfrac{1}{2}\rho v^2 C_L(\cos\phi - \tan\beta\sin\phi)\,b\,dr$$

$$= \tfrac{1}{2}\rho v^2 C_L[\{\cos(\phi + \beta)\}/\cos\beta]$$

where $\tan\beta = dD/dL = C_D/C_L$.

Now it can be seen from *Figure 9.14* that $v = V_a(1 + a)/\sin\phi$ so that

$$dT = \tfrac{1}{2}\rho C_L \times \frac{V_a^2(1+a)^2\cos(\phi+\beta)}{\sin^2\phi\cos\beta}\,bdr \qquad (9.10)$$

The total thrust produced by the propeller would then be

$$T = n\int_{d/2}^{D/2}\tfrac{1}{2}\rho C_L\frac{V_a^2(1+a)^2\cos(\phi+\beta)}{\sin^2\phi\cos\beta}\,bdr \qquad (9.11)$$

where *n* is the number of blades and the integration is taken from the root, i.e. the boss to the blade tips. The transverse force dM acting on the blade element at radius *r* from the axis is given by

$$dM = dL\sin\phi + dD\cos\phi = dL\{\sin\phi + (dD/dL)\cos\phi\}$$
$$= dL(\sin\phi + \tan\beta\cos\phi)$$
$$= \tfrac{1}{2}\rho v^2 C_L(\sin\phi + \tan\beta\cos\phi)\,bdr$$
$$= \tfrac{1}{2}\rho v^2 C_L\{\sin(\phi+\beta)/\cos\phi\}\,bdr$$

On substituting for v^2 and multiplying by the radius *r* to obtain the torque:

$$dQ = rdM = \tfrac{1}{2}\rho C_L r\frac{V_a^2(1+a)^2\sin(\phi+\beta)}{\sin^2\phi\cos\beta}\,bdr \qquad (9.12)$$

The total torque on the propeller becomes

$$Q = n\int_{d/2}^{D/2}\tfrac{1}{2}\rho C_L r\left\{\frac{V_a^2(1+a)^2\sin(\phi+\beta)}{\sin^2\phi\cos\beta}\right\}\,bdr \qquad (9.13)$$

From these two equations the thrust horsepower proportional to TV_a and the shaft horsepower proportional to NQ can be calculated and the propeller efficiency would be

$$\eta = TV_a/2\pi NQ$$

It is instructive to examine the efficiency of a blade element and it will be seen that this is given by

$$\eta = \frac{\tfrac{1}{2}\rho C_L\dfrac{V_a^3(1+a)^2\cos(\phi+\beta)bdr}{\sin^2\phi\cos\beta}}{2\pi N\tfrac{1}{2}\rho C_L rV_a^3(1+a)^2\sin(\phi+\beta)bdr/\sin^2\phi\cos\beta}$$

$$= \frac{V_a}{2\pi NR}\times\frac{1}{\tan(\phi+\beta)}$$

From *Figure 9.14* it will be seen that

$$\frac{V_a}{2\pi Nr} = \frac{V_a(1+a)}{2\pi Nr(1-a')}\times\left(\frac{1-a'}{1+a}\right) = \tan\phi\times\frac{1-a'}{1+a}$$

so that efficiency of the blade element is

$$\eta = \frac{1 - a'}{1 + a} \times \frac{\tan \phi}{\tan (\phi + \beta)} \tag{9.14}$$

It will be seen that there are two factors which affect the efficiency of a blade element, the first being bound up with momentum, i.e. the factor $(1 - a')/(1 + a)$, the second with the characteristics of the blade section, i.e. the angle β which is governed by the ratio of the drag and lift coefficients, and also with the angle ϕ. If β were zero, which is equivalent to saying that the drag of the section is zero, then the second term would be unity and the efficiency would be the ideal efficiency forecast by momentum considerations. There is therefore a further reduction in efficiency arising from the blade drag.

In order to be able to calculate the thrust and torque for a complete propeller it would be necessary to evaluate the quantities a and a' for each radial section, from which the angle ϕ could be calculated and hence the angle of incidence ϕ_1 obtained, from which C_D and C_L could be evaluated from published data. This cannot be done easily and will not be pursued further here.

There are many factors which have been left out of this simple theory. For example the factors a and a' are based on the assumption that the blades communicate a uniform increase in velocity to the annulus of area $2\pi r dr$. As there are gaps between the blades it would be expected that the velocity would be less than suggested by the simple theory. The theory would apply to a propeller with an infinite number of blades. The influence of a finite number of blades was worked out by J. Goldstein and a series of correction factors was introduced to take the effect into account.

The theory does not allow for interference between the blades. In marine propellers, because of the breadth of the blades the pressure field on one blade interferes with the adjacent blades which affects the lift and drag coefficients. This is known as the cascade effect and in developments of theory corrections are made to allow for it.

Other factors which should be taken into account include the fact that at the tips of the blades the flow spills over, producing a tip vortex which modifies the lift and drag of the section in that region; a complete theory should also allow for the drag of the boss.

It is not intended to pursue propeller theory any further as such developments are well outside the scope of this work. However, it should be noted that theory has developed very considerably in the last 30 years, based largely on considerations of the circulation around the blades, and it has now become possible to produce a quantitative theory which enables propellers to be designed for given conditions without recourse to experiment.

Law of comparison for propellers

As in the case of resistance a great deal of knowledge concerning the performance of propellers has been gained from experiments on models and it is important therefore to examine the relation between model and full-scale results. The theory of dimensions can be used to establish this relation and in what follows an expression will be obtained for the thrust obtained from a propeller.

It will be assumed that the thrust developed by a propeller depends upon a linear dimension, in this case the diameter D, the speed of advance V_a, the revolutions in unit time N, the density and viscosity of the water ρ and μ, the static pressure of the fluid p and the acceleration due to gravity g. Hence the thrust can be written

$$T \propto D^a V_a^b N^c \rho^d \mu^e p^f g^h$$

Therefore

$$ML/T^2 = L^a (L/T)^b (1/T)^c (M/L^3)^d (M/LT)^e (M/LT^2)^f (L/T^2)^h$$

Equating the indices of M, L and T on the two sides of the equation leads to

$$T = \rho V_a^2 D^2 \{F_1 (ND/V_a), F_2 (v/V_a D), F_3 (p/\rho V_a^2), F_4 (gD/V_a^2)\} \quad (9.15)$$

The quantity $\rho V_a^2 D^2$ has the dimensions of a force whilst all the terms in the brackets are functions of non-dimensional quantities. Considering these functions in turn it will be seen that $F_1 (ND/V_a)$ is a function of the advance coefficient J or if preferred is a function of slip. This is likely to be an important factor in the performance of a propeller. $F_2 (v/V_a D)$ is a function of Reynolds' number and in the propeller problem this is not likely to be nearly as important as in the ship resistance problem, because viscous forces are not such a large proportion of the total force involved. It would not be unreasonable therefore to neglect this term. The function $F_3 (p/\rho V_a^2)$ is concerned with the relation between the static pressure of the fluid and the dynamic pressure and is therefore bound up with cavitation. The importance of this term will be discussed later but under non-cavitating conditions it can be ignored. The final term $F_4 (gD/V_a^2)$ is of the form of a Froude number and is concerned with gravitational effects. The elementary theories discussed in previous sections have shown that gravitational considerations do not enter into the problem so that this term too can be ignored. However, in the case of a propeller operating near the surface wave making may occur, in which case the term might become important.

For deeply immersed propellers under non-cavitating conditions the expression for thrust reduces to

$$T = \rho V_a^2 D^2 \{F_1(ND/V_a)\} \qquad (9.16)$$

Consider now two geometrically similar propellers denoted by the suffices 1 and 2, then

$$T_1 = \rho_1 V_{a1}^2 D_1^2 \{F_1(N_1 D_1/V_{a1})\}$$

and

$$T_2 = \rho_2 V_{a2}^2 D_2^2 \{F_1(N_2 D_2/V_{a2})\}$$

If $N_1 D_1/V_{a1} = N_2 D_2/V_{a2}$, i.e. if both propellers work at the same advance coefficient then

$$\frac{T_1}{T_2} = \frac{\rho_1}{\rho_2} \frac{V_{a1}^2}{V_{a2}^2} \times \frac{D_1^2}{D_2^2} \qquad (9.17)$$

From equation 9.17 it is possible to obtain the thrust of a full size propeller from the results obtained on a model. It is not necessary to have both model and full size propeller working at the same Froude number, but it may be convenient to do so, particularly if the model propeller is working behind a hull model. In these circumstances

$$\frac{gD_1}{V_{a1}^2} = \frac{gD_2}{V_{a2}^2} \text{ or } \frac{V_{a1}^2}{V_{a2}^2} = \frac{D_1}{D_2}$$

and if this is substituted in equation 9.17

$$\frac{T_1}{T_2} = \frac{\rho_1}{\rho_2} \times \frac{D_1}{D_2} \times \frac{D_1^2}{D_2^2} = \frac{\rho_1}{\rho_2} \times \frac{D_1^3}{D_2^3} = \frac{\rho_1}{\rho_2} \times \lambda^3 \qquad (9.18)$$

where λ is the ratio of the linear dimensions of model and full size propellers. This is the same result as was obtained from Froude's law for wave making resistance.

Referring back to the condition for equality of advance coefficient in the two propellers

$$\frac{N_1}{N_2} = \frac{V_{a1}}{V_{a2}} \times \frac{D_2}{D_1} = \sqrt{\lambda} \times \frac{1}{\lambda} = \frac{1}{\sqrt{\lambda}}$$

If therefore $N_1 D_1$ refer to the ship propeller and $N_2 D_2$ to the model propeller then

$$N_2 = N_1 \sqrt{\lambda}$$

from which it will be seen that the revolutions of the model will be $\sqrt{\lambda}$ times those of the ship. In other words the model will rotate faster.

The relation between the power delivered by model and ship can now easily be determined. The thrust horsepower (T.H.P.) is proportional to $T \times V_a$ so that

$$\frac{T_1 V_{a1}}{T_2 V_{a2}} = \frac{\rho_1}{\rho_2} \times \lambda^3 \sqrt{\lambda} = \frac{\rho_1}{\rho_2} \times \lambda^{7/2}$$

at the same Froude number and the shaft horsepowers (S.H.P.) are related in the same way:

$$\frac{\text{S.H.P.}_1}{\text{S.H.P.}_2} = \frac{\rho_1}{\rho_2} \times \lambda^{7/2}$$

The torques of the two propellers will vary as $\rho_1/\rho_2 \times \lambda^4$ since torque is equal to a force multiplied by a linear dimension.

Coefficients for the presentation of propeller data

Dimensional analysis suggests how suitable coefficients can be derived for plotting propeller data. The basic expression for thrust has been seen to be

$$T = \rho V_a^2 D^2 F_3(J)$$

$$\text{where } J = V_a/ND$$

Now $V_a = NDJ$ and substituting this in the expression for T gives

$$T = \rho N^2 D^4 J^2 F_3(J)$$
$$= \rho N^2 D^4 \phi(J)$$

where J^2 has been absorbed in a new function $\phi(J)$. It follows that

$$T/\rho N^2 D^4 = \phi(J) \qquad (9.19)$$

This is a non-dimensional thrust coefficient and is denoted by k_T. It is a function of J only and for geometrically similar propellers has the same value if they work at the same advance coefficient. In like manner the torque Q can be expressed in non-dimensional form. Without further analysis it is apparent that

$$Q/\rho N^2 D^5 = \phi_1(J) = k_q \qquad (9.20)$$

Thrust and torque coefficients can be plotted to a base of advance coefficient and the type of curves obtained are shown in *Figure 9.15*. On this diagram propeller efficiency is also plotted. For a given propeller both thrust and torque coefficients fall off with increasing advance coefficient. Efficiency rises with increasing J, reaches a maximum and then falls off steeply.

Although this form of presentation of propeller data is suitable for analysis purposes it is not very suitable for design. The problem

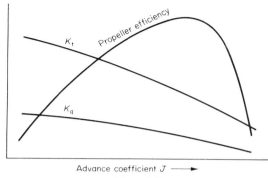

Advance coefficient J ⟶

Figure 9.15

here is to determine the principal dimensions of a propeller (diameter and pitch) to satisfy given values of power, revolutions and speed. Again referring back to the basic equation for thrust $T = \rho V_a^2 D^2 F_3(J)$ and eliminating D by writing $D = V_a/NJ$ the equation becomes

$$T = (\rho V_a^4/N^2 J^2) F_3(J) = (\rho V_a^4/N^2) \Phi(J)$$

where $1/J^2$ has been absorbed in the function $\Phi(J)$. From this it will be seen that

$$TN^2/\rho V_a^4 = \Phi(J)$$

and if top and bottom are multiplied by V_a

$$TV_a N^2/\rho V_a^5 = \Phi(J)$$

But since $TV_a =$ thrust horsepower $= U$ say, then $UN^2/\rho V_a^5 = \Phi(J)$. D. W. Taylor used this type of coefficient. He actually dropped the density term ρ and referred his results to salt water and took the square root of the other terms. He called the resulting coefficient

$$B_U = NU^{.5}/V_a^{2.5} \qquad (9.21)$$

Another coefficient used by Taylor referred to shaft horsepower P and this is given by

$$B_P = NP^{.5}/V_a^{2.5} \qquad (9.22)$$

For a series of model propellers in which the only feature varied was the pitch ratio Taylor plotted against B_U or B_P as abcissa and pitch ratio as ordinate contours of the reciprocal of the advance coefficient, which he called $\delta = ND/V_a$. These are shown in *Figure 9.16*, and also included are curves of propeller efficiency.

The method of using a diagram such as this is for a given power

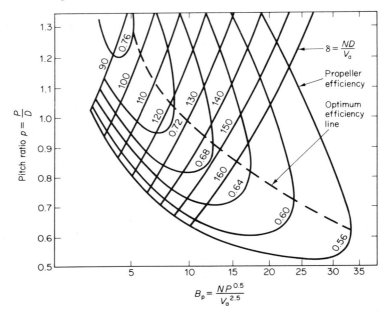

Figure 9.16

and speed of advance to decide upon the revolutions and this enables B_U or B_P to be calculated. If a vertical is erected on the B_U or $B_P - \delta$ diagram then it will be seen that the designer has the choice of a whole series of values of δ from which the diameter can be obtained, since $D = V_a\delta/N$. Associated with each of these diameters is a value of pitch ratio. The problem is to choose the propeller giving the maximum efficiency. This can be done easily, as it will be seen that it is possible to draw a line through the point where the tangents to the contours of efficiency are vertical. This line gives the maximum obtainable efficiency for any given value of B_U or B_P so that where the vertical from the particular value of B_U or B_P decided upon intersects this line it is possible to read off the pitch ratio and δ, from which the diameter and the pitch of the propeller can be obtained.

In the two coefficients B and δ used by Taylor, D is in feet, speed V_a is in knots, N is revolutions per minute and horsepower is in units of $33\,000$ ft lbf/min.

The Taylor presentation has found very wide acceptance as a suitable means of plotting propeller data determined from models for design purposes.

Open water tests on model propellers

A test on a model propeller which is run in a towing tank without a model hull in front of it is called an 'open water test'. The motor which rotates the propeller is attached to the tank carriage which is moved down the tank at the required speed of advance. The propeller shaft is geared to the motor and it is usual for the shafting and the associated gearing, or at least that part of the apparatus which is immersed in the water, to be in a streamlined housing. The propeller shaft projects forward beyond this housing, the propeller being attached to the end of the shaft. Various combinations of speed of advance and revolutions can be used to obtain results for different values of the advance coefficient *J*. Records of thrust, torque, revolutions and speed of advance are taken so that all the information is then available for calculating thrust and shaft horsepower and efficiency at a range of advance coefficients.

It is usual to carry out open water tests on standard series of propellers. It was seen in the early sections of this chapter that there are many features which could affect the performance of the propeller, such as number of blades, blade outline shape, blade area ratio, blade section shape, blade thickness fraction, boss diameter and pitch ratio. The procedure adopted in standard series work is to keep all these features except pitch ratio fixed and to carry out tests at a range of pitch ratios. If it is then intended to test propellers of different type, some feature such as blade area ratio would be altered and another set of tests carried out over a range of pitch ratio. When the data from these tests are plotted in some manner as described in the previous section they can be used for design purposes.

There have been many standard series tests carried out on propellers from time to time. One of the earliest of these was due to R. E. Froude.[3] The particulars of the screws tested by Froude were as follows:

Number of blades	3 and 4
Blade outline	elliptical 3 and 4 blades
	wide tipped 3 blades
Blade thickness fraction	0.035
Disc area ratio	0.3 to 0.525, 3 blade elliptical
	0.45 to 0.75, 3 blade wide tipped
	0.4 to 0.7, 4 blade elliptical
Pitch ratio	0.885 to 1.53
Rake and skew	nil
Shape of blade sections	flat face, circular back.

Froude plotted the results of his tests on these propellers in the form of two coefficients C_a and C_0, where

$$C_a = R^2 H/BV^5$$

and

$$C_0 = H/BD^2V^3$$

in which R = revolutions in hundreds per minute
$\quad\quad\;\; H$ = thrust horsepower
$\quad\quad\;\; D$ = propeller diameter
$\quad\quad\;\; V$ = speed
$\quad\quad\;\; B$ = blade factor determined from the experiments and depending upon the type of blade tested.

C_0 was plotted against C_a for different pitch ratios and also plotted on the diagrams were curves of efficiency for different pitch ratios. When revolutions, power and ship speed are known it is possible to calculate C_0 having decided on the blade type, i.e. on the value of B. It is then possible to lift off the curve the value of C_0 which contains diameter, choosing the pitch ratio which gives the highest efficiency. Hence diameter and pitch can be calculated. It is of interest to note that Froude's C_a is virtually the same as Taylor's B_U since $\sqrt{C_a} = RH^{.5}/B^{.5}V_a^{2.5}$. The only difference is that Froude included a blade factor B and Taylor did not. The probability is that Froude's coefficient pre-dates Taylor's.

At a later date (1937) Gawn carried out a similar set of experiments on propellers which were extensions of those done by Froude.[4] His experiments extended the disc area ratio up to 1.1 and he presented the data in the C_a—C_0 method.

Another important standard series of model propeller experiments is that due to D. W. Taylor, to which brief reference has already been made.[5] Taylor's tests were made on propellers of elliptical blade outline with flat face, circular back sections, the propellers having three or four blades. The blade thickness fraction was 0.05 and the mean width ratio (i.e. the mean width of the blades clear of the boss ÷ diameter) was 0.25. This with a boss diameter of $0.2D$ gave a disc area ratio of 0.4 for the three-bladed propellers, and 0.53 for the four-bladed propellers. Taylor plotted the results in terms of charts of B_U and B_P for the three-bladed and four-bladed series in the manner already described and the method of making use of these charts requires no further comment.

A much more up-to-date series of propeller experiments was carried out by Troost in Holland and his results have been recorded in three papers to the North East Coast Institution of Engineers and Shipbuilders in 1937–1938, 1939–1940 and 1950–1951,[6,7,8].

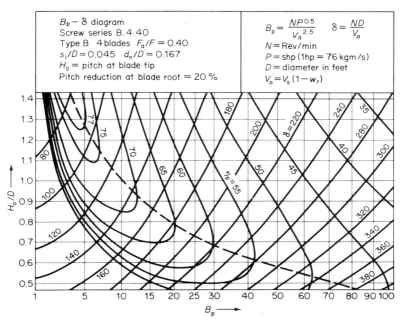

$B_p - \delta$ diagram
Screw series B.4.40
Type B   4 blades $F_a/F = 0.40$
$s_i/D = 0.045$ $d_n/D = 0.167$
H_o = pitch at blade tip
Pitch reduction at blade root = 20%

$$B_p = \frac{N P^{0.5}}{V_a^{2.5}} \qquad \delta = \frac{ND}{V_a}$$

N = Rev/min
P = shp (1hp = 76 kgm/s)
D = diameter in feet
$V_a = V_s (1 - w_t)$

Figure 9.17   (From *Trans. North East Coast Instn. of Engrs. & Shpbdrs.*, **54**, plate 10 (1937–1938))

These tests were made on two-,three-,four- and five-bladed propellers and they differ from the earlier series which have been discussed here, in that aerofoil sections were used over parts of the radius and appreciable departure was made from the elliptical blade outlines used in some of the earlier propellers. The data were presented in the Taylor $B_p - \delta$ form and the curves of thrust and torque coefficients K_t and K_Q and efficiency plotted to a base of advance coefficient were also included. A typical Troost $B_p - \delta$ diagram is shown in *Figure 9.17*.

Propeller and ship

So far in the study of the propeller and the resistance of the ship the two have been considered in isolation. However, the propeller has to work behind the ship and in consequence one has an interaction upon the other. The first problem is to consider how the ship affects the water in which the propeller is working. It has been seen that due to the form of the ship the velocity of the water around the hull varies, being less than the average at the ends and greater than the average at amidships. It follows that the water in the neighbourhood of the propeller at the stern has some forward velocity, and in consequence the speed of advance V_a of the propeller through the water in this region is less than the ship speed V_s. Also, because of viscosity the hull drags water along with it so that for this reason also the water at the stern will have a forward velocity. A third effect which contributes to the velocity of the water at the stern is wave making. In ships where there is heavy wave making at the stern the particles of water in the wave which are moving in circles move either forward or aft relative to the hull.

These three effects contribute to what is called the 'wake' and it is important to be able to determine the speed of the wake so as to find the speed of advance of the propeller relative to the water. The wake speed is not constant over the propeller disc but it will be found that it is highest near the hull and diminishes in going outwards and in going downwards from the surface. The distribution of the wake speed can be investigated by taking velocity measurements on models in a towing tank. It will be seen that a blade section at some radius r from the propeller axis in, say, a single-screw ship will be operating in a variety of water speeds as it rotates. The effect of this is that the section works at variable angles of incidence as it rotates. It is obviously not possible to adjust the blade pitch to give the best angle of incidence for each position in the revolution and what has to be done is to design for average conditions, that is to design for the mean value of the wake speed at any radius. Because the wake speed is greatest near the shaft axis, to avoid getting high angles of incidence of the blade sections at these positions and consequent flow breakdown, diminution in pitch is required towards the root. This is a common feature of modern propellers.

As one of the factors responsible for the wake is dependent upon viscosity it would be expected that the wake would be different on a model as compared with a ship, but it will be seen later that the designer is not entirely dependent upon model data for determining the wake speed for design purposes. It is possible to make estimates of wake speed from full scale trials.

In preliminary propeller design calculations the average speed of the wake over the whole of the screw disc is used and it is usual to express it as a fraction of the speed of advance or the ship speed. This is known as the 'wake fraction' so that if the wake speed is v_w then the wake fraction is either v_w/V_a or v_w/V_s. R. E. Froude used the former and D. W. Taylor the latter, so that

Froude wake fraction $w_f = v_w/V_a$ and
Taylor wake fraction $w_t = v_w/V_s$

These two are of course both expressing the same thing and are related to one another as follows:

$$w_t = w_f/(1 + w_f) \text{ or } w_f = w_t/(1 - w_t)$$

It has been found that the wake fraction is generally independent of speed, or in other words the wake speed is directly proportional to the ship speed. This may not always be true, especially in high speed ships where there is considerable wave making at the higher speeds and the wave making component of the wake becomes appreciable. In such cases the wake fraction may become negative, which means that due to the motion of the particles in the wave at the stern created by the ship there is a current moving aft instead of forward. As far as normal merchant ships are concerned the wake moves forward, reducing the speed of advance of the propeller through the water. The wake fraction varies with the position of the propeller relative to the hull and is greater in single-screw ships than in twin-screw ships. In the former the value of the Taylor wake fraction may be 0.25–0.30 but in the latter would be very considerably less than this.

A second factor which has to be taken into account is the influence of the propeller on the hull. Experiments on models have shown that if a model hull is run in a tank with a propeller working behind, the recorded resistance of the hull is greater than if the model were towed without the propeller. There is thus a resistance augment due to the presence of the propeller which could be accounted for by the suction or reduced pressure on the rear part of the hull arising from the propeller, and also by the increased rubbing velocity created by the velocity augment of the water entering the propeller. From this, if T is the thrust of the propeller and R is the towrope resistance of the hull at a given speed, then in order that the propeller shall propel the hull at this speed T must be greater than R. R. E. Froude expressed this in the inverse way, namely that R is less than T, and could be written

$$T(1 - t) = R \qquad (9.23)$$

The quantity $T \times t$ he called 'thrust deduction' and t he called the 'thrust deduction fraction' and $1 - t$ is the 'thrust deduction factor'.

Suppose now that both sides of equation 9.23 are multiplied by V_s, then $TV_s(1 - t) = RV_s = TV_a(1 + w_f)(1 - t)$ in the Froude notation. Now $TV_a =$ Thrust horsepower of the screw (T.H.P.) and $RV_s =$ Effective horsepower necessary to drive the ship at speed V_s (E.H.P.). Hence T.H.P. $(1 + w_f)(1 - t) =$ E.H.P. or

$$\text{T.H.P.} = \frac{\text{E.H.P.}}{(1 + w_f)(1 - t)} \qquad (9.24)$$

In Taylor's notation

$$\text{T.H.P.} = (1 - w_t/1 - t) \times \text{E.H.P.} \qquad (9.25)$$

The quantity $(1 + w_f)(1 - t)$ or $(1 - t)/(1 - w_t)$ is called the 'hull efficiency' and for normal ships is greater than unity. This means that the thrust horsepower which the propeller has to produce to drive a ship whose towrope resistance is R at a speed V_s is less than the effective horsepower at that speed. At first sight this seems an anomalous situation in that apparently something is being obtained for nothing. It can, however, be explained by the fact that the propeller is making use of the energy which is already in the wake because of its forward velocity and therefore for a given thrust the blades do not have to give the race as large an absolute velocity as when the propeller is working in open water.

One further factor enters into the interaction between hull and propeller, and this concerns the efficiency of the propeller itself. Since a propeller in open water is working in a uniform stream and a propeller behind a model or ship is working in a stream of variable velocity because of the wake, it is conceivable that the efficiency will be different in the two cases. Froude called the ratio Efficiency behind model or ship/Efficiency in open water the 'relative rotative efficiency'. This ratio may differ from zero by a very small percentage and often for design purposes it is taken as unity.

It is now possible to build up a complete picture of the relation between the E.H.P. necessary to drive the ship at a given speed and the shaft horsepower (S.H.P.) which must be supplied at the propeller. The ratio of the two is sometimes called the 'quasi-propulsive coefficient' (Q.P.C.). It will be seen that

$$\text{Q.P.C.} = \frac{\text{E.H.P.}}{\text{S.H.P.}} = \frac{\text{T.H.P.}}{\text{S.H.P.}} \times \text{Hull}\,\eta$$

$$= \text{Hull}\,\eta \times \text{Open water}\,\eta \times \text{Relative rotative}\,\eta \qquad (9.26)$$

Determination of wake and thrust deduction

Froude developed a method for finding wake and thrust deduction from model experiments. The model hull was towed in the usual way and the open water propeller apparatus was brought up behind the hull with the propeller in the correct position relative to the hull. Thus separate records were obtained of propeller thrust and model hull resistance. In present day work it is usual for the model to be internally propelled.

If a series of runs of the model hull with its propeller are made

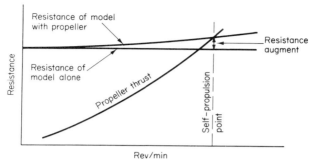

Figure 9.18

at different revolutions per minute and constant speed then it is possible to draw a curve of hull resistance and propeller thrust to a base of revolutions as in *Figure 9.18*. At the revolutions where thrust and resistance are equal is the self-propulsion point for the model. The amount by which the resistance is in excess of the resistance for the model hull without the propeller at the revolutions for self-propulsion is the thrust deduction or the resistance augment. Suppose now open water results are available for the propeller in the form of thrust or thrust coefficient plotted to a base of advance coefficient $J = v_a/nd$. The thrust being known, which is necessary to propel the model, it is possible to ascertain the value of J which would correspond to this value. Hence, knowing the revolutions n and the diameter d it is possible to determine v_a and knowing the speed of the model hull v_m, the difference between these speeds is the average wake speed. This method assumes of course that the thrust advance coefficient curve for the propeller in open water and for the propeller behind the model hull are the same. A similar method for determining the wake can be adopted using full scale trial information, as will be described below.

As far as the revolutions for self-propulsion are concerned by the method described these will be for the model, and in attempting to predict the revolutions for the ship a correction will be necessary because of the relatively greater resistance of the model as compared with the ship on account of the greater proportion of skin friction resistance in the model. One way of making this correction is to estimate the difference in resistance of ship and model when reduced to the model scale and apply this force to the dynamometer measuring resistance as a force tending to propel the model. In this way the revolutions at which the model is propelled will, when scaled up to the full size, according to the law of comparison give the correct revolutions for self-propulsion of the ship.

Values of wake fraction and thrust deduction

A primary factor affecting the wake fraction is the fullness of the form and early formulae giving values of this quantity have usually incorporated the block coefficient. A formula developed by D. W. Taylor based on results obtained by Luke gave the following expressions for single- and twin-screw ships:

$$w_t = -0.05 + 0.5 C_B \quad \text{single screws}$$
$$w_t = -0.20 + 0.55 C_B \quad \text{twin screws}$$

For a range of block coefficients these formulae would give the following results:

| | | w_t | |
C_B	*Single*		*Twin*
0.5	0.20		0.075
0.6	0.35		0.13
0.7	0.300		0.185
0.8	0.35		0.24

Data published by Burrill[9] give the following values for the Taylor wake fraction for single- and twin-screw ships:

| | | *Twin screws* | |
| | | *10°* | *30°* |
C_B	*Single*	*Bossing angle*	*Bossing angle*
0.50	0.215	0.074	0.01
0.60	0.242	0.130	0.069
0.70	0.273	0.187	0.130
0.80	0.313	0.254	0.206

These data give wake fractions for two different inclinations of twin-screw bossing to the horizontal and from these results this feature would appear to influence the wake fraction appreciably.

Formulae of the type quoted take into account fullness of form as the only variable. However, there are other factors which must influence the average wake fraction over the screw disc. For example, propeller diameter has an influence since a large diameter propeller could be expected to be working in a lower average wake than one of small diameter, because the wake diminishes in intensity in going outwards from the ship's hull.

Telfer analysed data presented by Bragg[10,11] and produced a formula for the wake fraction of single-screw ships which is as follows:

$$w_t = \frac{3}{1-F} \times \frac{BE}{LH} \left(1 - \frac{3D + 2R}{2B}\right) \tag{9.27}$$

where

$$F = \frac{\text{Prismatic coefficient}}{\text{Waterplane area coefficient}}$$

L = length of ship
B = breadth of ship
H = draught
E = height of shaft centre above keel
D = propeller diameter
R = tip rake plus skew.

Telfer also gave a formula for thrust deduction factor t which is

$$t = \frac{w_f}{2(1+u/V_a)} + \frac{SD(V_a+2u)^2}{2T} \qquad (9.28)$$

where

w_f = Froude wake fraction
V_a = speed of advance of propeller
T = total thrust of propeller
u = uniflow velocity at propeller
$u/V_a = (\sqrt{(1+C_T)}-1)/2$
$C_T = T/\pi D^2 V_a^2$
S = siding of the rudder post.

Recent work has been published by Parker[12] from results of the BSRA methodical series in which block coefficient C_b, speed/length ratio $V/\sqrt{L}$, position of longitudinal centre of buoyancy (L.C.B.) and propeller diameter are taken into account in assessing wake and thrust deduction for single-screw ships. A coefficient $D_w = D/\sqrt{\nabla^{\frac{1}{3}}D}$, where ∇ is the volume of displacement, was introduced into the formula for wake fraction and another coefficient $D_t = DB/\nabla^{\frac{1}{3}}$ was used in the thrust deduction formula. Taylor wake fraction w_t and thrust deduction factor t are given by

$$w_t = a_0 + a_1 C_B + a_2 C_B^2 + a_3 \frac{V}{\sqrt{L}} C_B + a_4 \left(\frac{V}{\sqrt{L}} C_B\right)^2$$

$$+ a_5 D_w C_B + a_6 \delta \text{L.C.B.} \qquad (9.29)$$

$$t = b_0 + b_1 C_B + b_2 C_B^2 + b_3 \frac{V}{\sqrt{L}} C_B + b_4 \left(\frac{V}{C_B \sqrt{L}}\right)^2$$

$$+ b_5 \left(\frac{V}{C_B \sqrt{L}}\right)^3 + b_6 \frac{V}{\sqrt{L}} + b_7 D_t + b_8 \delta \text{L.C.B.} + b_9 C_B \delta \text{L.C.B.} \qquad (9.30)$$

A formula is also given for the relative rotative efficiency η_R which is

$$\eta_R = C_0 + C_1 C_B + C_2 C_B^2 + C_3 \frac{V}{\sqrt{L}} C_B + C_4 \frac{D}{V^{\frac{1}{3}}} \qquad (9.31)$$

In these formulae δ L.C.B. is the distance of the actual L.C.B. from the basis position which is given as $20(C_B - 0.675)\%$ of length B.P. forward of amidships.

The values of the coefficients were obtained by regression analysis and are listed below:

$a_0 = -0.8715$, $a_1 = 2.490$, $a_2 = -1.475$, $a_3 = -0.3722$, $a_4 = 0.2525$, $a_5 = 0.2260$, $a_6 = -7.176 \times 10^{-3}$
$b_0 = -0.1158$, $b_1 = 8.859 \times 10^{-2}$, $b_2 = 0.3133$, $b_3 = 0.2758$, $b_4 = 5.432 \times 10^{-2}$, $b_5 = 2.419 \times 10^{-2}$, $b_6 = -0.4542$, $b_7 = 0.6044$, $b_8 = 5.171 \times 10^{-2}$, $b_9 = -8.622 \times 10^{-2}$
$C_0 = 1.716$, $C_1 = -2.378$, $C_2 = 1.742$, $C_3 = -3.080 \times 10^{-2}$, $C_4 = 0.6931$.

Cavitation

The phenomenon which has come to be known as cavitation was first experienced by Thornycroft and Barnaby on the destroyer *Daring*.[13] When on trial this ship achieved a speed of 24 knots instead of an expected speed of over 27 knots. Thornycroft and Barnaby cured the problem eventually by increasing the propeller blade area by 45%. They concluded that the loss of thrust at the higher speeds was due to the high loading of the screw and suggested that the phenomenon was likely to occur when the thrust per square inch of projected surface was in excess of 11.25 lbf (77.55 kN/m^2). The speed achieved eventually by the *Daring* with the modified propellers was over 29 knots. It was concluded that the high loading of the screws resulted in high negative pressure on the backs of the blades and consequently the development of cavities which were filled with air and water vapour. It was R. E. Froude who suggested to Thornycroft and Barnaby that the phenomenon should be called 'cavitation'. Around this time—the mid 1890s—Sir Charles Parsons was experimenting with his vessel the *Turbina* which was the first ship in which the steam turbine was used. He came across the same problem as had arisen in the *Daring*. Parsons solved the problem by using nine propellers on three shafts.

The experiments carried out by Thornycroft and Barnaby were made on the full scale by testing propellers of different blade area ratios. Parsons suggested another approach to the problem which was the basis of the development of modern methods of studying cavitation. He tested propellers on the model scale in a closed channel of water (or cavitation tunnel) in which the propeller was rotated without axial movement and the water was circulated in the tunnel. It was possible to reduce the pressure in the tunnel below atmospheric and thus create the necessary condition on the model scale for the occurrence of cavitation.

The need to reduce the static pressure on the model propeller can be seen by examining the pressure term in the dimensional equation. It was shown when developing the law of comparison for model and ship propellers that there was a term $F_3(p/\rho V_a^2)$ where p was the static pressure in the fluid. For this function to be the same for model and ship then

$$p_m/\rho_m V_{am}^2 = p_s/\rho_s V_{as}^2$$

Now if the propellers are working at the same Froude number then

$$V_{am} = V_{as}/\sqrt{\lambda}$$

where λ is the ratio of the linear dimensions so that

$$\frac{p_m}{\rho_m(V_{as}^2/\lambda)} = \frac{p_s}{\rho_s V_{as}^2}$$

or

$$p_m = \frac{\rho_m}{\rho_s} \times \frac{p_s}{\lambda}$$

Ignoring the difference in density between ship and model in the meantime, the pressure should be scaled down in proportion to the ratio of the linear dimensions. Consider now a full-size propeller and a geometrically similar model working in an ordinary towing tank, the surface of the water in which is open to the atmosphere and is therefore at atmospheric pressure. If h_m is the immersion of, say, the shaft axis in the model and h_s the immersion in the ship then it is quite easy to arrange that $p_m = h_s/\lambda$ so that the pressure due to the hydrostatic head will be reduced in the correct proportion. The atmospheric head H, say, will however be the same in model and ship and therefore

$$p_s \propto h_s + H \quad \text{and}$$
$$p_m \propto (h_s/\lambda) + H$$

In order, therefore, to have dynamical similarity as far as model and ship are concerned the atmospheric head should be scaled down to H/λ. In the ordinary open water test the propeller is working at a static pressure which is too high and cavitation which is associated with pressure reductions of the same order as the static pressure in the fluid at a particular point on the blade will not generally appear. Two courses are open for the study of propeller cavitation on the model scale: the method used by Parsons can be employed, in which the propeller is run in a closed channel or tunnel, the water being circulated and the static pressure reduced, or alternatively the ordinary towing tank can be used, the tank itself being enclosed in a room in which the pressure can be reduced. The former method has normally been adopted but in recent years the latter has been employed.[14]

The experience of Thornycroft and Barnaby showed that cavitation could result in loss of propeller efficiency which unltimately meant loss of ship speed for a given power. Another serious effect of cavitation is that the material of the propeller may suffer considerable damage. This can be appreciated by considering in a little more detail the mechanism by which cavities are formed in the fluid. It has been shown that the lift force on a blade section is generated by increased pressure on the face and by reduced pressure on the back, the latter contributing most to the total force. The pressure distribution on a blade section is shown in *Figure 9.19*. On this diagram a horizontal line is drawn which represents the static pressure on the back of the blade. If the reduction in pressure on the back should equal the static pressure at any point in the breadth of the blade then any tendency to reduce the pressure beyond this point will cause flow breakdown and the formation of cavities filled with air or water vapour. Now these cavities or bubbles will be swept along the blade and as they move towards the trailing edge they will come into regions of higher pressure and collapse. The collapse of the bubbles generates very high localised forces, sufficient to cause

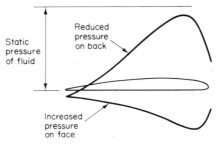

Figure 9.19

mechanical damage to the material of the blade, or 'erosion' as it is called. The magnitude of the reduced pressure on the back of the blade will depend upon the local velocity, the angle of incidence and the shape of the section, so that erosion damage on the back might be expected on sections towards the top where the rotational velocity is high. Cavitation can, however, occur at the root of a propeller blade on the face because of the high angles of incidence which exist there. Reduction of pitch at the root helps in this respect and the 'lifting' of the leading edge of the root sections also reduces the pressure peaks which can exist on the face.

Generally speaking, flat face circular back sections have lower pressure peaks than sections of aerofoil type and for this reason the former type was used in heavily loaded screws. Special sections have now been developed, however, which have a more uniform distribution of pressure with correspondingly less tendency towards cavitation. It can be seen that for a given thrust, greater blade area will reduce the average pressure on the blades and therefore the pressure peaks will be lower. The device of adopting greater blade area (which means greater blade breadth for a given diameter) adopted in the *Daring* is still sound and it will be found that heavily loaded propellers have much broader blades than lightly loaded ones.

Cavitation number

It is convenient to have some criterion which will indicate the conditions under which a propeller is working insofar as cavitation is concerned.

Let v_0 and p_0 be the velocity and pressure in the fluid at a point remote from the propeller blade and let v_1 and p_1 be the pressure at some point on the blade. Then

$$p_1 + \tfrac{1}{2}\rho v_1^2 = p_0 + \tfrac{1}{2}\rho v_0^2$$

At some point on the nose of the blade section the fluid is brought to rest and this point is called the 'stagnation point'. It follows that at this position

$$v_1 = 0 \quad \text{and} \quad p_1 = p_0 + \tfrac{1}{2}\rho v_0^2$$

or the increase in pressure at this point $\Delta p = p_1 - p_0 = \tfrac{1}{2}\rho v_0^2 = q$, say. At any other point on the blade the reduction in pressure will be

$$\Delta p = p_1 - p_0 = \tfrac{1}{2}\rho v_1^2 - \tfrac{1}{2}\rho v_0^2$$

$$\therefore \quad p_1 = p_0 + \Delta p$$

The pressure p_1 will become zero if

$$\Delta p = -p_0$$

This is the point beyond which flow breakdown is likely to take place, but in actual fact bubbles may occur in the water when the pressure has dropped to a value e which is the vapour pressure of water, i.e. when

$$\Delta p = -(p_0 - e)$$

If this is divided by $\frac{1}{2}\rho v_0^2$ then

$$-\Delta p / \frac{1}{2}\rho v_0^2 = (p_0 - e)/\frac{1}{2}\rho v_0^2$$

gives a non-dimensional quantity which is a measure of the onset of cavitation. This is called the 'cavitation number' and is usually denoted by σ so that

$$\sigma = (p_0 - e)/\frac{1}{2}\rho v_0^2 \tag{9.32}$$

It is clear that for a propeller v_0 would vary for different positions on the blade and also p_0 would vary with the angular position of the blade section concerned. It is, however, customary to take v_0 as the speed of advance of the propeller through the water and p_0 is taken as the pressure at the shaft axis.

If one wanted to derive a local cavitation number for a particular blade section then the actual velocity at that section would have to be used instead of the speed of advance in the expression for σ. This velocity would be $\sqrt{(V_a^2 + 4\pi^2 N^2 r^2)}$ where N is the revolutions in unit time and r is the radial distance from the shaft axis to the section concerned. In this, V_a should take into account the wake velocity which clearly would cause a variation in V_a round the screw disc. Also, as already pointed out, the static pressure would vary with the angular position of the blade section at any instant. The blade section would therefore be operating at different values of σ as it rotates and may therefore pass into and out of cavitating conditions.

The cavitation tunnel

The cavitation tunnel consists of a closed channel in which water is circulated by means of an impeller. The channel is in a vertical plane and is shown in *Figure 9.20*.

Water is circulated in a clockwise direction and the model propeller to be tested is placed in the upper horizontal section. As the water approaches the working section the cross section of the tunnel is reduced in order to steady the flow. The working section

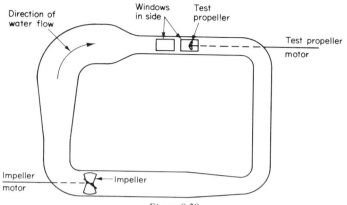

Figure 9.20

is of constant cross section for a considerable length so that uniform velocity can be obtained and windows are provided in the sides so that the test propeller can be observed. A vacuum pump is provided so that at the top of the section in front of the working section the static pressure can be reduced, and the static pressure in the working section is measured at the propeller shaft axis. The model propeller to be tested is stationary in the axial direction but can be rotated at any desired revolutions. Thrust, torque and rotational speed can be measured. When carrying out an experiment the impeller circulates the water at some predetermined speed, which therefore gives the speed of advance of the water relative to the propeller. By measuring thrust and torque at various revolutions the propeller can be tested at a whole series of J values. Such tests can be made at a series of different cavitation numbers by suitably adjusting the static pressure by means of the vacuum pump. In order that the flow conditions around the propeller can be observed, stroboscopic lighting is provided. This lighting gives a flash of very short duration once every revolution so that the propeller appears stationary and the flow around the blades can be examined. The position of the flash in the revolution can be altered so that each blade can be examined.

Figure 9.21 shows curves of thrust and torque coefficient to a base of J for various cavitation numbers. The curves can be compared with the results which would be obtained under non-cavitating conditions and it will be seen that both K_T and K_Q fall off at low J. The effect on the propeller efficiency can also be seen and it will be observed that at the higher cavitation numbers and low J the loss in efficiency is greatest.

If the revolutions of a propeller which is cavitating are increased

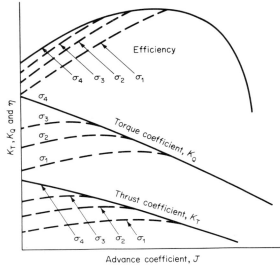

Advance coefficient, J

Figure 9.21

the cavity formed on the back of the blades will spread further and further across the breadth. Eventually a condition can be reached where the cavity covers the full breadth of the blade and the back is no longer in contact with the water. This is the 'fully cavitating condition' or the 'super-cavitating condition'. Propellers working under this condition will not experience erosion on the back. With a suitable type of blade section it is possible to design a propeller to work under these conditions. Such fully cavitating or super-cavitating propellers can be used in high speed craft such as motor boats.

The type of test described enables the propeller to be tested under cavitating conditions in a uniform stream. It is also possible to examine the influence of variable wake conditions on propeller cavitation. This can be done by using some device to give a non-uniform distribution of velocity over the propeller disc. If the tunnel section is large enough it is possible to introduce a model hull into the working section and thus produce the type of wake distribution which would exist at the after end of a ship.

The cavitation tunnel can be used to obtain 'open water' data for propellers similar to those obtained from tests in a towing tank. The tunnel has the advantage that it is possible to do these tests under various cavitation conditions, i.e. at different values of the cavitation number σ, and the regions where cavitation is likely to occur can be shown on the $B_P - \delta$ diagrams obtained from these tests. Results of tests carried out in this way have been published by Burrill and Emerson.[15]

Special types of propellers

Controllable pitch propellers

Most propellers are of fixed pitch and are designed for optimum efficiency at some operating condition. The adoption of a propeller in which the pitch can be altered to satisfy a range of operating conditions has some advantages in certain types of ships. For instance, in ships such as tugs and trawlers there is a great difference in the loading between the towing and trawling condition and the free running condition. In these ships the use of controllable pitch propellers enables the machinery to be run at constant revolutions in the different conditions so that full power can be developed when towing or trawling.

The pitch of the blades of a controllable pitch propeller is altered by means of a mechanism housed within the boss and this generally necessitates a larger boss than in an ordinary fixed pitch propeller. The mechanism alters the angle of the blade sections, each section having its pitch angle changed by the same amount. This means that if the propeller was originally of constant pitch it will no longer be so when a change is made from the basic condition. It is possible to reverse the pitch of the controllable pitch propeller so that astern thrust can be generated, and this has the advantage of simplifying the machinery since the ship can go astern with the same direction of rotation of the main engines. Manoeuvrability is thereby greatly increased.

Shrouded or ducted propellers

In this type the propeller is surrounded by a shroud or duct as shown in *Figure 9.22*. The object of the duct is to improve propeller

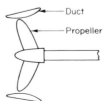

Figure 9.22

efficiency, the duct itself also generating a forward thrust. This propeller was first used in small ships such as tugs where a high bollard pull is desirable. It has been extended to larger ships and is considered to be suitable for use in large tankers. A discussion of the theory of the ducted propeller is beyond the scope of this book. An up-to-date theory has been developed by Ryan and Glover[16] and the reader is referred to this paper for a detailed study of the problem.

Contra-rotating propellers

One of the devices which has been considered for improving propulsive efficiency is to have two propellers rotating in opposite directions on the same shaft instead of a single propeller. The theory of contra-rotating propellers has been examined by Glover[17] and it is concluded that they can have a useful application to large tankers, where by the use of slow running contra-rotating propellers the Q.P.C. could be increased by up to 20%. The high speed dry cargo ship is another type where contra-rotating propellers could be used effectively and Glover suggests that in ships of this type, where propeller diameter may be restricted because of draught, propeller efficiency may be increased by as much as 12%.

As in the controllable pitch propeller there are mechanical problems with contra-rotating propellers.

Speed and power trials

It is customary for a number of reasons to run trials when a ship has been completed. Generally there will be a requirement in the contract for the ship to achieve a given speed under certain loading conditions. Penalties are often imposed for failure to obtain this speed, so that the builder has to satisfy himself that the ship has adequate power. On the other hand too great a margin of power might lead to the ship being too costly, so that the accurate powering of a ship is an important problem for the designer. The full scale trial, therefore, has the second purpose of enabling the designer to obtain data for estimates of power for future ships. It enables him to correlate the estimate of power obtained from a model experiment with the power actually obtained on the full scale. If the estimate was correct then these two powers should be the same. However, this ideal state of affairs rarely exists and the ratio of trial power to estimated power from a model experiment is often other than unity.

This ship/model correlation factor can therefore be very useful in future estimates of power.

Another reason for trials is to obtain the relation between ship speed engine revolutions and power, or in other words to calibrate the ship and engines. The data collected can be extremely useful in assessing the future performance of the ship under service conditions. For this purpose it is useful to have speed, engine revolutions and power data for a range of speeds, so that trials are run starting at low speed and working up to top speed. This type of trial is often called a 'progressive speed trial'.

Information collected on speed trials

From the ship designer's point of view the important information which is collected on a trial consists of the ship speeds, the engine revolutions at those speeds and the corresponding horsepowers. The engineer will of course collect much additional information, but this is directly concerned with the performance of the engines.

The revolutions per minute of the propeller shaft or shafts is quite readily obtained and does not require any discussion here. Regarding the question of power, this can be measured in several different ways. Originally, when ships were first mechanically propelled, the type of machinery was steam reciprocating and the power was measured by taking indicator cards of the pressures in the cylinders. From this information the indicated horsepower (I.H.P.) was obtained. To obtain the shaft horsepower, i.e. the power delivered to the propeller shaft, it was necessary to know the mechanical efficiency of the machinery, which is the ratio of the brake horsepower (B.H.P.) to the I.H.P. This efficiency can be obtained from test bed trials in the engine shop before the installation of the machinery in the ship, since brake horsepower cannot be measured on the sea trial. The same procedure can be adopted with diesel machinery. However, from the point of view of establishing the relation between ship speed, power and revolutions it is very desirable to have a direct measure of the shaft horsepower on the ship trial. What is required is some means of measuring the power transmitted to the propellers without absorbing power in the process. The means of achieving this is by using a torsionmeter. The torsionmeter measures the twist of a certain length of the shaft while it is transmitting power, the shaft having previously been calibrated statically by applying a known torque and measuring the angle of twist over a given length. It is thus possible to deduce the torque when the shaft is transmitting power on the sea trial, and knowing the corresponding revolutions per minute the shaft horsepower can be determined.

From the ship designer's point of view it would be desirable to know the effective horsepower of the ship at various speeds but it could only be achieved by towing the ship without propellers. Except for the purpose of special investigations, as was seen in the *Greyhound* and *Lucy Ashton* experiments for example, this is not a practical possibility and could not be undertaken for every ship. It is possible, however, to measure the thrust delivered by the screws by means of a thrustmeter or by the use of electrical resistance strain gauges attached to the propeller shaft. This is sometimes done, but not always. The resulting thrust will, as has been seen, exceed the towrope resistance because of the thrust deduction or resistance augment due to the propeller.

Measurement of speed

The speed required is the speed of the ship through the water. This can be measured directly by some form of log. The log formerly used consisted of what was virtually a small propeller (a windmill propeller really), towed some distance behind the ship to be clear of the influence of the propellers. The rotation of this propeller was communicated to a spindle on board the ship by means of the rotation of the towline and the number of revolutions in a given time was recorded. From this information, the log having previously been calibrated, it was possible to deduce the ship speed.

The modern type of log measures the speed of the ship directly by means of a pitot tube projecting from the bottom of the ship somewhere near amidships. With this type of equipment it is necessary that the pitot tube should extend beyond the boundary layer so as to obtain the true ship speed. Also, the instrument should be calibrated for the particular ship because of the variation in speed of the streamlines around the hull due to the ship form.

Although devices such as the log may be fitted in most ships and can be useful for determining the speed of the ship in service it is common practice for ship trial purposes to measure the speed first of all relative to the land. The 'ground' speed so determined will then require correction for the influence of currents and tides to obtain the true speed through the water. The ground speed is obtained by running the ship on a measured mile, the time taken for the ship to pass two objects one nautical mile apart being observed. How this is done is shown in *Figure 9.23*. Four posts A, B, C and D are erected on land so that AB and CD are two parallel lines one nautical mile apart. The ship course is adjusted to be at right angles to AB and CD. The time interval between the instant when the ship

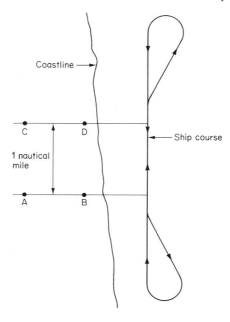

Figure 9.23

is in line with AB and the instant when it is in line with CD is then the time for it to travel one nautical mile, and this time in minutes divided into 60 gives the speed in knots over the ground.

Most measured miles are subject to currents and tidal effects, so that one single run will not give the speed of the ship through the water. If the speed of the water through which the ship is running was constant then the mean of two runs, one in each direction, would give the true speed. However, particularly where tidal effects are predominant the water speed is a function of time and to obtain the true speed more than two runs are necessary. The procedure adopted can be seen from *Figure 9.23*. The engine settings are fixed as the ship comes up to the mile for the first run and should not be altered during subsequent runs. Sufficient distance must be allowed before coming on to the mile to ensure that the ship is running at a steady speed. The first run on the mile is then made, the time being measured as previously described. After the completion of the first run the ship must turn to come back in the opposite direction. Turning involves loss of speed and this speed must be recovered before coming back on to the mile. It is therefore necessary for the ship to run some considerable distance past the end of the mile before turning, as shown in *Figure 9.23*. The distance will depend very much on the type of ship and may be as much as two or three miles. It will be seen that a considerable time may elapse between the first run on the mile and the return run, so that in that time the speed of the water could have changed appreciably with the result that the simple arithmetic mean of the speeds from two runs may not give the correct result. When making the return run on the mile the ship should if possible be the same distance from the land as on the initial run so as to be in the same conditions of current and tide. The procedure described is then continued for as many runs as are required at the particular engine setting.

Determination of speed through the water

Different methods exist for correcting the ground speed for current and tidal effects. One method which will be described makes certain assumptions regarding the variation of water speed with time. Suppose that tidal speed is the predominant factor in the speed of the water. The tidal speed curve to a base of time will have a cyclic character as shown in *Figure 9.24*. It may be approximately sinusoidal, one-half wave being of about six hours' duration. The time over which the runs on the mile are being made will probably lie between two points such as A and B and over this range of time it would not be unreasonable to suppose that the tidal speed, say y, is given by

$$y = a_0 + a_1 t + a_2 t^2 + a_3 t^3 + \ldots + a_n t^n \tag{9.33}$$

It will generally be sufficiently accurate to neglect all terms beyond the third so that

$$y = a_0 + a_1 t + a_2 t^2 \tag{9.34}$$

where a_0, a_1 and a_2 are constants.

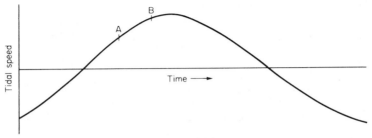

Figure 9.24

Suppose that four runs are made alternately with and against the tide, the zero of time being taken at the middle of the first run, the subsequent runs being made at equal time intervals T. Let the ground speeds measured be V_1, V_2, V_3 and V_4. Then if the true speed through the water is V

$$V_1 = V + a_0 \tag{9.35}$$

$$V_2 = V - a_0 - a_1 t - a_2 t^2 \tag{9.36}$$

$$V_3 = V + a_0 + 2a_1 t + 4a_2 t^2 \tag{9.37}$$

$$V_4 = V - a_0 - 3a_1 t - 9a_2 t^2 \tag{9.38}$$

Eliminating a_0 from each pair in turn it will be seen that

$$2V = V_1 + V_2 + a_1 T + a_2 T^2$$

$$2V = V_2 + V_3 - a_1 T - 3a_2 T^2$$

$$2V = V_3 + V_4 + a_1 T + 5a_2 T^2$$

If $a_1 T$ is eliminated from these equations

$$4V = V_1 + 2V_2 + V_3 - 2a_2 T^2$$

$$4V = V_2 + 2V_3 + V_4 + 2a_2 T^2$$

and finally if $2a_2 T^2$ is eliminated then

$$8V = V_1 + 3V_2 + 3V_3 + V_4$$

from which

$$V = \frac{V_1 + 3V_2 + 3V_3 + V_4}{8} \tag{9.39}$$

The same results would be obtained if the mean of each adjacent pair of speeds was taken and then the means of these means and finally the mean of the means so obtained. This is best shown as follows:

$$
\left.
\begin{array}{l}
V_1 \\[4pt]
V_2 \\[4pt]
V_3 \\[4pt]
V_4
\end{array}
\right\}
\quad
\left.
\begin{array}{l}
\dfrac{V_1 + V_2}{2} = M_1 \\[10pt]
\dfrac{V_2 + V_3}{2} = M_2 \\[10pt]
\dfrac{V_3 + V_4}{2} = M_3
\end{array}
\right\}
\quad
\left.
\begin{array}{l}
\dfrac{M_1 + M_2}{2} = M_{12} \\[14pt]
\dfrac{M_2 + M_3}{2} = M_{23}
\end{array}
\right\}
\quad
\dfrac{M_{12} + M_{23}}{2} = V
$$

If the tidal speed could be considered to be a linear function of the time, then only three speeds alternately with and against the tide would be required and the true speed V would be $(V_1 + 2V_2 + V_3)/4$.

Similar results can be obtained by retaining high terms in equation 9.33 for the tidal speed. Thus if the t^3 term is retained, five speeds would be required and

$$V = \frac{V_1 + 4V_2 + 6V_3 + 4V_4 + V_5}{16}$$

and if t^4 was retained six speeds would be required and

$$V = \frac{V_1 + 5V_2 + 10V_3 + 10V_4 + 5V_5 + V_6}{32}$$

A point of interest is that the multipliers in the expression for the mean speed are the numerical coefficients in the expansion of the binomial $(a+b)^{n+1}$ where n is the highest power of time in the expression for the tidal speed. In most cases it will be found that powers higher than the second need not be retained in the tidal speed equation, so that only four speeds are required at one particular engine setting.

Although a method has been developed for correcting ground speeds for variation in tidal speed it may often be found sufficiently accurate to take the simple arithmetic mean of two runs with and against the tide, but it must be remembered that this assumes that the tidal speed is constant. One aspect of this method is that the result is strictly speaking only true if the runs are made at equal time intervals, but the error involved if this condition is not absolutely observed should not be very great.

Another method of correcting the measured speeds for current and tidal effects is due to Telfer.[18] Suppose a series of runs is made on the measured mile and the time T taken for each run is noted. Then if the revolutions per minute are also known, say N, the revolutions to drive the ship forward one nautical mile can be calculated, i.e. $N \times T$. It is possible to divide the runs into two groups, those with the tide and those against the tide, so that two curves of revolutions per nautical mile can be drawn, one with and the other against the tide as shown in *Figure 9.25*, where the base is time of day. It is now possible to draw a curve giving the mean revolutions per nautical mile and calling this N_m, the speed of the ship in knots at any time being simply $(N/N_m) \times 60$, where N is the actual revolutions per minute at the time corresponding to N_m.

This method does not make any assumptions regarding tidal speed nor do the runs need to be made at equal time intervals. For that

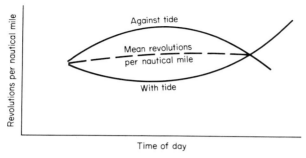

Figure 9.25

matter the runs need not be made at the same engine settings—the speed could be gradually increased.

Trial condition

Ideally it would be useful to have the results of speed trials for several conditions of loading of the ship, but as this is almost impossible from the point of view of cost it is necessary to agree some condition under which the ship can be tested. The most useful condition would probably be with the ship fully loaded, but in certain types of ship, particularly deadweight carriers, it may not be possible to get the necessary deadweight to load the ship down to the maximum draught. In these circumstances it is usual to specify some condition at a reduced displacement, and usually a higher speed is specified than would be expected if the ship were fully loaded.

In some ships such as oil tankers the necessary deadweight can be readily supplied in the form of water ballast so that it is possible to run fully laden trials.

In determining the trial speed it is usual to specify a higher speed than would be expected in service because of weather conditions. The trial will generally be run in weather conditions which are as near calm water as possible, but the ship in service will meet all sorts of conditions and on the average the resistance will be greater at a given speed than it would be in calm water. Consequently, if a certain service speed is envisaged the ship should be capable of achieving a higher speed under the relatively calm water conditions of the trial.

Before the ship is taken out on trial it should be docked and the immersed surface cleaned and painted to remove fouling organisms which it has been seen would increase the resistance. It is usual to note the number of days out of dock if the trial does not take place immediately after docking.

It is important to know the displacement of the ship in the trial condition, so that draughts forward and aft should be observed and the density of the water corresponding to these draughts should also be measured. In some high speed ships of low deadweight it may be necessary to correct the displacement for fuel consumption if the trials extend over a considerable period.

Ship trials should be carried out in deep water, as shallow water can have a marked effect on the resistance of the ship. This means that where trials are run on a measured mile the mile should be sited where the water is of sufficient depth without having to take the ship too far off shore.

Plotting and analysis of trial data

Having corrected the speed over the ground for tidal and current effects, power and revolutions per minute can be plotted against speed as shown in *Figure 9.26*. The resulting curves should be fair, the speed revolutions curve in many ships being practically a straight line. The power curve will increase quite rapidly as speed increases and in high speed ships it should be possible to detect humps and hollows in the curve if speed runs have been made at sufficiently close intervals.

If the speed revolutions curve is a straight line passing through the origin, this implies that the ratio V/N is constant, or in other words the propeller is working at constant J or constant slip

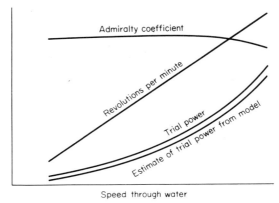

Speed through water

Figure 9.26

throughout the range. Now it has been seen that when a propeller is working at constant J, the thrust developed is proportional to the square of the ahead speed. It follows then that the resistance of the ship must also be proportional to the square of the speed. This situation often exists for merchant ships over a fairly wide range of speeds but if the speed is pushed higher then the square law will no longer hold and this can be detected by non-linearity in the speed revolutions curve. It can of course also be seen by plotting ⓒ to a base of speed. The value of ⓒ would be constant if resistance varied as the square of the speed, and if resistance did not follow this law at the higher speeds there would be a rise in the curve as shown in *Figure 8.12*.

Another useful coefficient which can be plotted from trial data is what is called the 'Admiralty coefficient'. The use of this coefficient goes back to the early days of mechanical propulsion. The coefficient which is still sometimes used is one of two which were formerly in use. They were the 'displacement coefficient' and the 'midship section coefficient'. These are given as follows:

$$\text{Displacement coefficient} = \Delta^{\frac{2}{3}}V^3/\text{H.P.} \qquad (9.40)$$

$$\text{Midship section coefficient} = AV^3/\text{H.P.} \qquad (9.41)$$

where

Δ = displacement in tons
A = area of the midship section
V = speed in knots
H.P. = horsepower.

The horsepower used was formerly the indicated horsepower, but if the coefficients are used nowadays they would be determined in terms of shaft horsepower. The midship section coefficient has fallen into disuse but the displacement coefficient is still sometimes used.

The Admiralty coefficient plotted to a base of speed would like ⓒ be constant over the speed range where resistance is proportional to the square of the speed, but when resistance varies as a higher power than the square the curve would drop as shown in *Figure 9.26*. The Admiralty coefficient is virtually the inverse of ⓒ.

If on the diagram showing the trial data the estimate of, say, shaft horsepower from model or standard series data is plotted then it is possible to compare the actual shaft horsepower required at a given speed with the estimate and so obtain a ship/model correlation factor. This factor will rarely be unity. The factor is useful for design purposes for future ships and can be applied to estimates from model experiments to obtain more realistic values of the power required in the ship.

Determination of wake fraction from ship trials

Most information concerning wake fraction is obtained from model experiments. It is, however, possible to estimate wake fraction from full scale trial results. Suppose at a given ship speed V_s on trial the torque Q and the revolutions per minute n have been determined, then the torque coefficient can be calculated, since $K_q = Q/\rho N^2 D^5$.

If torque coefficients for various values of J are available for the propeller from open water tests it is possible to find the value of J corresponding to the trial condition, as shown in *Figure 9.27*. Then

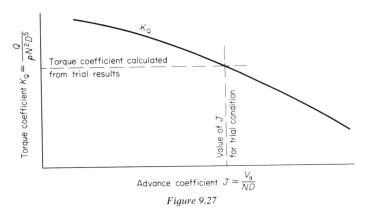

Figure 9.27

since $J = V_a/ND$ it follows that $V_a = J \times ND$, and since D is known, V_a can be calculated, so that the wake speed is obtainable since $v_w = V_s - V_a$ and hence the wake fraction v_w/V_s or v_w/V_a can be calculated. This procedure assumes that the open water curve for the propeller represents the performance of the full size propeller behind the ship. Where an open water curve is not available for the actual propeller a curve from a standard series can be used, but it should be pointed out that errors may arise here if the actual propeller departs very much from the standard series propeller. It may in this case be necessary to make corrections to the standard series curve to allow for this.

REFERENCES

1. Froude, R. E., 'On the part played in the operations of propulsion by differences in fluid pressure', *Trans. Instn. of Naval Architects*, 1889
2. Froude, W., 'On the elementary relation between pitch slip and propulsive efficiency', *ibid.*, 1878
3. Froude, R. E., 'Results on further model screw propeller experiments', *ibid.*, 1908

4. Gawn, R. W. L., 'Results of experiments on model screw propellers with wide blades', *ibid.*, 1937
5. Taylor, D. W., *Speed and power of ships*, United States Government Printing Office, Washington 1943
6. Troost, L., 'Open water tests series with modern propeller forms', *Trans. North East Coast Instn. of Engrs. & Shpbdrs.*, 1937–1938
7. Troost, L., *ibid.*, Part 2, 'Three bladed propellers', 1939–1940
8. Troost, L., *ibid.*, Part 3, 'Two and five bladed propellers', 1950–1951
9. Burrill, L. C., 'Developments in propeller design and manufacture for merchant ships', *Trans. Instn. of Marine Engineers*, 1943
10. Bragg, E. M., 'The wake and thrust deduction of self-propelled single screw models', *Trans. Society of Naval Architects and Marine Engineers, U.S.A.*, 1924
11. Telfer, E. V., 'The wake and thrust deduction of single screw ships', *Trans. North East Coast Instn. of Engrs. & Shpbdrs.*, 1935–1936
12. Parker, M. N., 'The B.S.R.A. Methodical series—an overall presentation. Propulsion factors', *Trans. Royal Instn. of Naval Architects*, 1966
13. Thornycroft, J. L., and Barnaby, S. W., 'Torpedo boat destroyers', *Proc. Instn. of Civil Engineers*, Part IV, 1894–1895
14. van Manen, J. D., 'The effect of cavitation on the interaction between propeller and ship's hull', *International Shipbuilding Progress*, Jan. 1972
15. Burrill, L. C., and Emerson, A., 'Propeller cavitation: further tests on 16 in propeller models in the King's College cavitation tunnel', *Trans. North East Coast Instn. of Engrs. & Shpbdrs.*, 1962–1963
16. Ryan, P. G., and Glover, E. J., 'A ducted propeller design method: a new approach using surface vorticity distribution technique and lifting line theory', *Trans. Royal Instn. of Naval Architects*, 1972
17. Glover, E. J., 'Contra rotating propellers, for high speed cargo vessels', *Trans. North East Coast Instn. of Engrs. & Shpbdrs.*, 1966–1967
18. Telfer, E. V., 'The practical analysis of merchant ship trials and service performance', *ibid.*, 1926–1927

10

Powering

In this chapter methods will be discussed for determining the amount of power which is required to propel a ship. From the ship designer's point of view it is important first of all to determine the power required to achieve the contract speed. The power also affects the weight of machinery and the weight of fuel which must be allowed for when determining the total displacement of the ship, which in turn influences the principal dimensions. Methods of determining the effective or towrope horsepower have already been discussed, but from a machinery point of view it is the shaft horsepower which is required.

In the early design stage an approximate estimate of power would perhaps suffice and this can be checked at a later stage when more detailed information for the ship becomes available. One of these approximate methods makes use of the Admiralty coefficient. As with C so with the Admiralty coefficient: it should be constant for two geometrically similar ships at corresponding speeds if scale effect is neglected. This is justified if two ships not differing greatly in size are being considered.

Suppose an Admiralty coefficient curve is available from full scale trials for a range of speeds for a ship whose length and displacement are L_1 and Δ_1 and suppose an approximately similar ship is being designed whose speed, length and displacement are V_2, L_2 and Δ_2, then the corresponding speed V_1 for the basis ship would be determined from

$$V_1/\sqrt{L_1} = V_2/\sqrt{L_2}$$

If then a value for the Admiralty coefficient at the speed V_1 is obtained for the basis ship this should apply to the design at speed V_2. Hence calling this value C it will be seen that

$$\text{S.H.P. for design} = \Delta^{\frac{2}{3}} V_2^3 / C \qquad (10.1)$$

The merit of this procedure is that it makes use of full scale results and should therefore form a suitable basis for estimating power for a ship. The defect of the method is that unless the two ships are exactly geometrically similar, error may arise unless correction factors have been developed for small changes in form and proportions between the two ships. Another factor which has not been taken into account is the possibility of differences in the propulsive efficiency of the ships. However, used with intelligence the method can still prove useful for at least first estimates of power.

Estimation of power from model experiments

In Chapter 8 it was shown that the effective horsepower necessary to drive a ship can be determined from a model experiment for the individual ship or from a standard series. Depending upon the conditions under which these tests have been made it may be necessary to make additions to the power obtained. For example, if the models are tested without appendages then a small percentage addition would have to be made for the resistance of these items. Also, if extrapolation from model to ship is carried out using, say, the Schoenherr or the I.T.T.C. 1957 lines, the power will be that for a smooth ship and would require an addition for the roughness of the ship. A further addition may also be necessary due to errors in the extrapolation method to bring the estimate of power from the model in line with the ship. These two latter corrections could be considered to be included in a ship/model correlation allowance, say, x, so that if E.H.P.$_m$ and E.H.P.$_s$ represent the powers estimated for the ship from the model and the actual power for the ship respectively, then

$$\text{E.H.P.}_s = (1+x)\text{E.H.P.}_m \qquad (10.2)$$

The determination of x is a difficult matter. It could only be made at the E.H.P. stage by towing the ship without propellers and comparing the result with the estimate from the model. This procedure could be adopted only in special circumstances. It is, however, possible to introduce this addition at the S.H.P. stage so that

$$\text{S.H.P.}_s = (1+x)\text{S.H.P.}_m \qquad (10.3)$$

in which S.H.P. on trial is compared with estimates from the model experiment. The disadvantage is that other factors are introduced such as variations of wake, thrust deduction and propeller efficiency between ship and model. This cannot be avoided and the comparison of shaft horsepower is the most suitable way of determining this factor.

However the factor is determined, the power obtained will represent that required to drive the ship in calm water. To this must be added allowances for increases in power due to rough weather at sea. Such allowances can only be obtained by the analysis of the service performance of ships. It is not customary to allow sufficient power to maintain speed in extreme weather conditions but it is usual to legislate for average conditions. To do this properly, analysis of power, revolutions and speed have to be made over a long period, from which it is possible to ascertain the average increase in power required. The analysis is complicated by the fact that with the passage of time the resistance of the ship will increase due to fouling so that part of the increase in power under particular weather conditions may be due to this cause. If, however, the ship is travelling in smooth or nearly smooth water at certain times it should be possible to assess the increase in resistance due to fouling at these times, from which it would be possible to deduce the increased resistance due to weather. The average increase in power for weather might be of the order of 20% but the exact figure should be determined from analysis of actual ships.

When additions have been made to the estimate of E.H.P. for the ship for all the factors considered above, the E.H.P. for the ship in service will be obtained. In order that the shaft horsepower for the machinery can be obtained it is necessary to assume some value for the quasi propulsive coefficient (Q.P.C.) from which S.H.P. = E.H.P./Q.P.C.

The Q.P.C. can be obtained from self-propelled model experiments. The shaft horsepower determined in this way is that at the propeller or the tail shaft. It is usual to allow for shaft losses in order to find the S.H.P. at the engine. This generally amounts to about 3% for machinery amidships and 2% for machinery aft.

Having decided upon the shaft horsepower it is then possible to choose a suitable engine. Machinery and particularly diesel machinery is usually manufactured in standard sizes so that in choosing an engine for a particular ship it may not be possible to obtain one which would give the exact power required for a given ship speed. In such cases it may be necessary to use a slightly larger engine than would actually be required, which would mean that a slightly greater ship speed would be obtained.

In order to design a suitable propeller to fit to a given engine and ship, standard series data can be used to obtain a first approximation to the principal dimensions of the propeller, i.e. the diameter and pitch. The blade area required may have to be determined from cavitation consideration in heavily loaded propellers. To use standard series data the revolutions per minute and power required would first of all have to be decided upon and from what has already been said these would be governed by the engine chosen. The speed of advance would also have to be calculated from the ship speed and data for the wake fraction. The coefficient B_p can then be calculated and the optimum propeller chosen from a $B_p - \delta$ diagram as described previously. The detailed design of the propeller goes very much beyond this simple process and beyond the scope of the present work.

In the next section an arithmetical example is worked, showing how the ship power can be arrived at and how the propeller particulars can be determined.

Example of determination of power for a ship and the use of B_p-δ diagrams

For the purpose of this calculation the ship on page 242 for which the E.H.P. was calculated is considered.

The resistance of the ship at 15 knots was calculated as 60 240 lbf (267 960 N) using the Schoenherr formula for a smooth ship. (See page 245.) From this

E.H.P. of smooth ship	= 2774 h.p.	2068 kW
Allowance for appendages, say 5%	139	103
	2913	2171
Roughness allowance (equivalent to $\Delta Cf = 0.0003$)	339	253
E.H.P. in smooth water	3252	2424.
Weather allowance (20%)	650	485
	3902	2909
say,	3900	2910

It is assumed that the ship/model correlation allowance is made up of a roughness allowance only. It will further be assumed that the wake fraction hull efficiency and the quasi propulsive coefficient can be obtained from available data. These will be taken as follows:

Taylor wake fraction	= 0.27
Hull efficiency	= 1.15
Q.P.C.	= 0.75
Relative rotative efficiency	= 1.00

The required shaft horsepower is

$$\frac{3902}{0.75} = 5200$$

Shaft losses 2% $\frac{104}{5304}$

say,   5310

This would give the power at the engine and as stated in the last section it may be necessary to go to a higher power if a standard engine is to be fitted. It will, however, be assumed that an engine of 5310 horsepower is available which would give 5200 horsepower at the propeller. The speed of the ship is 15 knots so that the speed of advance $V_a = V_s(1 - w_t) = 15(1 - 0.27) = 10.95$ knots.

At this stage it is necessary to decide upon the revolutions per minute for the machinery. In direct drive engines a compromise has to be reached between the best revolutions for the machinery and the best revolutions for the propeller. The former should generally be high and the latter low. With high speed diesel engines and turbines both can work much more nearly at their best values by the introduction of reduction gearing between engine and propeller. Fast running machinery means lighter weight, although allowance has to be made for the introduction of the reduction gearing.

In the present example a fairly low figure of 110 rev/min will be assumed and the influence of increasing the revolutions will be investigated. If $N = 110$ then

$$B_{P4} = N P^{0.5} / V_a^{2.5} = (110 \times 5200^{0.5})/10.95^{2.5} = 19.99$$

From the $B_p - \delta$ chart (*Figure 9.17*) which is for a four-bladed propeller of 0.4 blade area ratio it will be found that

$$\delta = ND/V_a = 181$$
$$\text{Pitch ratio } p = 0.78$$
$$\text{Efficiency } \eta = 0.652$$

The values of δ and p are those for optimum efficiency.

Now $D = (\delta \times V)/N = (181 \times 10.95)/110 = 18.01$ ft (5.49 m), say 18 ft (5.5 m),

and pitch $= pD = 0.78 \times 18.01 = 14.05$ ft (4.28 m).

Q.P.C. = hull efficiency × propeller open efficiency
× relative rotative efficiency
$= 1.15 \times 0.652 \times 1.00 = 0.75$

This happens to be the same as was assumed in the calculation of the shaft horsepower. If there had been any large difference then the shaft horsepower would have had to be recalculated and the process repeated.

Consider now the effect of increasing the revolutions to 120 per minute. Then

$$B_{P4} = (120 \times 5200^{0.5})/10.95^{2.5} = 21.8$$

From the $B_p - \delta$ chart the values of δ, p and η are

$$\delta = 188 \qquad p = 0.765 \qquad \eta = 0.642$$

Hence

$$\text{Diameter} = (188 \times 10.95)/120 = 17.15 \, \text{ft} \, (5.23 \, \text{m})$$

$$\text{Pitch} = 17.15 \times 0.765 \qquad = 13.12 \, \text{ft} \, (4.00 \, \text{m})$$

$$\text{Q.P.C.} = 1.15 \times 0.642 \times 1.00 = 0.738$$

It will be seen that the faster running propeller will have a smaller diameter for the given power and ship speed and the efficiency will be lower.

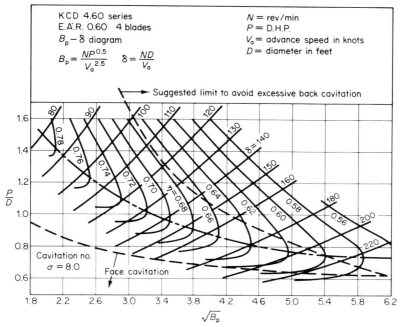

Figure 10.1 (From *Trans. North East Coast Instn. of Engrs. & Shpbdrs.*, **79**, p. 304 (1962–1963))

Repeating this process for revolutions of 130 per minute the results obtained are:

$$\text{Diameter} = 16.5\,\text{ft}$$
$$\text{Pitch} = 12.29\,\text{ft}$$
$$\eta = 0.630$$

$$\text{Q.P.C.} = 1.15 \times 0.63 \times 1.00 = 0.724$$

It will be seen that again the diameter is smaller and the efficiency has dropped still further. This conclusion was one that was arrived at from the simple momentum theory and it is clear that for high propeller efficiency the propeller should have as large a diameter as possible. However, this means low revolutions and hence the need for compromise where direct coupled machinery is concerned.

It is unlikely that cavitation would be a serious problem for the propeller considered here, but this could be checked by using $B_p - \delta$ diagrams obtained from the results of tests in the cavitation tunnel at the appropriate cavitation number. Such a diagram due to Burrill and Emerson[1] is shown in *Figure 10.1*, which gives the limits beyond which face or back cavitation are likely to occur.

Appendix

Note on the use of SI Units

In the chapters dealing with resistance, propulsion and powering it will have been seen that Imperial units have been used in many places. This is because a great deal of data in these branches of naval architecture have been published in terms of these units and it would seem that the designer will for a long time to come have to think in these terms. It is, however, fitting that some reference should be made to the use of SI Units for many of the coefficients used in these subjects. It will be found that in other chapters the SI system has been adopted.

In the SI system the unit of length is the metre and for small lengths the millimetre is used. The unit of time is the second and the unit of mass is the kilogram. A mass of 10 000 kg is called a tonne. The unit of force is called the newton (N) and is defined as the force necessary to cause an acceleration of one metre per second per second on a mass of one kilogram. One pound force is therefore equivalent to 4.448 22 N. The density of fresh water is taken as 999 kg/m^3, and of sea water as 1026 kg/m^3, both at a temperature of 15° C.

Power, instead of being measured in units of 33 000 ft lb/min is measured in kilowatts.

The relations between British horsepower and kilowatts is 1 British horsepower = 745.7 W or 0.7457 kW.

It has been suggested that the powers which have been discussed in the preceding chapters and referred to as shaft horsepower (S.H.P.), brake horsepower (B.H.P.), delivered horsepower (D.H.P.) and effective horsepower (E.H.P.) be denoted by P_S, P_B, P_D, and P_E respectively. This system has been adopted by the International Towing Tank Conference (I.T.T.C.).

With regard to the displacement of a ship, it has been proposed that it be considered in terms of mass rather than weight. A new symbol $\sum$ has been suggested by Paffett for ship mass.

As far as speed is concerned, for scientific purposes it would be in metres per second, but the knot is still likely to be used and it is considered that the international nautical mile of 1852 m be used in defining it instead of the British nautical mile of 6080 ft.

Coefficients such as C_t, C_f and C_w used in connection with resistance are truly non-dimensional so that as long as they are calculated in a consistent system of units the change from imperial to SI Units

Quantity	Formula giving value in salt water	
	SI Units	*Imperial units*
	$\rho SW = 1026 \, \text{kg/m}^3$ $1 \text{ knot} = 1852 \text{ m/h}$	$\rho SW = 1/35 \, \text{ton/ft}^3$ $1 \text{ knot} = 6080 \text{ ft/h}$
Length constant Ⓜ	$Ⓜ = 1.0083 \dfrac{L_m}{(\sum(\text{tonne}))^{\frac{1}{3}}}$	$Ⓜ = 0.3057 \dfrac{L_{ft}}{(\Delta(\text{ton}))^{\frac{1}{3}}}$
Skin constant Ⓢ	$Ⓢ = 1.0167 \dfrac{S(\text{m}^2)}{(\sum(\text{tonne}))^{\frac{2}{3}}}$	$Ⓢ = 0.0935 \dfrac{S(\text{ft}^2)}{(\Delta(\text{ton}))^{\frac{2}{3}}}$
Speed constant Ⓚ	$Ⓚ = 0.5848 \dfrac{V(\text{knot})}{(\sum(\text{tonne}))^{\frac{1}{6}}}$	$Ⓚ = 0.5834 \dfrac{V(\text{knot})}{(\Delta(\text{ton}))^{\frac{1}{6}}}$
Resistance constant Ⓒ	$Ⓒ = 579.7 \dfrac{P_E}{(\sum(\text{tonne}))^{\frac{2}{3}}(V(\text{knot}))^3}$	$Ⓒ = 427.1 \dfrac{P_E(\text{h.p.})}{(\Delta(\text{ton}))^{\frac{2}{3}}(V(\text{knot}))^3}$
Advance coefficient δ	$\delta = 3.2808 \dfrac{N(\text{rev/min}) D(\text{m})}{V_a(\text{knot})}$	$\delta = \dfrac{N(\text{rev/min}) D(\text{ft})}{V_a(\text{knot})}$
Power coefficients B_U	$B_U = 1.158 \dfrac{N(\text{rev/min}) P_T^{0.5}(\text{kW})}{(V_a(\text{knot}))^{2.5}}$	$B_U = \dfrac{N(\text{rev/min}) P_T^{0.5}(\text{h.p.})}{(V_a(\text{knot}))^{2.5}}$
B_p	$B_p = 1.158 \dfrac{N(\text{rev/min}) P_D^{0.5}(\text{kW})}{(V_a(\text{knot}))^{2.5}}$	$B_p = \dfrac{N(\text{rev/min}) P_D^{0.5}(\text{h.p.})}{(V_a(\text{knot}))^{2.5}}$

makes no difference to their values. However, the Froude 'constants', while still being non-dimensional, were developed in what may be called a mixed system of units so that the numerical coefficients in these require modification when the SI system is employed. The same remark applies to the Taylor B_p and δ coefficients used in propeller work. Paffett in a paper entitled 'Metrication in Ship Research and Design' given to the Royal Institution of Naval Architects in 1971 has worked out the values for the numerical parts of all these coefficients. A comparison of these coefficients with the formulae which were derived in the chapters on resistance, propulsion, and powering is shown in the table on page 321. It should be noted that the actual numerical value of the coefficients will be the same whether worked out in imperial or SI Units.

This enables existing data for $\copyright$, for example, which have been calculated in terms of imperial units to be used directly when SI Units are employed.

REFERENCE

1. Burrill, L. C., and Emerson, A., 'Propeller cavitation: further tests on 16 in propeller models in the King's College cavitation tunnel', *Trans. North East Coast Instn. of Engrs. & Shpbdrs.*, 1962–1963

11

Vibration

When a body which is in stable equilibrium under a system of forces is displaced from the equilibrium position by an external force it will return to that position when the displacing force is removed. It will, however, overshoot the equilibrium position and in the absence of damping forces take up a displacement of equal magnitude but of opposite sign. It will then return to the original position and once again overshoot the equilibrium position and achieve a displacement of the same magnitude and sign as the original displacement. This procedure will be repeated indefinitely if there are no forces to drain energy from the system. A typical example of this type of system is the simple pendulum.

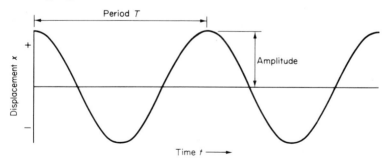

Figure 11.1

The type of motion described is called 'oscillatory' or 'vibratory' motion and the body is said to oscillate or vibrate. The motion can be represented graphically as in *Figure 11.1*, where the ordinate represents displacement of the body from the equilibrium position and the abscissa is time. The maximum displacement is called the 'amplitude' of the motion and the time interval between two successive positions of the body having the same displacement is called the 'period'. In certain systems the period of one complete vibration or oscillation is very small and it is often more convenient to consider the number of vibrations in unit time. This is referred to as the 'frequency' of vibration, so that if T is the period and n the frequency then $n = 1/T$.

The vibrations referred to above are called 'free vibrations' and associated with any system there is a definite frequency which is known as the 'natural frequency' of free vibration. Some systems have more than one natural frequency and the vibrating string is a good example of this. Again in some systems the natural frequency depends upon the magnitude of the original displacement, but where one is dealing with small vibrations it is sufficiently accurate to consider that the natural frequency is independent of the amplitude.

All known engineering systems are subject to damping forces to a greater or lesser degree; that is to say there are forces such as those due to viscosity which are absorbing energy from the system. In such cases the amplitude of the free vibration will gradually diminish and instead of the type of time displacement curve shown in *Figure 11.1*, that shown in *Figure 11.2* will be obtained. This can easily be demonstrated by the example of the pendulum. The original amplitude will diminish steadily until the pendulum comes to rest.

The damping force not only affects the amplitude of vibration but it also has an effect on frequency. However, it is only when the damping force is large relative to the mass of the system that there is any appreciable effect and for all practical purposes the natural frequency of undamped oscillations can be taken as the frequency of the real system with damping.

Before dealing with the specific problem of the vibration of a ship it is appropriate to consider some fundamental principles concerning vibrations and for this purpose the simple mass-spring system will be examined.

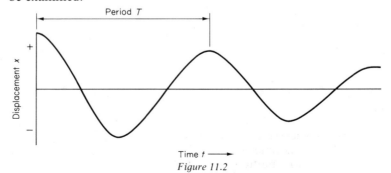

Figure 11.2

Vibration of undamped mass-spring system

Suppose a mass M is suspended by means of a spring S as shown in *Figure 11.3*. The mass of the spring can be ignored as being small relative to the mass M. The position of static equilibrium is at A. Let x be the displacement of the mass from this position at some time t.

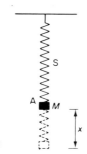

Figure 11.3

If k is the spring constant then the force generated by the displacement x is kx, on the assumption that there is a linear relation between the force and the displacement. The force kx is directed towards A and is the only force acting in the absence of damping, except the constant force due to gravity which can be ignored.

Applying Newton's second law, the relation between the mass, the acceleration and the displacement is given by

$$M(\mathrm{d}^2x/\mathrm{d}t^2) = -kx$$

or

$$\mathrm{d}^2x/\mathrm{d}t^2 = -(k/M)x \tag{11.1}$$

The solution to this differential equation is

$$x = B\sin\left(t\sqrt{\frac{k}{M}} + \delta\right) \tag{11.2}$$

where B is an arbitrary constant and would depend upon the initial displacement; δ is a phase angle dependent upon the time at which the displacement is measured.

The type of motion represented by equation 11.2 is called 'simple harmonic' and is the type which always exists when there is a linear relation between the displacement of the system and the restoring force generated. The motion is represented by the curve in *Figure 11.1* already referred to.

It will be seen that if t is increased by $2\pi\sqrt{(M/k)}$ then x will have the same value. Therefore the period of the motion T which is defined as the lapse of time between the occurrence of displacements of equal magnitude is given by $T = 2\pi\sqrt{(M/k)}$ and the frequency n is given by

$$n = 1/2\pi\sqrt{(k/M)} \qquad (11.3)$$

The first point to be noted here is that the frequency of vibration depends upon the ratio of some factor (in this case k) affecting the stiffness of the system to the mass M of the system. It will be found that this is true generally for all kinds of systems, although it may appear in a much more complicated form than shown in the above expression.

Another point to be noted is that the period and the frequency are independent of the amplitude of the oscillation and this will always be true so long as a linear relation exists between the displacement and the force generated by the spring. Should conditions be such that the linear relation breaks down, then the period and frequency would no longer be independent of the amplitude.

T and n are the natural period and the natural frequency of free vibration respectively.

Damped vibrations

As stated earlier, all known systems are subject to damping forces to a greater or lesser degree. It is therefore necessary to consider the influence of damping on a vibrating system. The simplest type of damping is where the force generated is proportional to the velocity, so that if μ is some damping coefficient then

$$\text{Damping force} = \mu(\mathrm{d}x/\mathrm{d}t)$$

This is what is called 'linear damping' and under this type of damping the differential equation of motion for the mass spring system can be written

$$M(\mathrm{d}^2x/\mathrm{d}t^2) + \mu(\mathrm{d}x/\mathrm{d}t) + kx = 0 \qquad (11.4)$$

There are several solutions to equation 11.4 depending upon the degree of damping in relation to the mass of the system. For example, if μ is very large then vibration will not exist, but if the mass is displaced from its equilibrium position and released it will return to the equilibrium position without the displacement changing sign. This is illustrated in *Figure 11.4(a)*. Another case is where the mass will return to and pass the equilibrium position once as shown in *Figure 11.4(b)* and then sink to rest.

In a great many systems the damping is not so great and a series

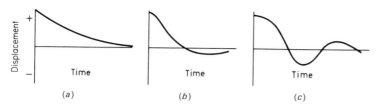

Figure 11.4 (a) *Heavy damping* (b) *Lighter damping* (c) *Normal damping*

of complete oscillations of steadily diminishing amplitude will take place as shown in *Figure 11.4(c)* until the mass eventually comes to rest. This is the case of particular interest as far as ship vribration is concerned. Under these circumstances the motion can be represented by

$$x = B\exp(-\alpha t)\sin(pt + \delta) \tag{11.5}$$

where as before B is the initial amplitude of the motion, δ is a phase angle and α and p are constants yet to be determined. Various methods can be used for determining α and p. The following is a simple approach which involves inserting equation 11.5 in equation 11.4 and comparing the coefficients of sin and cos on each side of the equation. Hence

$$\frac{dx}{dt} = -\alpha B\exp(-\alpha t)\sin(pt + \delta) + pB\exp(-\alpha t)\cos(pt + \delta)$$

$$\frac{d^2x}{dt^2} = +\alpha^2 B\exp(-\alpha t)\sin(pt + \delta) - \alpha pB\exp(-\alpha t)\cos(pt + \delta)$$

$$-\alpha pB\exp(-\alpha t)\cos(pt + \delta) - p^2 B\exp(-\alpha t)\sin(pt + \delta)$$

Therefore, equation 11.4 becomes

$$\alpha^2 B\exp(-\alpha t)\sin(pt + \delta) - 2\alpha pB\exp(-\alpha t)\cos(pt + \delta)$$

$$- p^2 B\exp(-\alpha t)\sin(pt + \delta)$$

$$+ \frac{\mu}{M}(-\alpha B\exp(-\alpha t)\sin(pt + \delta) + pB\exp(-\alpha t)\cos(pt + \delta))$$

$$+ \frac{k}{M}B\exp(-\alpha t)\sin(pt + \delta) = 0$$

Dropping out the common $B\exp(-\alpha t)$ and equating coefficients of sin and cos on each side of the equation gives

$$\alpha^2 - p^2 + (\mu/M)\alpha + k/M = 0 \tag{11.6}$$

and

$$2\alpha p - (\mu/M)p = 0 \tag{11.7}$$

From equation 11.7 $\alpha = \mu/2M$ and substituting this in equation 11.6 gives

$$(\mu^2/4M^2) - p^2 - (\mu^2/2M^2) + (k/M) = 0$$

$$p^2 = (k/M) - (\mu^2/4M^2)$$

The value of x is then given by

$$x = B \exp(-(\mu/2M)t) \sin\left\{ t \sqrt{\left(\frac{k}{M} - \frac{\mu^2}{4M^2} \right)} + \delta \right\} \qquad (11.8)$$

It will be noted that the rapidity with which the vibrations die away depends upon the ratio of the damping coefficient to the mass of the system, i.e. upon $\mu/2M$. Also, since the frequency $n = p/2\pi$ then

$$n = \frac{1}{2\pi} \sqrt{\left(\frac{k}{M} - \frac{\mu^2}{4M^2} \right)} \qquad (11.9)$$

From this it will be noted that the damping has an influence on the frequency of free vibration. Once again, however, unless μ is large relative to M the effect is small and in many but not all cases the influence of the damping force on the vibration frequency can be neglected, so that it is often sufficiently accurate to take the natural frequency of free vibration as $1/2\pi\sqrt{(k/M)}$.

Forced vibrations

So far only free vibrations of a system have been considered, i.e. the system has been imagined to be displaced from its equilibrium position and then released and allowed to oscillate freely. This situation can occur in practice where, for example, a structural member is struck by an instantaneous blow when the resulting vibrations will be of the type which have been discussed. More often, however, the disturbing forces applied to the system will be continuous over a long period and will fluctuate in magnitude. One of the simplest types of oscillating disturbing force can be written in the form

$$F = F_0 \sin \omega t \qquad (11.10)$$

This is a sinusoidal force with maximum value F_0 and with frequency $\omega/2\pi$.

It represents a force of constant maximum value applied for an indefinitely long time. Often it will be an over-simplification to represent the disturbing force by this expression but it is instructive to investigate the consequences of imagining that this force is applied to the damped mass-spring system. The differential equation of motion becomes

$$M(d^2x/dt^2) + \mu(dx/dt) + kx = F_0 \sin \omega t \qquad (11.11)$$

The solution to this equation is in two parts, the complete solution being their sum. The first is the solution of equation 11.11 with zero on the right-hand side and is therefore the damped solution as previously given, i.e.

$$x_1 = B \exp(-(\mu/2M)t) \sin \left\{ t \sqrt{\left(\frac{k}{M} - \frac{\mu^2}{4M^2} \right)} + \delta \right\}$$

The second part consists of an oscillation in the same frequency as the applied disturbing force and is

$$x_2 = C \sin (\omega t - \gamma)$$

The complete solution is

$$x = x_1 + x_2 = B \exp(-(\mu/2M)t) \sin \left\{ t \sqrt{\left(\frac{k}{M} - \frac{\mu^2}{4M^2} \right)} + \delta \right\}$$
$$+ C \sin (\omega t - \gamma) \qquad (11.12)$$

It is clear that after a sufficient lapse of time the damped oscillation will disappear and the oscillation in the frequency of the disturbing force only will be left. This oscillation is called a 'forced oscillation' and it is important to be able to determine its amplitude, i.e. to determine the value of C in equation 11.12. Also the evaluation of γ, the phase angle, is important. The values of C and γ can be obtained in various ways and once again the method of substituting the solution x_2 in the original differential equation can be adopted. Carrying out this process leads to the following results.

$$\tan \gamma = \frac{\mu \Lambda / \sqrt{Mk}}{1 - \Lambda^2} \qquad (11.13)$$

$$C = \frac{F_0/k}{\sqrt{\{(1 - \Lambda^2)^2 + (\mu^2 \Lambda^2 / Mk)\}}} \qquad (11.14)$$

where Λ is the tuning factor $= \omega / \sqrt{(k/M)}$, i.e. is the ratio of the frequency of the applied force to the natural frequency of the system. The forced vibration is then given by

$$x = \frac{F_0/k}{\sqrt{\{(1 - \Lambda^2)^2 + (\mu^2 \Lambda^2 / Mk)\}}} \sin (\omega t - \gamma) \qquad (11.15)$$

Now since k is the spring constant, F_0/k represents the displacement which would be caused by a static force F_0. If this is called X then

$$\frac{x}{X} = \frac{1}{\sqrt{\{(1-\Lambda^2)^2 + (\mu^2\Lambda^2/Mk)\}}}\sin(\omega t - \gamma) \qquad (11.16)$$

and this then represents the ratio of the dynamic displacement to the static displacement. This is called the 'magnification factor' and represents the ratio of the response of the system under the dynamic force to the response obtained if the same maximum force were applied statically. It will be seen that it is a function of Λ, the tuning factor, i.e. the ratio of the frequency of the disturbing force to the natural frequency of the mass-spring system. The magnification factor is also a function of the ratio of the square of the damping coefficient μ to the product of the mass and the spring constant. For a given system this means that it is a function of Λ and μ. If response curves are drawn for different values of μ a diagram such as that shown in *Figure 11.5* is obtained. For small values of Λ, x/X tends

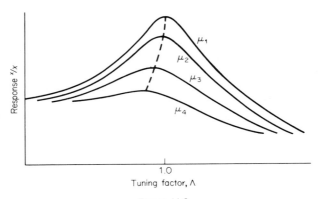

Figure 11.5

to unity and for large values it tends to zero. In between these extreme values the response builds up to a maximum which can be obtained by differentiating x/X with respect to Λ and equating to zero. If this is done the value of Λ for maximum response is given by

$$\Lambda = \sqrt{\{1 - (\mu^2/2Mk)\}} \qquad (11.17)$$

The maximum value of x/X corresponding to this is

$$\frac{x}{X} = \frac{1}{\sqrt{[(\mu^2/Mk)\{1 - (\mu^4/4M^2k^2)\}]}} \qquad (11.18)$$

Although the maximum value of x/X really occurs at the value of the tuning factor given by equation 11.17, in lightly damped systems it is often sufficiently accurate to take this value as being unity. This is the situation known as 'resonance', where the frequency of the applied force is the same as the natural frequency of free vibration of the system. This is equivalent to ignoring the second term in the bracket under the root sign in equation 11.18 when calculating the amplitude of vibration. Whether or not this approximation is made it is evident from equation 11.18 that if μ were zero the amplitude would become infinite. When dealing with practical systems, however, damping is always present and the amplitudes at or near resonance will be large but finite. It is important, then, if large amplitudes of vibration are to be avoided, that the frequency of any disturbing forces should be as far as possible away from the frequency of free vibration of the system.

Phase angle at resonance

The phase angle γ by which the forced vibration lags behind the applied force is given by equation 11.13 and if in this expression $\Lambda = 1$, i.e. the condition for resonance is inserted, then $\tan \gamma = \infty$ and $\gamma = \pi/2$. This means that there is a phase lag of one-quarter of a period between the displacement of the system and the force causing that displacement. In other words the system achieves its maximum displacement one-quarter of a period after the disturbing force has achieved its maximum value. The curves representing the force and the displacement would then be as shown in *Figure 11.6*.

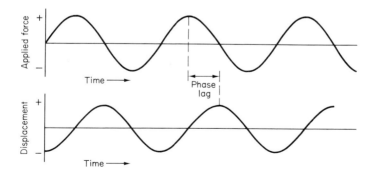

Figure 11.6

Systems with several natural frequencies

Many systems have more than one natural frequency of free vibration. A typical example is a beam undergoing flexural vibration as shown in *Figure 11.7*, where several 'modes' of vibration are possible with different numbers of nodes. Each of these modes of vibration has its own natural frequency and an applied oscillating force will excite all these modes simultaneously.

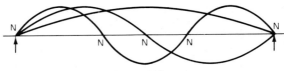

Figure 11.7

Expressions similar to equation 11.18 can then be obtained for each mode of vibration and the response of the system to the applied oscillating force will be the sum of these responses. It follows that if the frequency of the applied force is imagined to increase then several resonance peaks could be obtained as the frequencies coincide with the various natural frequencies in turn.

Irregular exciting force

In the above discussion it has been assumed that the exciting force was simple harmonic, but in practice it may be that the exciting force, although having a constant frequency, is of quite irregular form and cannot therefore be correctly represented by a single sine term. Where that is the case it is possible to analyse the actual disturbing force into a Fourier series which consists of an infinite series of terms which can be written as follows:

$$F = \sum_{n=1}^{n=\infty} F_n \sin(n\omega t + \varepsilon_n) \qquad (11.19)$$

where n is any integer. Written out in full this is

$$F = F_1 \sin(\omega t + \varepsilon_1)$$
$$+ F_2 \sin(2\omega t + \varepsilon_2) + \dots$$
$$+ F_n \sin(n\omega t + \varepsilon_n) \qquad (11.20)$$

If the values of F_1, etc., and ε_1, etc., can be calculated then the response of the system to which this irregular exciting force is applied can be calculated for each of the components in the series. By the principle of linear superposition the total response is then equal to the sum of the responses to the individual components.

For each term in the series an expression for the response of the system similar to equation 11.16 will be obtained and the total response will therefore be

$$x = \sum_{n=1}^{n=\infty} \frac{X_n}{\sqrt{\{(1-\Lambda_n^2)^2 + (\mu^2/Mk)\Lambda_n^2\}}} \sin(n\omega t - \gamma_n) \quad (11.21)$$

In this Λ_n is the tuning factor applying to the nth term in the series for the disturbing force.

Although the analysis of the exciting force is shown as an infinite series, in practice a limited number of terms only need be considered provided that they represent with sufficient accuracy the true form of the exciting force.

Ship vibration

In the previous sections the general problem of vibrating systems has been considered. It is now necessary to turn to the particular problem of vibration in ships.

The ship is an elastic structure and when it is subjected to oscillating forces vibration can result. The disturbing forces which might cause vibration in a ship can originate from within the ship itself or may be due to some external cause. In the former category can be included unbalanced forces in the main and auxiliary machinery. Such forces occur particularly with reciprocating machinery. The reciprocating masses present in diesel engines, for example, produce forces which if not balanced completely will transmit forces to the hull and can thus excite vibration. These forces will generally have frequencies of the order of the revolutions per minute of the machinery and as will be seen later, since the natural frequency of the hull is of the same order as the revolutions, serious vibration may result due to resonance. There may also be unbalanced couples created by the reciprocating forces in the machinery which could generate vibration. With turbine machinery, since there are no reciprocating parts there should not be any out-of-balance forces or couples.

Forces due to reciprocating masses may also exist in auxiliaries such as diesel generators. As they usually run at much higher speeds the frequencies of the unbalanced forces will be very much higher and in consequence they are unlikely to excite the lowest mode of vibration of the hull. Their effect will generally be to excite the higher modes of the hull or to create local vibration, i.e. the vibration of some part of the structure such as a flat or deck or a bulkhead.

As far as external forces are concerned there are two types which have to be considered, firstly those due to the propeller, and secondly those due to the environment of the ship, i.e. the sea. Propeller forces arise because of the pressure field around the blades. The pressures on the blades are communicated to the surfaces of the hull which are in close proximity. Thus, the shell plating in the neighbourhood of the propeller and other surfaces such as the bossing enclosing the shaft and shaft brackets will have forces exerted on them due to the pressure field of the propeller. As the propeller rotates the position of any particular blade relative to the hull changes and consequently the pressure on the hull changes, resulting in periodic forces being communicated to the hull. Such forces will have a frequency equal to the revolutions per unit time of the propeller shaft multiplied by the number of blades. Apart from these forces, which would be present even with a propeller working in a uniform wake, there will also be forces generated because of the variable wake conditions which exist in the neighbourhood of the propeller. The water close to the hull moves forward at a higher speed than the water at some distance from the hull and in consequence the speed of advance of a propeller blade section varies with its circumferential position. This means that the angle of incidence of a section changes as it moves round and hence the forces generated on the blade will also vary. The result is that the circumferential (and of course also the axial) forces on the blades will not all be equal at any instant, but there will be some resultant out-of-balance force transmitted to the shaft which will vary in magnitude as the propeller rotates. Once again the frequency of this varying force will be equal to the revolutions per unit time multiplied by the number of blades.

The forces generated by the propeller will be seen to be of much higher frequency than the revolutions per unit time of the shaft, so that it is most unlikely that resonance would occur between this force and the lowest mode of vibration. It might be expected therefore that such forces will excite the higher modes of vibration.

Two types of forces can be generated by the external environment of the ship. There are impulsive forces due to the sea, and continuously applied oscillating forces due to the same cause. Impulsive forces can arise due to severe pitching motion of the ship. If pitching is sufficiently heavy the bow of the ship may leave the water and when it plunges into the water again the phenomenon known as 'slamming' occurs, i.e. a large instantaneous force is generated near the bow and the situation is very similar to that of a bar being struck by a hammer. The bar would vibrate in its natural frequency of vibration and would eventually come to rest due to damping. In like manner the ship when struck by the slamming force would vibrate in its natural frequency and eventually come to rest. High stresses can result from this whipping action and the stresses are often referred to as 'whipping' stresses. The phenomenon can often be detected on records of longitudinal strain taken on ships at sea. Superimposed on the normal record of strain approximately in the frequency of encounter of the waves with the ship can be seen variations of strain of a very much higher frequency, which die away very rapidly.

The second category of wave-applied forces which can excite hull vibration is that due to the steadily applied forces arising from the passage of waves past the ship. If a ship is considered to be moving through regular waves of relatively small length compared with the ship length, as shown in *Figure 11.8*, then from purely static considerations it will be seen that, where there is a crest passing the ship,

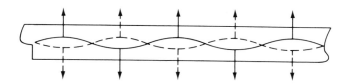

Figure 11.8

buoyancy will be increased and where a trough passes the ship buoyancy will be reduced, so that this is equivalent to the ship being subjected to a series of alternate upward and downward forces. These forces create couples tending to bend the ship in the longitudinal plane. After one-half wave length has passed the ship where there were crests, troughs will appear and the forces will be reversed in sign, so that the situation indicated by the dotted lines in *Figure 11.8* will occur. The couples will then be reversed in sign and the ship will be bent in the opposite direction. After another half wave length has passed the original situation will be established. It follows that one complete cycle of the disturbing couple will have a period equal to the period of encounter of the waves with the ship.

Consider waves of length l approaching head on a ship whose speed is V. The time taken for one complete wave to pass a fixed point on the ship is given by

$$T_E = l/(V + v)$$

where v is the speed of the wave.

$$T_E = \frac{l/v}{1 + V/v} = \frac{T_W}{1 + V/v}$$

in which T_W is the wave period.

From the theory of waves $v = \sqrt{(gl/2\pi)}$ and $T_W = \sqrt{(2\pi l/g)}$

$$\therefore \quad T_E = \frac{\sqrt{(2\pi l/g)}}{1 + V/\sqrt{(gl/2\pi)}} = \frac{0.8\sqrt{l}}{1 + 0.8(V/\sqrt{l})}$$

$$T_E(1 + 0.8(V/\sqrt{l})) = 0.8\sqrt{l}$$

$$T_E(\sqrt{l} + 0.8V) = 0.8l$$

$$l - 1.25\,T_E\sqrt{l} - T_E V = 0$$

If V is in knots then

$$l - 1.25\,T_E\sqrt{l} - 0.51\,T_E V = 0 \qquad (11.22)$$

This gives the length of wave l in metres which would be required to give a period of encounter T_E with a ship speed V in knots.

In a modern large tanker or bulk carrier of say 300 m length and speed 15 knots the frequency of the lowest mode of vibration might be 60 per minute, hence $T_E = 1$ s.

$$\therefore \quad l - 1.25\sqrt{l} - 0.51 \times 1 \times 15 = 0$$

$$l - 1.25\sqrt{l} - 7.65 = 0$$

$$\therefore \quad \sqrt{l} = \frac{1.25 \pm \sqrt{(1.25^2 + 4 \times 7.65)}}{2} = 3.46$$

$$l = 11.97 \text{ m}$$

If the frequency of vibration was as low as 30 per minute then $T_E = 2$ s and $l = 28.7$ m.

These numerical examples show that with relatively short waves, frequencies of encounter can be obtained which could give resonance with the lower frequencies of vibration of a ship.

This brief discussion of the forces which can generate vibration shows that it is important to know and to be able to estimate the frequency or frequencies of vibration of the hull. The accuracy with which this can be done will determine the accuracy with which severe vibration can be avoided.

Types of vibration

Before considering the calculation of frequencies of vibration of the ship it is important to examine the various ways in which the ship may vibrate. It is first of all necessary to distinguish between what may be called local vibration and main hull vibration.

Local vibration is concerned with some small portion of the structure vibrating, for example a deck, a bulkhead or a flat. This type of vibration is usually of much higher frequency although of smaller amplitude than main hull vibration. If it occurs in accommodation it can be most unpleasant and probably more troublesome to passengers and crew than main hull vibration. In practice investigations of the possibility of local vibration are seldom undertaken in so far as the calculation of frequencies are concerned unless there is some special reason for doing so. This is largely because there is an almost infinite number of possible parts of the structure which may vibrate in many different frequencies.

Main hull vibration is concerned with the whole of the structure vibrating and it is necessary first to distinguish between two types, namely flexural vibration, where the hull bends like a beam, and torsional vibration, the structure twisting about a longitudinal axis. A third type is also theoretically possible: longitudinal vibration where the structure is alternately compressed and extended, the displacement of the structure being parallel to the length of the ship. An alternating axial force would be required to generate this type of vibration and this could be provided by the variable axial force on the propeller operating in a non-uniform wake.

Of the types of vibration mentioned, flexural vibration is the type which has been given most consideration in the past and is probably the most important.

Flexural vibration

The ship structure can flex in two different planes, the vertical and the horizontal, and the vibrations associated with these two planes are called 'vertical' and 'horizontal' vibrations respectively. In either of these flexural types the structure has an infinite number of degrees of freedom and the mode of vibration is usually described by the number of nodes which exist in the length. The fundamental mode is the two-node mode as shown in *Figure 11.9(a)*. A one-node mode is impossible with flexural vibration as the ship would not be in dynamic equilibrium without in addition some rigid body motion. The mode shown in *Figure 11.9(a)* yields displacement at the ends since there is no rigid support at the ends of the ship. This is sometimes referred to as a 'free-free' mode of vibration and it therefore differs from the mode which would be taken up by a beam in a structure where there would be zero displacement at one end at least.

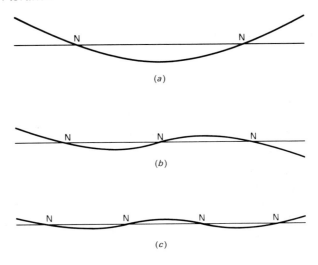

Figure 11.9 (a) *Two-node* (b) *Three-node* (c) *Four-node*

The next mode of vibration is one in which there are three nodes as in *Figure 11.9(b)* and the higher mode having four nodes is shown in *Figure 11.9(c)*. All these modes are free-free and can exist either in vertical or horizontal vibration. Associated with each mode there is a natural frequency of free vibration, the frequency being higher for the higher modes. If the ship were of uniform rigidity and uniform mass distribution throughout its length and was supported at its ends then the frequencies of the higher modes of vibration would be simple multiples of the fundamental frequency. As, however, none of these conditions apply the frequencies of the higher modes do not bear any simple relation to the fundamental frequency.

Torsional vibration

With this type of vibration a one-node mode is possible, since dynamic equilibrium can be maintained because only twisting of the structure is involved. Some of the possible modes of vibration are then as shown in *Figure 11.10*, which indicates the first three. In the figure the displacement shown is angular displacement.

Coupling

It is usually assumed that the various modes of vibration are independent of one another, for example vertical and horizontal

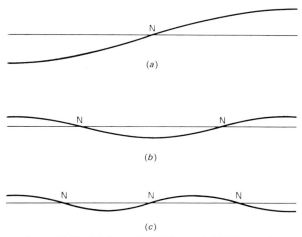

Figure 11.10   (a) *One-node* (b) *Two-node* (c) *Three-node*

vibration can be treated separately. Sometimes, however, if a structural member (in this case the ship) is vibrating in one mode it can
generate vibration in another mode. Under these circumstances the
modes are said to be 'coupled'. The horizontal and torsional modes
in the ship problem come into this category. Because of the non-
uniform distribution of mass in the vertical plane, if a ship is vibrating
in the horizontal plane it is possible for a twisting moment to be
generated which will excite torsional vibration.

Flexural vibration of a beam

Before considering the ship problem it is instructive to examine the
vibration of a simple beam of uniform section with a concentrated
mass at the centre of length. *Figure 11.11* shows a beam of length l
and moment of inertia I with a mass M at mid length. The beam
is assumed to be vibrating freely and its own mass is neglected.

Under a static load W at mid span the deflection δ would be
$Wl^3/48EI$ where E is Young's modulus of elasticity. Consequently if
y is the dynamic deflection at mid span then the relation between

Figure 11.11

the force and deflection is

$$\text{Dynamic force} = \frac{48EI}{l^3} y$$

so that $48EI/l^3$ represents the spring constraint k in the simple theory of vibrating systems which was discussed earlier. The equation of motion becomes

$$M\frac{d^2y}{dt^2} + \frac{48EI}{l^3} y = 0$$

or

$$\frac{d^2y}{dt^2} + \frac{48EI}{Ml^3} y = 0 \tag{11.23}$$

The solution of this equation is

$$y = A \sin\left(t\sqrt{\left(\frac{48EI}{Ml^3}\right)} + \gamma\right)$$

and the frequency is

$$N = \frac{1}{2\pi}\sqrt{\frac{48EI}{Ml^3}} \tag{11.24}$$

This gives the frequency of the lowest mode of vibration which is the only one possible with a concentrated mass when the mass of the beam is neglected.

It can be shown that if the mass M, instead of being concentrated at mid length, was distributed uniformly over the whole length of the beam, the mass per unit length being m, the frequency would be given by

$$N = n^2 \frac{\pi}{2}\sqrt{\frac{EI}{Ml^3}} \tag{11.25}$$

where n is any whole number. The lowest mode of vibration is associated with $n = 1$. If the value of the numerical constant in equation 11.24 is worked out then

$$N = 1.103 \sqrt{\frac{EI}{Ml^3}}$$

for a concentrated mass at mid span and

$$N = 1.57 \sqrt{\frac{EI}{Ml^3}}$$

when the mass M is distributed uniformly over the length of the beam. This comparison brings out the fact that, for the same total mass, how the mass is distributed over the length has an important influence on the frequency. Another important factor governing the frequency is the manner in which the ends of the beam are supported. It can be shown, for example, that for the beam with uniformly distributed mass the frequency of vibration in the free-free mode is given by

$$N = \frac{n^2}{2\pi} \sqrt{\frac{EI}{Ml^3}}$$

where for two-, three- and four-node vibration n has the values 4.73, 7.85 and 10.99 respectively. This would give a frequency for the lowest mode of vibration of $N = 3.56 \sqrt{(EI/Ml^3)}$.

Formulae for frequency of vibration of a ship

The formulae referred to in the previous section for the frequency of vibration for a beam suggest the form which simple formulae for the frequency of a ship might take. The difference here is of course that neither the mass nor the stiffness are constant along the length of the ship. However, if the total mass is considered and the stiffness is characterised by the product of the moment of inertia of the midship section and Young's modulus for the material of construction a suitable formula would have the form

$$N \propto \sqrt{\frac{EI_\otimes}{Ml^3}}$$

Now since the material of construction in most modern ships is steel, E can be regarded as constant. If the structure consists of materials having different elastic moduli then they can be referred to steel with a suitable correction to I. Since in the past, at any rate, naval architects have been used to thinking in terms of displacement Δ of the ship instead of mass then with the length being denoted by L a suitable formula for frequency would be

$$N = \text{CONST.} \sqrt{\frac{I_\otimes}{\Delta L^3}} \tag{11.26}$$

This type of formula was one of the earliest ever used in connection with the determination of ship vibration frequency. It has been associated with Otto Schlick and is often called the Schlick formula.[1] The constant in the formula obviously depends in a complicated way on mass and stiffness distribution and can only be evaluated empirically. The method of obtaining values for the constant is to observe frequencies of vibration for a given loading condition of a ship and thus knowing Δ and L and calculating $I_\otimes$ it is possible by substitution in the formula to find a value for the constant. One method of finding the vibration frequency of the ship experimentally is to run the main machinery through a range of revolutions. If the natural frequency lies in that range there will be some position in the range where synchronism exists with whatever unbalanced forces there are in the machinery, resulting in large amplitudes of vibration. Another more direct method is to fit a vibrator on the ship. This consists of a reciprocating mass whose frequency can be varied until synchronism is obtained with the hull frequency.

A formula of the Schlick type can only be of use if sufficient data have been collected from actual ships for it to be possible to have values of the constant for different types of ships in different conditions of loading. Mass and stiffness distribution are two of the major factors which would affect the values of the constant in the formula. There are, however, others which have not yet been discussed, which are:

(1) Departure from ordinary bending theory
    (a) shear deflection
    (b) structural discontinuities
(2) Added virtual mass of the water surrounding the ship
(3) Rotary inertia.

All these factors have an important bearing on the determination of frequencies by direct calculation and they will now be discussed in some detail.

Shear deflection

Ordinary bending theory which is the basis of the calculation of frequencies of flexural vibration is based on the assumption that when a beam is bent sections which were plane and perpendicular to the neutral axis before bending remain so after bending. It follows from this that there can be no shear stress in the beam. In fact, however, there are few if any cases where this condition exists in engineering structures and it certainly does not exist as far as the longitudinal flexing of the ship's structure is concerned.

The influence of the presence of shear stress in a beam is twofold: firstly the linear theory of bending stress is modified so that the stress is no longer simply proportional to the distance from the neutral axis, and secondly additional deflection of the beam occurs which is appropriately referred to as 'shear deflection'. The vibration frequency is very much tied up with deflection so that if shear deflection is appreciable it will have an important influence on the frequency.

The influence of shear deflection on the frequency can be illustrated by a simply supported beam with a mass M at mid span. It was stated that for a static force W the deflection based on bending theory is simply $\delta_m = Wl^3/48EI$. It can be shown that for a beam of solid rectangular cross section the deflection due to shear alone is $\delta_S = 3Wl/10bdG$ where b and d are the breadth and depth of the beam and G is the modulus of rigidity of the material. Hence total deflection is given by

$$\delta_T = \delta_m + \delta_S = \frac{Wl^3}{48EI} + \frac{3Wl}{10bdG}$$

$$= \frac{Wl^3}{48EI}\left(1 + \frac{3Wl/10bdG}{Wl^3/48EI}\right)$$

$$= \frac{Wl^3}{48EI}\left(1 + \frac{6}{5}\frac{E}{G}\frac{d^2}{l^2}\right)$$

It is possible therefore to write the total deflection as $\delta_m(1 + r_s)$ where r_s is the ratio of the shear to the bending deflections and is in the case quoted $(6/5)(E/G)(d^2/l^2)$.

The quantity r_s depends on the ratio of the two constants for the material, E and G, the properties of the structure, and will also depend upon how the structure is loaded. For the present purpose it is sufficient to say that it is a ratio which can be determined so that the total deflection is simply

$$\delta_T = \frac{Wl^3}{48EI}(1 + r_s)$$

so that the static force generated by a deflection δ_T is

$$W = \frac{48EI}{l^3(1 + r_s)} \times \delta_T$$

Consider now the vibrating beam. If the dynamic deflection is y then the force generated on the mass is $(48EI/l^3(1 + r_s)) \times y$ and the differential equation for the beam is

$$M\frac{d^2y}{dt^2} + \frac{48EI}{l^3(1 + r_s)} \times y = 0$$

giving the solution

$$y = A\sin\left(t\sqrt{\left[\frac{48EI}{Ml^3(1 + r_s)}\right]} + \gamma\right)$$

and the frequency will be

$$N = \frac{1}{2\pi} \sqrt{\frac{48EI}{Ml^3(1+r_s)}} \qquad (11.27)$$

Now if N_m is the frequency based on bending theory only the modified frequency N when shear deflection is taken into account is

$$N = \frac{N_m}{\sqrt{(1+r_s)}} \qquad (11.28)$$

The influence of shear deflection is then to reduce the frequency. It was seen from the example for the solid rectangular beam that shear deflection was dependent among other things on the ratio of the depth of the beam to the length, so that for shallow beams it may be relatively unimportant. However, in the case of the ship it is not unimportant. The length/depth ratio may be as high as ten and also the ship structure is more correctly represented by a box girder, which influences greatly the value of the shear correction factor. J. Lockwood Taylor examined the problem of the ship's structure[2,3] on the basis of a modified bending theory and developed the following formula for the shear deflection factor r_s:

$$r_s = C \times 100 \times \frac{D^2}{L^2}$$

where

D = depth of ship
L = length of ship
C = a coefficient which depends upon the ratio of breadth B to depth D.

The value of C is shown plotted in *Figure 11.12*. In Taylor's original paper C is

$$0.035 \left(\frac{3a^3 + 9a^2 + 12a + 1.2}{3a+1} \right)$$

where $a = B/D$.

Johnson[4] in a paper on ship vibration has, however, shown that the value of the numerical coefficient in the above expression should be 0.025, so that with this value the shear correction factor would be

$$r_s = 2.5 \frac{D^2}{L^2} \left(\frac{3a^3 + 9a^2 + 12a + 1.2}{3a+1} \right)$$

It is of interest to ascertain what influence shear deflection has on

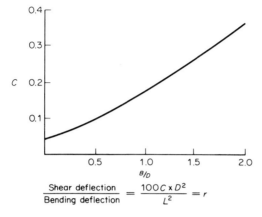

$$\frac{\text{Shear deflection}}{\text{Bending deflection}} = \frac{100C \times D^2}{L^2} = r$$

Figure 11.12

the frequency of vibration by considering a numerical example. Suppose $D/L = 1/10$ and $a = 1.7$, then

$$r_s = 2.5 \times \frac{1}{10^2} \left(\frac{3 \times 1.7^3 + 9 \times 1.7^2 + 12 \times 1.7 + 1.2}{3 \times 1.7 + 1} \right)$$

$$= 0.255$$

The factor affecting the frequency is $1/\sqrt{(1 + r_s)}$

$$= \frac{1}{\sqrt{(1 + 0.255)}} = 0.892$$

The reduction in the frequency as calculated using ordinary bending theory would therefore amount to something over 10%. This is for two-node vibration. It can be shown that for the higher modes the influence of shear deflection on the vibration frequency is very much greater. In fact for these higher modes the influence of shear becomes so important that flexure could be a secondary factor and shear the predominant one.

Other approaches have been made to the influence of shear deflection on vibration and some of these are quoted by Johnson. The Taylor approach has, however, given quite satisfactory results in practice and is probably as good as any.

Structural discontinuities

In many ships the structure resisting longitudinal bending is virtually continuous over the entire length, but there are some ships where the depth of the structure changes fairly abruptly in certain places, as for example where there are large superstructures particularly in passenger ships. Many full scale experiments have shown that where the structure is continuous the distribution of stress over the depth is reasonably linear apart from the shear effect which has already been discussed in the previous section. When there are abrupt changes in section the stress distribution departs from linearity and it is not possible to determine in any simple way how the stress varies.

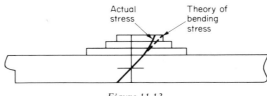

Figure 11.13

Figure 11.13 shows a superstructure on the uppermost continuous deck of a ship. When the structure bends in the longitudinal vertical plane, because the ends must be free from longitudinal stress shear stresses are generated in the vertical sides of the superstructure. This results in distortion of the section. The effect of this is that plane sections no longer remain plane as postulated by ordinary bending theory and thus there is a reduction in the longitudinal stress which would be determined from that theory. The effect is greatest at the ends where, as has already been stated, the stress must be zero. Proceeding further from the ends of the superstructure the effect diminishes and if the superstructure were sufficiently long, near the centre the stress might approach very closely to that forecast by bending theory. It follows that the superstructure is less effective than it would be if the stress distribution were linear. In other words the effective value of I, the second moment of area of a section through the structure in way of the superstructure, is less than it would be if plane sections remained plane.

Another factor which reduces the efficiency of a superstructure is what may be called deck flexibility. A superstructure is caused to bend by the shear stress at the connection between it and the main hull and also because of the vertical forces in the sides which are attempting to pull the superstructure into the same radius of curvature as the hull. Where the sides of the superstructure are continuations of the side shell of the ship it is likely that the two will bend to the same radius of curvature. But if the sides of the superstructure are set in from the ship's side the vertical forces in the sides tend to

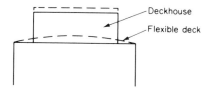

Figure 11.14

cause transverse deformation of the deck structure (as shown in *Figure 11.14*), with the result that the radii of curvature of hull and superstructure are different and this further relieves the stress in the superstructure.

It is most difficult to quantify these two effects as far as a super-structure is concerned. There has been a considerable amount of research aimed at determining stresses in these discontinuous members,[5,6,7,8,9,10] but it is probable that modern approaches to structural analysis can greatly assist in resolving this problem.[11] It is one which cannot be ignored in attempting to calculate ship vibration frequencies. Whilst as far as structural strength is concerned superstructures might be ignored in strength calculations, this is not possible in vibration calculations since it might be equally dangerous to underestimate the frequency as to overestimate it. Corrections for the effectiveness of superstructures when estimating the value of I to be used in frequency calculations are often made on a semi-empirical basis. Reference will be made again to this problem when considering the calculation of frequencies of vibration from first principles.

Influence of entrained water on frequency

The vibration of the main hull of a ship differs in one important aspect from the vibration of a simple beam in that the latter is normally vibrating in air whilst the former is partially immersed in water. When a member vibrates in any fluid the displacement of the member causes a movement of the fluid, so the kinetic energy of the system is the sum of the kinetic energies of the masses and the mass of the fluid set in motion. For the member vibrating in air the kinetic energy of the air is of small amount because of the low density, and consequently there is a negligible effect on the frequency. Frequencies are calculated in these circumstances as though the member in question were in a vacuum. When the fluid is water, however, the density is so very much larger that the kinetic energy of the water must be taken into account when considering vibration.

How the vibrating ship affects the water in its vicinity can be seen readily by considering a section of the ship. As the section moves into the water it pushes fluid aside and as it moves out of the water the fluid returns to fill up the space. The water is thus in a constant state of oscillation and this effect is communicated to the whole of the fluid out to infinity. The phenomenon is often called an 'entrained water effect' and it is equivalent to an increase in mass of the system. This increased mass is called the 'added virtual mass' and it is important to be able to calculate its value when dealing with the vibrating ship.

To obtain some idea of the magnitude of the added virtual mass consider a cylinder of radius a moving with velocity V through a fluid as in *Figure 11.15*. The motion is assumed to be two-dimensional so that the length of the cylinder can be taken as unity.

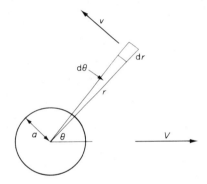

Figure 11.15

It can be shown that as the cylinder moves, the surrounding fluid is set in motion and an element at radius r from the centre of the cylinder has a resultant velocity $v = Va^2/r^2$.

Kinetic energy of element $= \frac{1}{2}\rho v^2 r \, d\theta \, dr$

$$= \frac{1}{2}\rho \frac{V^2 a^4}{r^4} r \, d\theta \, dr$$

The fluid is assumed to extend out to infinity so that

$$\text{Total kinetic energy of fluid} = \int_0^{2\pi} \int_a^{\infty} \frac{1}{2}\rho \frac{V^4 a^4}{r^3} \, d\theta \, dr$$

$$= \frac{\pi}{2}\rho V^2 a^2$$

Now the mass of the fluid displacement by the cylinder per unit length is $\pi \rho a^2 = M'$. Therefore kinetic energy of the water surrounding the cylinder

$$= \tfrac{1}{2}M'V^2 \tag{11.29}$$

The quantity M' is the added virtual mass due to the entrained water and if the mass of the cylinder itself were M per unit length then the total energy of the system would be $\tfrac{1}{2}(M+M')V^2$. The sum $M + M'$ is the total virtual mass of the system.

This result is for a totally immersed cylinder. The problem of the ship is different in that there is a free surface, the ship section being only partly immersed. If in the first place a cylinder is imagined to be floating with its diameter in the waterline and it is assumed to be executing small vertical oscillations then the assumption is made that the added virtual mass for the floating cylinder is half that for the totally immersed cylinder. Hence

Added virtual mass $= \tfrac{1}{2}\pi\rho a^2$ per unit length

Now $\tfrac{1}{2}\pi a^2$ is the immersed area of the half-cylinder so that the added virtual mass $= \rho A$ where $A = \tfrac{1}{2}\pi a^2$.

It follows that for a cylinder length l the added virtual mass would be $\rho A l$ and since the cylinder is floating freely this is also the mass of the fluid displaced and is therefore the mass of the cylinder itself. This brings out an important point, namely that the influence of the entrained water is equivalent to increasing the mass of the system by 100%.

Calculation of added virtual mass for ship forms The theoretical result for a cylinder suggests a simple approximate method for finding the added virtual mass for a ship form. It is reasonable to suppose that for vertical motion the added virtual mass would depend upon the slenderness of the section in the direction of motion, i.e. the breadth/draught ratio. If therefore the local breadth of a section on the waterline is b and the draught is d then a slenderness factor $b/2d$ would measure roughly the influence of this on the added virtual mass. Hence

Added virtual mass per unit length $= \rho A(b/2d)$

where A is the immersed area of the ship section. This will give the correct value for a semicircular section and gives a rough measure of the dependence of the mass on the shape of the section. For the entire ship

$$\text{Added virtual mass} = \int_0^L \rho A \frac{b}{2d}\,\mathrm{d}x \tag{11.30}$$

The value of the added virtual mass for an actual ship calculated from this formula could well be of the order of the displacement at the load draught and for the light draught might be well in excess of this since at the lower draught b/d for the various sections can be in excess of two.

The added virtual mass is dependent on the ratio of breadth to draught for the various sections so that taking an overall picture it could be said that it is dependent upon B/d where B is the moulded breadth of the ship. It would be reasonable to say, therefore, that it is equal to $k(B/d)$ multiplied by the mass of the ship. This would suggest that

$$\text{Total virtual mass } M' = M\left(1 + \frac{kB}{d}\right)$$

where M is the mass of the ship.

Burrill[12] has suggested that $k = \frac{1}{2}$ so that

$$M' = M\left(1 + \frac{B}{2d}\right) \tag{11.31}$$

On the other hand Todd[13] from the analysis of various data has suggested

$$M' = M\left(1.2 + \frac{1}{3}\frac{B}{d}\right) \tag{11.32}$$

The following table gives a comparison of these two formulae.

B/d	Burrill	Todd
2.0	2.00	1.86
2.25	2.12	1.95
2.50	2.25	2.03
2.75	2.37	2.11
3.00	2.50	2.20
3.25	2.62	2.28
3.50	2.75	2.36

The Todd formula consistently gives a value for the total virtual mass factor less than that of Burrill.

Alternative method of calculating added virtual mass   The formulae discussed in the last section, although based on the theoretical result for a cylinder, are nevertheless semi-empirical and can therefore only give an approximate result. A much more detailed study of added virtual mass of near ship shape sections was made by Lewis.[14] Lewis took the results for a cylinder totally immersed and by a process known as 'conformal transformation' derived from the cylinder a series of prismatic sections which represented ship sections plus their images.

If a cylinder whose radius is unity, taken with its centre at the origin in the x,y plane, has applied to it the transformation

$$Z = z + \frac{a}{z} + \frac{b}{z^3}$$

where

$Z = X + iV$
$z = x + iy$
a and b are constants

then a closed figure will be obtained whose co-ordinates are X and Y and which is symmetrical about these axes. Such a figure is shown in *Figure 11.16*. The co-ordinates of the transformed section are related to those of the cylinder by

$$\begin{aligned} X &= (1+a)\cos\theta + b\cos 3\theta \text{ and} \\ Y &= (1-a)\sin\theta + b\sin 3\theta \end{aligned} \right\} \tag{11.33}$$

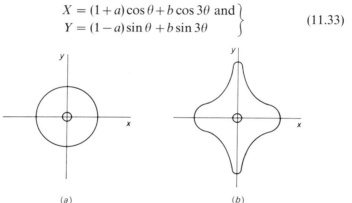

(a)   (b)

Figure 11.16   (a) *Circular cylinder* (b) *Transformed section*

It will be seen that if only the portion of the figure below OX is considered a section is obtained which reasonably represents a stern section for a ship.

By adjusting the values of the coefficients a and b different shapes can be obtained.

The second stage in the Lewis process was to transform the flow around the cylinder to that around the new section and to calculate the energy of the entrained water for the section. This was then compared with the energy of the entrained water for the cylinder. In both cases the assumption made in the previous section was adopted, namely that the energy for the semi-immersed section was one-half of that for the full section.

Lewis showed that for a transformed section having a half waterline breadth B the energy was

$$T' = \frac{\pi}{2}\rho B^2 V^2 \left\{ \frac{(1-a)^2 + 3b^2}{(1-a+b)^2} \right\} \tag{11.34}$$

The result for a cylinder of the same half waterline breadth is

$$T = \frac{\pi}{2}\rho B^2 V^2$$

The ratio of the two is

$$\frac{T'}{T} = \frac{(1-a)^2 + 3b^2}{(1-a+b)^2} \tag{11.35}$$

This ratio Lewis called C and it can be shown to be a function of the fullness of the section and the ratio of the half breadth B to the draught. It follows that the kinetic energy of the surrounding water is equal to

$$T' = \frac{\pi}{2}\rho B^2 V^2 C$$

and therefore the added virtual mass per unit length is

$$M' = \tfrac{1}{2}\pi\rho B^2 C \tag{11.36}$$

for the half of the section which is immersed. Lewis plotted the values of C for various breadth/draught ratios and for sections of various fullnesses. A typical diagram is shown in *Figure 11.17*.

In applying these results for the calculation of the added virtual mass for a ship the intention was to compare the sections for the

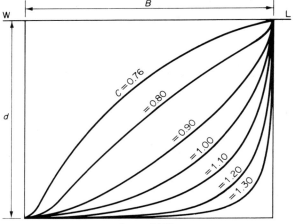

Figure 11.17

actual ship with the Lewis standard sections and lift off values of C and then substitute in equation 11.36 to find the added virtual mass per unit length at any position in the length of the ship. This was not very easy to apply because not only had interpolation to be made between the various C curves for a particular value of B/d but interpolation was also necessary between two values of B/d.

A more convenient way is to plot C against a cross-sectional area coefficient for a particular breadth/draught ratio and plot similar curves for other breadth/draught ratios. Landweber and Macagno[15] in a development of Lewis's work plotted the results in this way and extended the Lewis results to give added virtual mass for horizontal vibration as well as vertical vibration. These curves are shown in *Figures 11.18* and *11.19* and they give added virtual mass coefficients C_V and C_H in terms of sectional area coefficient σ and half breadth/draught ratio λ where these are defined as follows:

Added virtual mass A_V per unit
length for vertical vibration $= \frac{1}{2}\pi\rho B^2 C_V$ (11.37)

Added virtual mass A_H per unit
length for horizontal vibration $= \frac{1}{2}\pi\rho B^2 C_H$ (11.38)

Breadth/draught ratio $\lambda = B/d$

Sectional area coefficient $\sigma =$ Area of immersed section$/2Bd$

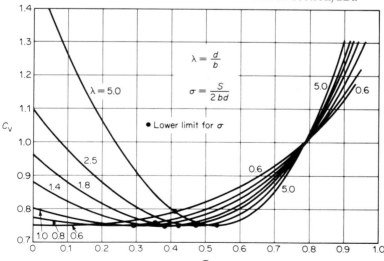

Figure 11.18 (From *Journal of Ship Research*, Society of Naval Architects and Marine Engineers, U.S.A., **1**, 3, p. 28 (Nov. 1957))

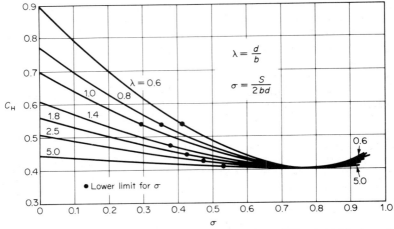

Figure 11.19   (From *Journal of Ship Research,* Society of Naval Architects and Marine Engineers, U.S.A., **1**, 3, p. 29 (Nov. 1957))

Influence of three-dimensional flow   The simple formula developed earlier and the theoretical work of Lewis and Landweber and Macagno are all based on the assumption of two-dimensional flow around the sections of the ship. It will be clear, however, that the problem is really a three-dimensional one.

In *Figure 11.20* is shown the vertical vibration profile of a ship in the two extreme positions A and B. In moving from A to B it will

Figure 11.20

be seen that at amidships the ship sections are moving down and thus pushing water aside. Towards the ends the sections are emerging and tending to form a cavity in the water. Water thus flows in the axial direction to fill this cavity and when the ship moves from B to A the situation is reversed. This situation would exist even if the ship were infinitely long. Because the ship is of finite length there will also be flow in an axial direction round the ends.

The effect of this three-dimensional flow around the ship is to reduce the added virtual mass per unit length. The effect is usually taken account of by the introduction of a three-dimensional added virtual mass factor J which is defined as

$$J = \frac{\text{Kinetic energy of water in three dimensions}}{\text{Kinetic energy of water in two dimensions}} \quad (11.39)$$

Lewis[14] worked out the values of J for the two- and three-node vibration of ellipsoids of revolution of varying length/breadth ratios and the results are as follows:

Length ÷ breadth	5	6	7	8	9	10	11	12	13	14	15
J_2 (two-node)	0.700	.752	.787	.818	.840	.858	.872	.887	.900	.910	.919
J_3 (three-node)	0.621	.675	.720	.758	.787	.811	.830	.845	.860	.872	.883

These are overall coefficients in the sense that the total added virtual mass for a ship would be reduced by multiplying the values of J obtained from the above table by the added virtual mass calculated for two-dimensional flow. This means that the added virtual mass at every section throughout the length is reduced in the same proportion. This cannot really be true, since at amidships for example the flow is probably two-dimensional whereas at the nodes and the ends of the ship considerable three-dimensional flow exists. Because of the difficulty of assessing correctly the J value for each section, however, it is customary to apply these overall coefficients to each section so that in the Lewis formula C is replaced by CJ.

Taylor's analysis of J coefficients Taylor[16] investigated theoretically the value of J and the values given by him for two-node vibration are

L/B	22.22	10.00	7.09	4.97
J	0.946	0.825	0.729	0.615

These values will be seen to be consistently lower than those given by Lewis.

Experimental investigations of added virtual mass A considerable amount of experimental work has been carried out on the influence of added virtual mass on the frequency of vibration on both rigid bodies and bodies subjected to flexure. Reference will only be made, however, to some of the work carried out in the Department of Naval Architecture and Shipbuilding at the University of Newcastle upon Tyne. This has been summarised by Townsin[17] and a relatively simple formula derived for assessing the value of J in determining the added mass for a ship.

It is assumed in this work that the two-dimensional added mass can be obtained correctly by the use of the Lewis method so that the problem resolves itself into finding the value of J. To determine the value of added virtual mass experimentally it is necessary to vibrate the structure in both air and in water. It is obviously impossible to vibrate a full size ship in air so that experiments of this nature have to be done on the model scale. In the work carried out by Townsin the models consisted of light metal ellipsoids and models of actual ship-shape section. Calculations were made for the frequency of vibration in air and they were compared with the actual observed results. A very small difference was found for the two-node vibration but the two differed by a greater percentage for three- and four-node modes. By comparing the frequencies obtained for the models in water with the results in air it was possible to derive a formula for the value of J. It was found that J could be written as a function of the length/breadth ratio and the number of nodes in the length.

The formula finally derived was:

$$J_n = 1.02 - 3\left(1.2 - \frac{1}{n}\right)\frac{B}{L} \tag{11.40}$$

where

L = length of ship
B = breadth of ship
n = number of nodes in the length
J_n = J value for n nodes.

These J values associated with the Lewis results for two-dimensional flow should give a sufficiently accurate estimation of added virtual mass for ship forms in various modes of vibration.

Rotary inertia

The simple formula for the frequency of vibration of a beam with a concentrated mass which was developed above assumed that the mass executed linear oscillation only. Because of the depth of the ship girder, however, the rotation of the mass about a transverse axis is also important, as can be seen from *Figure 11.21*. If the vibration profile for the ship is $y = f(x)\sin pt$ then the slope at any point in the length is $\partial y/\partial x = f'(x)\sin pt$ and consequently the angular

Figure 11.21

velocity is

$$\omega = \frac{\partial}{\partial t}\left(\frac{\partial y}{\partial x}\right) = pf'(x)\cos pt$$

If the mass per unit length is m at any point and its radius of gyration is k then the kinetic energy of rotation is $\frac{1}{2}mk^2\omega^2$ per unit length

$$= \frac{1}{2}mk^2\left(\frac{\partial^2 y}{\partial x \partial t}\right)^2 = \frac{1}{2}mk^2p^2f'(x)^2\cos^2 pt$$

The kinetic energy of translation is $\frac{1}{2}m(\partial y/\partial t)^2 = \frac{1}{2}mp^2f(x)^2\cos^2 pt$ so that the total energy is the sum of these two.

Depending upon the magnitude of k this correction to the energy of the system due to what is called rotary inertia may be appreciable. The correction will be dealt with in more detail later but it is sufficient to say in the mean time that it leads to a reduction in frequency and can be made as an adjustment to the frequency obtained when this factor is ignored as in the case of shear deflection.

If r_r is the ratio of rotational energy to that of translation then the correction factor to the frequency, ignoring rotational energy, is $1/\sqrt{(1 + r_r)}$. The application of this and other correction factors to the frequencies of vibration obtained by applying ordinary bending theory will be discussed in a later section when considering empirical formulae for the estimation of vibration frequency.

Frequencies of vibration by direct calculation

It has been shown that empirical formulae of the Schlick type can give only a very rough estimate of the frequency of vibration of a ship. The empirical constant in such a formula has to take into account distribution of mass and stiffness of the hull girder, shear deflection, added virtual mass and rotary inertia as well as the problems arising with large partial superstructures. Unless, therefore, a large amount of data are available for ships of all types in various conditions of loading it is not possible to obtain really accurate values for the frequency.

Whilst empirical formulae do permit a first shot being made at the frequency, for a detailed study of the problem much more accurate methods are required. This has led investigators to study how frequencies may be determined by direct calculation. There are basically two approaches to this problem which have been employed, namely the deflection method (often nowadays called the 'full integral method') and the energy method. In the following sections they will be discussed in detail.

The deflection or full integral method

This method has often been associated with the name of Morrow[18] and has been developed by others such as Taylor[3] and Todd.[19,20]

Consider a beam or the hull of a ship vibrating with some deflection which can be expressed as

$$y = f(x)\sin pt \qquad (11.41)$$

This assumes the motion to be simple harmonic of frequency $N = p/2\pi$ and that at any instant the deflection at any position in the length is given by $f(x)$. The function $f(x)$ for non-uniform mass and stiffness distribution is an unknown quantity but as will be seen later this can be assumed in the first place.

Differentiating equation 11.41 twice with respect to t it will be seen that

$$\frac{\partial^2 y}{\partial t^2} = -p^2 f(x)\sin pt = -p^2 y \qquad (11.42)$$

This shows that the acceleration is proportional to the displacement y at any point in the length since all points vibrate with the same frequency. It follows that the dynamic load on the structure is given by $m(\partial^2 y/\partial t^2) = -mp^2 y$. Integrating this expression for the load the dynamic shearing force is obtained so that

$$\text{Shearing force } F = -p^2 \int my\,dx + A$$

and

$$\text{Bending moment } M = -p^2 \iint my\,dx\,dx + Ax + B$$

where A and B are constants of integration. In the case of the vibrating ship with free-free ends, both the bending moment and shear force must be zero at $x = 0$ so that A and B are both zero.

Now the relation between the bending moment and deflection is given by $-EI(\partial^2 y/\partial x^2) = M$ so that

$$\frac{\partial^2 y}{\partial x^2} = -\frac{M}{EI} = \frac{p^2}{EI}\iint my\,dx\,dx$$

in which E is Young's modulus for the material and I is the second moment of area at any point of the cross sections resisting bending. In the case of the ship I varies along the length so that at this stage in the calculation a curve of M/I would have to be plotted. Carrying out two further integrations gives

$$\text{Slope} = \frac{\partial y}{\partial x} = -\int \frac{M}{EI}\,dx = \int\left(\frac{p^2}{EI}\iint my\,dx\,dx\right)dx + C$$

$$\text{Deflection} = y = -\iint \frac{M}{EI}\,dx\,dx = \iint\left(\frac{p^2}{EI}\iint my\,dx\,dx\right)dx\,dx$$

$$+ Cx + D$$

The values of the two constants C and D can be obtained from the end conditions. If for convenience the deflection is measured from the chord joining the two ends of the deflection curve then at $x = 0$ and $x = L$, $y = 0$. The first condition gives $D = 0$ and the second gives

$$\left[\iint\left(\frac{p^2}{EI}\iint my\,dx\,dx\right)dx\,dx\right]_L + CL = 0$$

Hence

$$C = -\frac{1}{L}\left[\iint\left(\frac{p^2}{EI}\iint my\,dx\,dx\right)dx\,dx\right]_L$$

It is now possible to write the deflection at any point as

$$y = p^2\left\{\left(\iint\frac{1}{EI}\iint my\,dx\,dx\right)dx\,dx\right.$$

$$\left. -\frac{1}{L}\left[\left(\iint\frac{1}{EI}\iint my\,dx\,dx\right)dx\,dx\right]_L x\right\} \qquad (11.43)$$

After four integrations a value of y has been obtained which should be the same as the original assumed y in the calculation. The only unknown in equation 11.43 is p^2, so that comparing the original value of y with this derived value it is possible to solve for p and hence the frequency. If the correct value of y as given by $f(x)$ had been assumed then the assumed and derived values of y would agree at every point in the length. In other words if both curves were drawn to the same scale they would coincide at every point. As, however, the correct value of $f(x)$ is unknown at the beginning of the calculation, this is not possible and it is customary to equate the deflections at some selected point in the length, usually at the centre of length. Thus the two curves shown in *Figure 11.22* would be obtained and the divergence of the derived curve from the assumed curve will depend upon the accuracy of the original assumption.

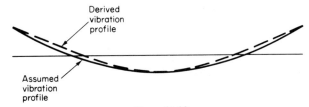

Figure 11.22

Let the value of

$$\left(\iint \frac{1}{EI} \iint my \, dx \, dx \right) dx \, dx - \frac{1}{L}\left[\iint \frac{1}{EI} \iint my \, dx \, dx \, dx \, dx \right]_{L} x$$

in equation 11.43 at $x = L/2$ be δ then

$$y_{L/2} = p^2\delta$$

and let the assumed value of y at $L/2$ be y_0 then

$$y_0 = p^2\delta$$
$$p^2 = y_0/\delta$$

and the frequency is given by

$$N = \frac{1}{2\pi}\sqrt{\frac{y_0}{\delta}} \qquad (11.44)$$

It remains to decide what form the assumed deflection should take and for ship hull vibration calculations this is usually taken as the free-free vibration curve for a uniform beam. The theory of vibration of such beams gives the deflection as

$$f(x) = A \cos mx + B \sin mx + C \cosh mx + D \sinh mx$$

and when the appropriate end conditions are inserted for the free-free case the following ordinates of the vibration profile are obtained, assuming the end ordinate to be unity for two-node vibration:

x/L	0	0.05	0.10	0.15	0.20	0.25	0.30	0.35	0.40	0.45	0.50
$f(x)$	1.000	0.708	0.537	0.313	0.097	−0.099	−0.272	−0.414	−0.521	−0.586	−0.608

In making use of this vibration profile for a ship there is another important matter still to be considered. In the ship the mass is not uniformly distributed, nor is it generally symmetrically distributed about amidships. This means that in carrying out the integrals $\int my \, dx$ and $\iint my \, dx \, dx$ using the values of y for the uniform bar, these integrals would produce curves which if plotted would not close at $x = L$ and the consequence of this is that there would be some force acting on the ship and a moment, both of which would cause rigid body motions, one of translation and one of rotation. To avoid this the ordinates for the uniform bar have to be corrected. A bodily shift of the base line would make the shearing force curve close and a tilt of the base line would close the bending moment curve. This is equivalent to adding a quantity $a + bx/L$ to the original assumed deflection curve at any point where a and b are constants yet to be determined. If y is the deflection curve ordinate for the uniform bar then the conditions to be satisfied are

$$\int_0^L m\left(y + a + \frac{bx}{L}\right) dx = 0 \tag{11.45}$$

and

$$\iint_L m\left(y + a + \frac{bx}{L}\right) dx\, dx = 0 \tag{11.46}$$

It will be seen that these are two simultaneous equations in the two unknowns a and b from which their values can be obtained.

The base line correction referred to here only arises because of the free-free condition of support which exists in the case of the floating ship. If the deflection method were applied to a beam with supports at the ends this correction would be unnecessary since the forces which would have to be applied to maintain equilibrium without rigid body motion would be automatically supplied by the reactions at the supports.

If the basic uniform bar profile is corrected in the way described for the ship it will be found that the frequency obtained from the deflection or full integral method is quite accurate for the two-node mode at least. It is, however, possible to make a second approximation to the frequency by repeating the calculation, starting with the derived deflection curve from the first calculation and going through the process already described. It is usually considered that for two-node vibration this second approximation is not necesssary and that sufficient accuracy can be obtained from the first calculation.

When making the calculation the mass per unit length m includes the mass of the entrained water calculated by a method such as one of those described in the section dealing with added virtual mass.

The method for determining the frequency is based on the application of bending theory, which ignores shear deflection and rotary inertia effects. These are usually made as final corrections of the type already described so that if N is the frequency from the basic calculation the final corrected frequency would be

$$N' = \frac{N}{\sqrt{\{(1+r_s)(1+r_r)\}}} \tag{11.47}$$

where r_s and r_r are the shear deflection and rotary inertia correction factors already discussed.

The integrations required in the actual calculations can be done in either of two ways, i.e. graphically or by a tabular method. In the graphical method a curve of the mass per unit length of the ship must first of all be plotted and to this must be added the added virtual mass of the entrained water. The latter can be done fairly easily but the former, which includes the mass of the structure of the ship, its machinery and fittings, as well as the mass of all the deadweight items, is much more difficult and often approximate methods have to be employed to reduce the amount of calculation. The mass per unit length m is then multiplied by the ordinates of the assumed y curve to give the dynamic load my. The y values have of course to be corrected as described earlier in order to make the shearing force and bending curve close at $x = L$.

The curve of dynamic load my is then plotted and the various integrations carried out by simple graphic integration or by the use of instruments such as the planimeter, integrator or integraph.

If the tabular method is employed the length of the ship is first of all divided into a number of equal parts, say 20 or 40. The mass of the ship plus the added virtual mass of the entrained water is then calculated and this is assumed to be constant over each of the divisions of length. This if plotted would give a stepped type of mass distribution curve. The mass of each division of length is then multiplied by the mean value of y for that division to give the dynamic load. This is then integrated in tabular fashion to give the dynamic shear force and the other integrations are carried out in the same way.

Example of full integral method for calculating frequency of two-node vertical vibration   The particulars of a ship are as follows:

> Length 220 m
> Breadth 31 m
> Depth 15.25 m
> Displacement (mass)   60 000 t
> Added virtual mass   65 000 t
> Total virtual mass   125 000 t

Second moment of area of the midship section is $195\,\text{m}^4$ and is assumed to be constant over the length. The total virtual mass is assumed to be distributed parabolically with its maximum ordinate at amidships and zero ordinates at the ends. This gives a symmetrical distribution of mass about amidships.

The calculation is shown set out in *Table 11.1*. For simplicity only ten sections have been taken in the length although in actual calculations 20 or even 40 sections might be necessary. The first column lists the sections, which in this case are spaced 22 m apart, i.e. $220 \div 10$. In the second column the average mass over each division is given. The third column gives the assumed amplitude curve which is taken to be that for a free-free uniform beam. The ordinates given in this column are those at the centre of each division of length. Column 4 gives the product my which is a measure of the dynamic load. This, when summed from 0 to 220 m, gives an excess value of -985.2 units. A correction is therefore necessary to the assumed amplitude curve. This is obtained by dividing the excess by the total mass and works out at 0.172. This is shown in column 6. In the example chosen there is no correction required for excess moment, i.e. the baseline of the amplitude curve does not have to be tilted. The corrected amplitude curve is shown in column 7. Column 8 gives the product my for the corrected amplitude. This is summed in column 9. The total at the bottom should be zero but there is a small excess due to rounding off the values of the amplitude. The figures in column 9 represent the shearing force and when integrated give the bending moment. This integration is done by taking the arithmetic mean values of the shearing forces over the intervals of length and summing them to give the bending moment in column 10b. The figures in columns 9 and 10b require to be multiplied by the length of the divisions and the length squared respectively to give shearing force and bending moment, i.e. by 22 m and $(22 \text{ m})^2$. However, as all intervals have the same length this can be left until the end of the calculation.

The bending moment M in column 10b is divided by I, which in this case is constant but in practice would vary over the length of the ship. The values of M/I are shown in column 11. The integration of M/I is carried out in column 12 by summing the mean values of M/I for each division of length. A similar procedure is adopted to obtain $\iint (M/I) \, dx \, dx$ as shown in column 13. At this stage the constant of integration A referred to in the discussion of the full integral method has to be calculated. This is given by

$$-\frac{\iint_L \frac{M}{I} \, dx \, dx}{L}$$

Column 14 gives the value of $-Ax$ and when this quantity is deducted from the figure in column 13b a series of values representing deflection are obtained which are shown in column 15. To obtain the actual deflection these figures will require to be multiplied by 22^4 and by p^2 and divided by E (Young's modulus).

It will be seen that

$$\text{Central deflection} = -\frac{p^2}{E}\iint_{x=L/2}\frac{M}{I}\,dx\,dx$$

$$= -\frac{p^2}{E} \times 56.943 \times 22^4$$

Taking $E = 208\,000\,\text{MN/m}^2$ and remembering that the mass is in tonnes, the deflection becomes

$$-\frac{p^2 \times 56.943 \times 22^4 \times 10^3}{208\,000 \times 10^6} = -\frac{56.943 \times 22^4}{208\,000 \times 10^3}$$

$$= -0.064\,13\,p^2$$

Now this value of the deflection must be the same as the original assumed deflection measured from the chord joining the two ends of the curve. This will be seen to be -1.608.
Therefore

$$-0.064\,13\,p^2 = -1.608$$

or $$p^2 = 1.608/0.064\,13 = 25.07$$

Now

$$\text{Frequency } N = p/2\pi = (1/2\pi)\sqrt{25.07}$$

$$= 0.797 \text{ cycles per second}$$

$$= 47.8 \text{ cycles per minute}$$

There remains the correction for shear deflection. Using the Lockwood Taylor formula the modified frequency $N' = N/\sqrt{(1+r_s)}$ where $r_s = 100\,C(D^2/L^2)$.
From *Figure 11.12* it will be seen that for

$$\frac{B}{D} = 31/15.25 = 2.03 \qquad C = 0.36$$

so that $r_s = 100 \times .36 \times (15.5/220)^2 = 0.178$. Therefore corrected frequency $= 47.8/\sqrt{(1+0.178)} = 44$ cycles per minute.
If desired a further correction could be made for rotary inertia but this is likely to be small for the ship considered.

Energy method

The energy method for finding frequencies of vibration is based on the principle of conservation of energy, i.e. that in the absence of

Table 11.1 CALCULATION FOR FREQUENCY OF TWO-NODE VERTICAL VIBRATION BY FULL INTEGRAL METHOD

1	2	3	4	5	6	7	8	9	10		11	12		13		14	15
Section	Mass of ship + added virtual mass (m) (tonne/ metre)	Assumed amplitude (y) (metres)	$m \times y$	$\int my\,dx$	Correction to y (metres)	Corrected y (metres)	my	$\int my\,dx$ = Shearing force F	$\iint my\,dx\,dx$ a	b Bending moment	M/I	$\int \frac{M}{I}dx$ a	b Sum α slope	$\iint \frac{M}{I}dx\,dx$ a	b Sum	−Ax	$\iint \frac{M}{I}dx\,dx$ −Ax Sum α Deflection
0–1	162	+0.708	+124.4	+124.4	+0.172	+0.940	+152.3	+152.3	+76.6	+76.6	0.392	0.196	0.196	0.098	0.098	−15.957	−15.859
1–2	435	+0.313	+136.1	+260.5	+0.172	+0.485	+211.0	+363.3	+257.8	+334.4	1.715	1.053	1.249	0.723	0.821	−31.914	−31.093
2–3	639	+0.099	+63.3	+323.8	+0.172	+0.271	+173.2	+536.5	+449.9	+784.3	4.022	2.868	4.117	2.683	3.504	−47.871	−44.367
3–4	775	−0.414	−320.8	+3.0	+0.172	−0.242	−187.5	+349.0	+442.7	+1227.0	6.292	5.157	9.274	6.695	10.199	−63.828	−53.629
4–5	844	−0.586	−494.6	−491.6	+0.172	−0.414	−349.4	−0.4	+174.3	+1401.3	7.186	6.739	16.013	12.643	22.842	−79.785	−56.943
5–6	844	−0.586	−494.6	−986.2	+0.172	−0.414	−349.4	−349.8	−174.3	+1226.6	6.290	6.738	22.751	19.382	42.224	−95.742	−53.518
6–7	775	−0.414	−320.8	−1307.0	+0.172	−0.242	−187.5	−537.3	−443.5	+783.1	4.016	5.153	27.904	25.327	67.551	−111.699	−44.148
7–8	639	+0.099	+62.3	−1244.7	+0.172	+0.271	+173.2	−364.1	−450.7	+332.4	1.705	2.861	30.765	29.335	96.886	−127.656	−30.770
8–9	435	+0.313	+136.1	−1108.6	+0.172	+0.485	+211.0	−153.1	−258.6	+73.8	0.378	1.041	31.806	30.785	127.671	−143.613	−15.942
9–10	162	+0.768	+124.4	−984.2	+0.172	+0.940	+152.3	−0.8	−76.9	−3.1	−0.016	0.181	31.987	31.896	159.567	−159.570	—

damping the total energy of a vibrating system is constant. It has been seen that damping does exist in any real system but it is of sufficiently small amount to justify the assumption that the energy is constant. Hence it can be stated that the sum of the kinetic energy and the potential energy in a vibrating system is constant.

If a vibrating beam is considered then the kinetic energy is that of the moving masses and in the first place it will be assumed that this is due to linear motion only.

Let the dynamic deflection be $y = f(x)\sin pt$, then at any instant the velocity is $\partial y/\partial t = pf(x)\sin pt$ and the kinetic energy of a small element dx whose mass per unit length is m will be

$$\tfrac{1}{2}m(\partial y/\partial t)^2\,dx = \tfrac{1}{2}mp^2 f(x)^2\cos^2 pt\,dx$$

$$\text{Total kinetic energy} = \int_0^L \tfrac{1}{2}mp^2 f(x)^2\cos^2 pt\,dx \qquad (11.48)$$

The potential energy is the strain energy of bending and this can be shown to be equal to

$$\frac{EI}{2}(\partial^2 y/\partial x^2)^2\,dx = (EI/2)(\partial^2 f(x)/\partial x^2)^2\sin^2 pt\,dx$$

for a small length dx

$$\text{Total potential energy} = \int_0^L \frac{EI}{2}\left(\frac{\partial^2 f(x)}{\partial x^2}\right)^2\sin^2 pt\,dx \qquad (11.49)$$

When the beam is passing through rest the velocity will be a maximum and there will be no bending of the beam at that instant. It is therefore reasonable to conclude that in this case the potential energy will be zero and all the energy will be kinetic. Similarly, when the beam is deflected to its extreme position and is about to return to its original position the velocity is instantaneously zero so that there will be no kinetic energy in the system and all the energy will be potential. As the total energy of the system is constant this means that

P.E. in extreme position = K.E. in passing through rest

The kinetic energy in passing through rest is obtained by putting $\cos pt = 1$ in equation 11.48, and the potential energy in the extreme position by putting $\sin pt = 1$ in equation 11.49. Hence

$$\int_0^L \tfrac{1}{2}mp^2 f(x)^2\,dx = \int_0^L \frac{EI}{2}\left(\frac{\partial^2 f(x)}{\partial x^2}\right)^2\,dx$$

From this the factor p can be determined and hence

$$N = \frac{p}{2\pi} = \frac{1}{2\pi} \sqrt{\frac{\int_0^L EI(\partial^2 f(x)/\partial x^2)^2 \, \mathrm{d}x}{\int_0^L mf(x)^2 \, \mathrm{d}x}} \qquad (11.50)$$

As in the deflection method it is necessary to assume some function $f(x)$ and for the vibrating free-free ship this would be taken to be the deflection profile for the uniform bar. In the case of the ship it will be necessary to ensure equilibrium by making the baseline correction $a + (bx/L)$ in the same way as was done with the deflection method. If this is done the frequency of vibration should be the same when calculated by the two methods. As before, allowance can be made for shear deflection at the end of the calculation by multiplying the result obtained from equation 11.50 by $1/\sqrt{(1 + r_s)}$.

The energy method has been developed in the case of a ship by Tobin[21] and Nicholls.[22]

Rotational inertia   It was shown above that when the rotation of an element was taken into account the kinetic energy was increased by the addition of an amount $\frac{1}{2}mk^2p^2 f'(x)^2 \cos^2 pt \, \mathrm{d}x$ so that the total energy contributed by rotation is

$$\int_0^L \frac{1}{2}mk^2p^2 f'(x)^2 \cos^2 pt \, \mathrm{d}x = \int_0^L \frac{1}{2}mk^2p^2 (\partial f(x)/\partial x)^2 \cos^2 pt \, \mathrm{d}x$$

and equating the kinetic energy in passing through rest to the potential energy in the extreme position gives

$$\int_0^L \frac{1}{2}mp^2 f(x)^2 \, \mathrm{d}x + \int_0^L \frac{1}{2}mk^2p^2 (\partial f(x)/\partial x)^2 \, \mathrm{d}x$$

$$= \int_0^L \frac{EI}{2}(\partial^2 f(x)/\partial x^2)^2 \, \mathrm{d}x$$

The frequency then is

$$N' = \frac{p}{2\pi} = \frac{1}{2\pi} \sqrt{\frac{\int_0^L EI(\partial^2 f(x)/\partial x^2)^2 \, \mathrm{d}x}{\int_0^L mf(x)^2 \, \mathrm{d}x + \int_0^L mk^2 (\partial f(x)/\partial x)^2 \, \mathrm{d}x}} \qquad (11.51)$$

This can be re-written as

$$N' = \frac{1}{2\pi} \sqrt{\frac{\int_0^L EI(\partial^2 f(x)/\partial x^2)^2 \, \mathrm{d}x}{\int_0^L mf(x)^2 \, \mathrm{d}x}} \times \sqrt{\frac{1}{1 + \dfrac{\int_0^L mk^2 \, (\partial f(x)/\partial x)^2 \, \mathrm{d}x}{\int_0^L mf(x)^2 \, \mathrm{d}x}}}$$

$$= \frac{N}{\sqrt{(1 + r_r)}}$$

where N is the frequency, neglecting rotational energy, and

$$r_r = \frac{\int_0^L mk^2(\partial f(x)/\partial x) \, \mathrm{d}x}{\int_0^L mf(x)^2 \, \mathrm{d}x} \qquad (11.52)$$

Energy method applied to beam with concentrated mass   The case of a simply supported uniform section beam with a concentrated mass M at mid span has already been considered by forming the differential equation for the motion. It is instructive to examine this problem using the energy method. In this case a suitable form for the function $f(x)$ is $\sin \pi x/l$. This satisfies the end conditions of simple support. The frequency in this case will be given by

$$N = \frac{1}{2\pi} \sqrt{\frac{\int_0^L EI\{-(\pi^2/l^2)\sin^2(\pi x/l)\}^2 \, \mathrm{d}x}{\int_0^L Mf(x)^2 \, \mathrm{d}x}} = \frac{1}{2\pi} \sqrt{\frac{EI(\pi^4/l^4) \times l/2}{M \times 1}}$$

since the integral on the bottom simply consists of the mass M multiplied by the value of $\sin^2(\pi x/l)$ at mid span

$$\therefore \quad N = \frac{1}{2\pi} \sqrt{\frac{\pi^4 EI}{2Ml^3}} \qquad (11.53)$$

The result previously obtained gave the value of the constant under the root sign as 48 and it will be found that the value of $\pi^4/2$ is 48.7. The difference between the exact formula and that obtained by the approximation $f(x) = \sin(\pi x/l)$ for the deflection profile is quite small. The ratio of the frequencies so calculated is 1.0073 which shows that there is an error of less than 1%.

This example is instructive in showing that so long as the correct end conditions are satisfied there is considerable latitude in the choice of the form of the deflection profile. The assumed profile $f(x) = \sin(\pi x/l)$ is the correct one for a simply supported uniform beam with a uniformly distributed load so that no great error is involved in applying this to the most severe case of concentration of load.

Uniform section simply supported beam with uniformly distributed load

Consider a simply supported beam length l of uniform section and uniform mass distribution. An appropriate function $f(x)$ would be $\sin(n\pi x/l)$ where n is any whole number. If this expression is substituted in equation 11.50 the frequency is

$$N = \frac{1}{2\pi} \sqrt{\frac{\displaystyle\int_0^l EI(n^4\pi^4/l^4)\sin^2(n\pi x/l) \times \mathrm{d}x}{\displaystyle\int_0^l m\sin^2(n\pi x/l)\,\mathrm{d}x}}$$

$$= 1/2\pi \sqrt{(EIn^4\pi^4/ml^4)}$$

$$= (\pi/2)n^2 \sqrt{(EI/ml^4)} \tag{11.54}$$

This expression gives the frequencies of all modes of vibration depending upon the value assigned to n. It is in fact the exact formula for the frequency of vibration of a beam based on bending theory and ignoring shear deflection and rotational inertia.

This particular example is instructive in showing the factors which affect the rotational inertia correction.

From equation 11.52:

$$r_r = \frac{\displaystyle\int_0^l mk^2(\partial^2 f(x)/\partial x)^2\,\mathrm{d}x}{\displaystyle\int_0^l mf(x)^2\,\mathrm{d}x}$$

$$= \frac{\displaystyle\int_0^l mk^2\{(n\pi h/l)\sin(n\pi x/l)\}^2\,\mathrm{d}x}{\displaystyle\int_0^l m\sin^2(n\pi x/l)\,\mathrm{d}x}$$

$$= k^2n^2(\pi^2/l^2)$$

The value of r_r then depends upon the ratio of k to l and also upon n, i.e. the mode of vibration. Consider the case of a solid rectangular section, breadth b and depth d. Then $k^2 = d^2/12$ and

$$r_r = n^2 \times d^2/12 \times \pi^2/l^2 = n^2 \times 0.822\, d^2/l^2$$

The important factors affecting the value of r_r are the ratio of depth to length and the mode of vibration as represented by n. If, for example, $d/l = 1/10$ and $n = 1$, i.e. the fundamental mode of vibration is considered, then $r_r = 1 \times 0.822 \times 1/100 = 0.008\,22$. The correction factor $1/\sqrt{(1 + r_r)} = 1/\sqrt{(1 + 0.008\,22)} = 0.996$. This shows that the correction is negligible for proportions of depth to length below $1/10$ and for the lowest mode of vibration. The value of r_r does, however, go up as n^2 so that for the next two modes r_r would be 0.0328 and 0.0739 respectively, giving values of the correction factor of 0.983 and 0.965.

It is not easy to interpret these results as far as an actual ship is concerned, but it is probable that the influence of rotational inertia may be somewhat larger than for the simple beam. Johnson[4] gives the following values of $1 + r_r$ for two ships, the *Clan Alpine* and the *Ocean Vulcan*:

Mode		Light ship	Loaded ship
2	Node vertical	1.029	1.023
3	Node vertical	1.063	1.048
4	Node vertical	1.108	1.081
2	Node horizontal	1.031	1.022
3	Node horizontal	1.069	1.048
4	Node horizontal	1.118	1.082

The correction factor $1/\sqrt{(1 + r_r)}$ ranges from 0.988 for the two-node vertical to 0.945 for four-node horizontal. In two-node vibration the correction is therefore only a little over 1% and in view of the other uncertainties in the calculation of frequencies it could be safely neglected in ships of average proportions. When dealing with the higher modes of vibration, however, the correction becomes important and like the correction for shear deflection, it cannot be ignored.

Calculation of frequency of higher modes of flexural vibrations

At first sight it would seem that the frequency of vibration of a ship in the higher modes could be obtained by assuming the free-free profile for a uniform bar for the appropriate mode and following out the process described, for example, for the deflection or full integral method. It has been found, however, that if this process is followed for, say, the three-node mode and the necessary baseline corrections are made to satisfy the conditions of equilibrium, the vibration profile degenerates into that for the two-node mode as successive approximations are made. In other words instead of the assumed deflection profile converging to the correct one for the particular mode with successive iterations it actually diverges further and futher from the original assumed profile.

This can be explained by the fact that the assumed profile contains a component of the two-node profile which becomes predominant as the calculation proceeds. To avoid this it is necessary in calculating the frequencies of the higher modes to 'sweep' the assumed profile of all the lower components before proceeding with the calculation.

How this sweeping can be carried out has been described by Den Hartog,[23] Townsin[24] and others. The profile for some mode of vibration n can be expressed as

$$Y_n = \sum_{q=2}^{q=\infty} a_q y_q$$

where $a_q y_q$ is the exact profile for the qth mode.

If this summation is written out in full the deflection profile is given by

$$Y_n = a_2 y_2 + a_3 y_3 + a_4 y_4 + \ldots + a_q y_q$$

where y_2, y_3, etc., are the exact profiles for two-, three-node, etc., vibration.

It is necessary to remove all components below the qth node so nat if the three-node mode was being considered, then neglecting the components contributed by the higher modes

$$Y_3 = a_2 y_2 + a_3 y_3$$

so that the correct value of y_3 is

$$y_3 = (Y_3 - a_2 y_2)/a_3 \tag{11.55}$$

where in this case Y_3 is an assumed profile for the three-node vibration.

If the mass per unit length is m and both sides of this expression are multiplied by m and y_2 and integrated from 0 to L the following result is obtained:

$$\int_0^L m y_3 y_2 \, dx = \frac{\displaystyle\int_0^L m Y_3 y_2 \, dx - \int_0^L m a_2 y_2^2 \, dx}{a_3}$$

Since y_2 and y_3 are the correct profiles for the two-node and three-node modes respectively then

$$\int_0^L m y_2 y_3 \, dx = 0$$

and hence the value of a_2 is determined from

$$a_2 = \frac{\displaystyle\int_0^L m Y_3 y_2 \, dx}{\displaystyle\int_0^L m y_2^2 \, dx} \tag{11.56}$$

This enables the correct profile to be determined for the three-node vibration from equation 11.55. It involves first of all finding the correct two-node profile and the assumption of some reasonable function for Y_3 which could be the three-node profile for the uniform bar.

The same procedure can be adopted for the higher modes. In this way a corrected profile can be obtained for any mode of vibration and the frequency determined by applying the full integral method.

An alternative method for dealing with the problem of the frequencies of the higher modes of vibration has been suggested by Richards,[25] but this work is outside the scope of the present discussion.

In recent years the finite element method has been applied to ship vibration problems and future development will probably be along this line.

Frequencies of torsional vibration

It has been seen that torsional modes of vibration are possible in the case of a ship. The calculation of the frequencies of these modes of vibration, however, present many more difficulties than do the flexural modes.

Consider first the torsional vibration of a uniform rod. It will be assumed that a relation exists between the applied torque T and the angle of twist θ at any point, similar to the relation which exists for flexure. The relation is

$$T = JG\theta/l$$

where

J = polar second moment of area
l = length over which the angle θ is measured
G = modulus of rigidity of the material.

Consider now a small element of a shaft length δx, then the moments on the two ends will be T and $T + \delta T$, the net twisting moment being δT.

The mass of the element is $\rho a \delta x$ where a is the cross-sectional area. Hence the equation of motion for the element is

$$\rho a \delta x k^2 (\partial^2 \theta / \partial t^2) = \delta T = \partial / \partial x \{ JG(\partial \theta / \partial x) \} \delta x$$

$$\rho a k^2 (\partial^2 \theta / \partial t^2) = JG(\partial^2 \theta / \partial x^2) \qquad (11.57)$$

If it is assumed that the angle θ at any point x from one end is $\theta = f(x) \cos pt$ then a suitable function is $f(x) = A \cos(n\pi x / l)$. This gives $\partial \theta / \partial x = 0$ at $x = 0$ and $x = l$ for a free-free rod. If this function is substituted in equation 11.57 then

$$- \rho a k^2 p^2 A \cos(n\pi x / l) \cos pt = (- JGn^2 \pi^2 A / l^2) \cos(n\pi x / l) \cos pt$$

from which

$$\rho a k^2 p^2 = JGn^2 \pi^2 / l^2$$

or

$$N = \frac{p}{2\pi} = \frac{1}{2\pi} \sqrt{\frac{JGn^2 \pi^2}{\rho a k^2 l^2}}$$

$$= \frac{1}{2} \sqrt{\frac{JGn^2}{\rho a k^2 l^2}} \qquad (11.58)$$

The lowest mode of vibration is associated with $n = 1$ and if N is given in cycles per minute then

$$N = 30 \sqrt{(JG / \rho a k^2 l^2)} \qquad (11.59)$$

This shows the factors upon which the torsional frequency depends and it suggests in the case of the ship a formula of the Schlick type where if J refers to, say, the midship section of the ship and M is the total mass, the frequency could be written simply as

$$N = \text{CONST.} \sqrt{(J / Mk^2 L^2)} \qquad (11.60)$$

in which k could be a mean radius of gyration for the whole ship or the value of this quantity for the ship at amidships.

One important problem here is that while J is the polar second moment of area for a solid or hollow circular shaft, when a box structure like that of a ship is being dealt with J cannot be calculated in this way. This is because in the simple theory of torsion for a circular section the shear stress is proportional to the distance from the axis of rotation, leading to the result that the shear stress at any radius is constant. Such an assumption is not valid for box type sections. The problem has been tackled for an elliptical section and

Vedeler[26] was one of the first to deal with the box section as applied to the ship. He showed that the effective polar moment of inertia of a box section could be written

$$J_e = 2kA^2/\sum(l/t)$$

where k is an experimental constant derived from tests on box girders and is given by $k = 2-(b/a)$, b being the smaller dimension and a the larger. A is the total area enclosed by the section and l is the width of plating at any section whose thickness is t.

Prior to this work by Vedeler little interest was shown in the torsion of ships, but with the advent of the container ship more extensive investigations have been carried out, such as the work of De Wilde.[27] This considers the torsion problem in much more detail than earlier work.

When considering the ship the same problem exists as for flexural vibration, namely that the mass distribution and hence the mass moment of inertia distribution is not constant along the length, nor is the effective stiffness as represented by J_e. It is, however, possible to calculate the torsional frequency in the same way as in the case of flexure, as has been demonstrated by Taylor.[3]

If as was stated earlier $\theta = f(x)\cos pt$ then

$$\partial^2\theta/\partial t^2 = -p^2 f(x)\cos pt = -p^2\theta$$

It follows that the angular acceleration is proportional to the angular displacement. Now

$$\partial^2\theta/\partial x^2 = (\rho ak^2/J_e G)(\partial^2\theta/\partial t^2) = -(I/J_e G)p^2\theta$$

where

I = mass moment of inertia at any section = ρak^2
J_e = effective stiffness as already mentioned.

Integrating this expression twice will give

$$\theta = -p^2\iint\frac{I}{J_e G}\theta\,dx\,dx \qquad (11.61)$$

In this, θ inside the integral sign is the assumed value of the angular displacement and may be taken as that for the uniform shaft, namely $A\cos(\pi x/L)$ for one-node vibration. The value of θ obtained by equation 11.61 is the derived angular deflection. Suppose this has some value $-\theta_0$ at $x = L$ and suppose that $\iint(I/J_e G)\,dx\,dx$ at $x = L$

is δ, then

$$\theta_0 = p^2\delta$$

$$N = (p/2\pi = (1/2\pi)\sqrt{\theta_0/\delta} \qquad (11.62)$$

This is a similar expression to that derived for the case of flexural vibration.

The first integral of $(I/J_e\theta)p^2\theta$ must close at $x = L$ and because of the non-uniformity of inertia and stiffness this will not necessarily be so. It will therefore be necessary to make a baseline correction to the deflection $A\cos(\cos \pi x/L)$ by a parallel shift as in the case of flexure. Should the final derived curve differ greatly from the assumed curve the calculation should be repeated using the first derived curve as an assumed curve for the second approximation.

In calculating the mass moment of inertia for use in the calculation the moment of inertia of the entrained water should be included, but this is not likely to have the same importance as it has in vertical and horizontal vibration.

Approximate formulae for vibration frequency

It was stated above that formulae of the Schlick type had severe limitations, which had led to a more detailed study of the problem of calculating vibration frequencies. However, it will be obvious from the preceding sections that to calculate frequencies of the hull of a ship by 'direct' methods necessitates the availability of a considerable amount of information about the ship in question.

Two factors will have been seen to be of great importance, namely the distribution of mass throughout the length of the ship, and the distribution of stiffness as represented by the second moment of area of the structural sections of the ship. These would generally not be available until an advanced stage in the design of the ship had been reached. It is usually necessary, however, for a reasonable estimate of vibration frequency to be made at a much earlier stage in the design before much of this detailed information is available. There is thus a need to develop relatively simple empirical formulae to attempt a first shot at the frequency of vibration. Several investigators have suggested modifications to the basic Schlick type formula in an attempt to remove some of the uncertainty of the global constant present in this formula. Burrill[12] produced a formula by introducing factors dealing with added virtual mass and shear deflection. His formula gave the frequency of vertical two-node vibration in cycles per minute as

$$N = \frac{\phi_2}{\sqrt{\{1 + (B/2d)\}}\,(1 + r_s)} \sqrt{\frac{I}{\Delta L^3}} \qquad (11.63)$$

where

Δ = displacement of the ship
B and d = respectively the breadth and the draught
r_s = the Taylor shear deflection correction factor
L = length of the ship
I has the usual meaning.

From an analysis of the actual vibration frequencies for a number of ships of different types it was concluded that ϕ_2 should have a value of 200 000 when used in conjunction with L equal to the length between perpendiculars. If the relation between length overall and length between perpendiculars is unusual it was suggested that ϕ_2 should be 220 000 used in conjunction with the length overall. The values of ϕ_2 given were based on pre 1935 data and consequently if this formula was used for present day ships of the large tanker or bulk carrier types, modified values would probably have to be used.

In the Burrill formula, as in the original Schlick formula, Δ is in tons, L is in feet and I is in $\text{in}^2 \times \text{ft}^2$. The value of ϕ_2 would have to be modified if these quantities were given in SI Units.

Some idea of the frequencies which might be expected in ships using Burrill's formula can be obtained from the following example of a tanker which had the following particulars:

$$L = 700\,\text{ft}$$
$$B = 102\,\text{ft}$$
$$d = 39.4\,\text{ft}$$
$$\Delta = 45\,560\,\text{tons}$$
$$I = 3\,254\,500\,\text{in}^2\,\text{ft}^2$$
$$r_s \text{ from curves} = 0.06$$

$$N = \frac{200\,000}{\sqrt{\{1 + (102/2 \times 39.4)\}}\,(1 + 0.06)} \sqrt{\frac{3\,254\,500}{45\,560 \times 700^3}}$$

$$= 58.5 \text{ cycles per minute}$$

Todd has suggested a modification to the Schlick formula to include the effect of added virtual mass. It was shown earlier that Todd had suggested that the total virtual mass could be given by $M' = M\{(B/3d) + 1.2\}$ or in terms of displacement $\Delta_1 = \Delta\{(B/3d) + 1.2\}$. Taking this into account in the Schlick formula would give

$$N = \phi\sqrt{(I/\Delta_1 L^3)}$$

Todd investigated the influence of superstructures of various lengths and configurations on the value of I to be used with this formula and plotted actual observed results for two-node frequencies against $\sqrt{(I/\Delta_1 L^3)}$. He concluded that the best results would be obtained if I included superstructures which were in excess of 40% of the length of the ship. For ships both with and without superstructure most of the observed results lay on a straight line given by

$$N = 110\,000 \sqrt{(I/\Delta_1 L^3)} + 29 \qquad (11.64)$$

If this formula is applied to the ship for which the frequency has already been obtained from the Burrill formula the following result is obtained:

$$\Delta_1 = 45\,560 \left(\frac{102}{3 \times 39.4} + 1.2 \right) = 93\,987 \text{ tons}$$

$$N = 110\,000 \sqrt{\left(\frac{3\,254\,500}{93\,987 \times 700^3} \right)} + 29 = 60.9 \text{ cycles per minute}$$

Todd has produced a further modification to the Schlick formula to avoid the necessity of calculating the value of I in the early stages of a design. For a solid section, breadth B and depth D, $I = \frac{1}{12}BD^3$ so that it would be reasonable to suppose that for the box type section which would more nearly represent the structure of a ship it can be assumed that $I = CBD^3$ where C is a coefficient which would vary with length/depth ratio and type of ship. If C is imagined to be merged in the constant outside the root in the Schlick formula then the expression for frequency would be

$$N = \beta \sqrt{(BD^3/\Delta L^3)}$$

and if the Todd formula for added virtual weight is used, namely $\Delta_1 = \Delta\{(B/3d) + 1.2\}$, then

$$N = \beta \sqrt{(BD^3/\Delta_1 L^3)} \qquad (11.65)$$

In applying this formula to ships with superstructures it was difficult to decide what depth should be included. Todd developed a method whereby an effective depth D_E was calculated depending upon the extent of the superstructures. If as in *Figure 11.23* the length between perpendiculars of a ship is L and there is a first tier of superstructure length $L_1 = x_1 L$ and a second tier length $L_2 = x_2 L$ and the depth of the ship to the uppermost continuous deck is D and to the two superstructure decks the depths are D_1 and D_2, then

$$D_E = \sqrt[3]{\{D^3(1-x_1) + D_1^3(x_1 - x_2) + D_2^3 x_2\}}$$

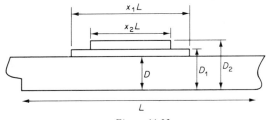

Figure 11.23

Todd found that for the ships for which he had data the two-node vertical frequency could be given by

$$N = 52\,000 \sqrt{(BD_E^3/\Delta_1 L^3)} + 28 \text{ for tankers and}$$

$$N = 46\,750 \sqrt{(BD_E^3/\Delta_1 L^3)} + 25 \text{ for cargo and passenger ships}$$

Applying the first of these formulae to the same example gives

$$N = 52\,000 \sqrt{(102 \times 50^3/93\,987 \times 700^3)} + 28 = 60.7 \text{ cycles per minute}$$

In Todd's formula the correction for the presence of deckhouses has been made purely on the influence which the length of the house has on the frequency. Work carried out by Johnson and Ayling[28] has also endeavoured to take account of how the stress distribution varies in the various sections of a superstructure and has attempted to assess the influence of this on the frequency. The theory determines an overall correction factor which is to be applied to the second moment of area to be used in a Schlick type formula for the calculation of two-node vibration frequency. Observed frequencies were plotted against the factor $\sqrt{Ik/\Delta_1 L^3}$ where k is the correction factor and Δ_1 is the total virtual mass as given by the formula $\Delta_1 = \Delta\{(B/3d) + 1.2\}$. Values of k can be derived from a series of graphs.

General remarks on approximate formulae

Only a few of the approximate formulae which have been developed for determining ship hull frequencies have been considered here and whilst values of empirical coefficients have been quoted as given by the authors of these formulae they should be treated with caution. Every investigator should derive his own data by actually observing frequencies and endeavouring to fit the results into some type of formula which correctly takes into account the fundamentals of the problem. The Schlick type formula does this in that it assesses the frequency as being proportional to the square root of the quantity $I/\Delta L^3$.

This type of formula could be used in its simplest form for geometrically similar ships loaded in the same way. For example, if for a certain ship of length L_1, second moment of area I_1 and displacement Δ_1 the frequency has been observed to be N_1, then for a geometrically similar ship whose particulars are L_2, I_2 and Δ_2 it follows that its frequency N_2 will be given by

$$\frac{N_1}{N_2} = \sqrt{\frac{I_1/\Delta_1 L_1^3}{I_2/\Delta_2 L_2^3}}$$

so that

$$N_2 = \sqrt{\left(\frac{I_2}{I_1} \times \frac{\Delta_1}{\Delta_2} \times \frac{L_1^3}{L_2^3}\right)} \qquad (11.66)$$

What is being assumed here is that the constant in the basic formula is the same for both ships, which is only true if the ships are geometrically similar and loaded in exactly the same way. Much depends upon judgement as to whether these conditions are fulfilled. If sufficient data have been collected for ships of various types in various conditions of loading then it should be possible to choose one which is close enough to a particular new design to get a good estimate of the frequency. However, after such first approximations have been made a detailed calculation should be carried out at the earliest opportunity.

Calculation of amplitudes of vibration

The theory of vibration shows that whenever there is a periodic force applied to an elastic structure there will be a forced vibration in the frequency of the applied force. In the case of the ship this means that if there are out-of-balance forces in the working parts or forces induced by the sea there will be vibration. It is important, therefore, to know what the amplitude of the vibration is from three points of view: firstly from that of habitability of crew and passengers, secondly from that of disturbance of instruments, and finally from that of structural strength.

Concerning habitability it is obvious that continuous vibration can be extremely unpleasant. Acceleration has often been used as a measure of what is tolerable as far as vibration is concerned, and this is a function of both frequency and amplitude. In simple harmonic motion, if the amplitude is y then

$$y = A \sin pt = A \sin 2\pi nt$$

where n is the frequency. $d^2y/dt^2 = -A4\pi^2 n^2 \sin 2\pi n^2 t = -4\pi^2 n^2 y$ so that the acceleration is proportional to the amplitude y and the square of the frequency.

Lewis[29] quotes the 'threshold of perception' based on acceleration as about $0.8 \, \text{in/s}^2$ for horizontal vibration and $1.6 \, \text{in/s}^2$ for vertical vibration. The corresponding figures for the tolerable limit are given as 20 and $40 \, \text{in/s}^2$. Taking this highest value then $40 = 4\pi^2 N^2 y$ and $y = 40/4\pi^2 N^2$.

In a ship where the natural frequency of the hull is of the order of 100 per minute then the limiting amplitude would be

$$\frac{40}{4\pi^2(100/60)^2} = 0.36 \, \text{in} \, (9.1 \, \text{mm})$$

Vibrations of this amplitude are in most cases not likely to occur, but nevertheless sustained vibration can still be quite unpleasant. Usually, however, it will be the higher frequency vibrations of very small amplitude which will be most unpleasant from the human point of view and they may be associated with local structure such as panels. Of course if the frequencies of such structural parts should happen to be in the audible range then greater discomfort still could be experienced.

Vibration of appreciable amount can also affect the performance of instruments on board the ship. This may be more important in naval vessels than in merchant ships, but severe vibration could adversely affect navigation instruments in all types of ships. What would constitute undesirable vibration under these circumstances would have to be determined for each individual case.

The importance of limiting vibration as far as the structure of the ship is concerned arises because of the possibility of high stresses being created in the ship which is already highly stressed. These stresses, which are cyclic, could result in fatigue failure, at least in local parts of the structure, which might ultimately lead to more general failure. It has been noted for example by Goodman[30] that the stresses induced in the structure by the passage of waves can be high. The importance of being able to make some estimate of these stresses is obvious and as a step towards this the determination of vibration amplitudes requires consideration.

The determination of the amplitudes of the vibration of a simple mass-spring system having one degree of freedom is a relatively simple matter and it was shown that the ratio of the dynamic deflection y to the static deflection Y produced by a load F_0 was

$$\frac{y}{Y} = \frac{1}{\sqrt{\{(1-\Lambda^2)^2 + \mu^2 \Lambda^2/Mk\}}}$$

This is seen to depend mainly on the turning factor Λ, but is also dependent upon the damping factor μ, especially near resonance.

The calculation of the amplitude of a beam in flexural vibration is much more complicated and the case of the ship is even more so. It has been shown that if a force $F \sin \omega t$ acts on a beam, and in the present case the hull of the ship, the resulting amplitude y_{it} of the ith mode of vibration at time t and at any position x in the length is given by

$$y_{ix,t} = \frac{FX_i(x_0)\sin(\omega t - \phi_i)X_i(x)}{\omega_i^2\left[\sqrt{\{1-(\omega/\omega_i)^2\}^2 + \{(\mu/m)(\omega/\omega_i^2)\}^2}\right]\int_0^L mX_i^2(x)\,dx}$$

(11.67)

where

X_i = vibration profile of the ith mode of vibration
x_0 = point of application of the load $F \sin \omega t$
ω_i = circular frequency of the ith mode
ϕ_i = phase angle for the ith mode
μ = damping coefficient
m = mass per unit length.

The phase angle is given by

$$\tan\phi_i = \frac{(\omega/\omega_i^2) \times (\mu/m)}{1-(\omega/\omega_i)^2}$$

(11.68)

Equation 11.67 for the amplitude is for one mode of vibration only. Since there is an infinite number of modes of vibration in flexure, all of which may be excited by the disturbing force, the total response of the structure would have to be summed for all values of i and would be given by

$$y_{(x\,t)} = \sum_{i=1}^{i=\infty} \frac{FX_i(x_0)\sin(\omega t - \phi_i)X_i(x)}{\omega_i^2\sqrt{[1-(\omega/\omega_i)^2]^2 + \{(\mu/m)(\omega/\omega_i^2)\}^2]}\int_0^L mX_i^2(x)\,dx}$$

(11.69)

In practice it would be sufficient to make this summation for the first three or four modes only, as the contribution of the higher modes to the total amplitude is very small.

To carry out calculations of this sort requires the determination of the frequencies of vibration of the hull in its several modes and also a correct assessment of the damping coefficients for the different modes, as well as a determination of the force F. The latter, instead of being a concentrated force, may be distributed as would be the case in the force generated due to waves, in which case to find the total response it would be necessary to make a further summation over the length of the ship.

When the amplitude y has been obtained, the bending moment corresponding to this can be obtained from $M = -EI(\mathrm{d}^2y/\mathrm{d}x^2)$, which would involve differentiating the deflection curve twice. The stress in the structure can then be obtained from the ordinary beam formula.

Prevention of vibration

The importance of minimising or eliminating vibration altogether has been demonstrated and in this final section some comments will be made as to how this can be brought about. The courses open to the designer are:

(1) Balance all forces in the machinery, auxiliaries and propeller.

(2) Avoid synchronism by adjusting the frequency of vibration of the structure or the frequency of the distributing force so that synchronism does not exist.

(3) Use some form of vibration damper.

It is extremely difficult to balance out all the forces and moments in machinery, particularly of the reciprocating type, but this is the object which the marine engineer tries to achieve. Where there are no reciprocating parts there should be no unbalanced forces and this applies in turbine machinery. With regard to the propeller, because of the fact that it always works in a variable wake, it is impossible to obtain a complete balance of all the forces acting on the blades. It is, however, possible to minimise these out-of-balance forces by giving adequate tip clearance between the blades and the hull and the appendages such as bossing and shaft brackets.

It should be noted that balancing all the disturbing forces completely is the only way in which to eliminate vibration in a ship altogether. It has been shown that if there is any periodic force acting on the ship there must of necessity be a forced vibration generated even although the frequency of the force is far away from the natural frequencies of the hull.

It follows from this that the second method of avoiding vibration, i.e. that of avoiding synchronism, is to some extent only a partial solution to the problem in that it reduces vibration rather than eliminating it completely. Synchronism or resonance can be avoided by assessing correctly the frequencies of the hull in its various modes of vibration and arranging if possible that the frequency of the disturbing forces does not coincide with these. This in general means choosing the revolutions of the main machinery so that they do not coincide with the hull frequencies. In cases where propeller excited vibration is suspected, this means choosing the revolutions multiplied by the number of blades so as not to coincide with the frequencies of the higher modes.

The alternative to altering machinery revolutions is to alter the frequency of the hull. This can be done in either of two ways. An adjustment in load distribution will alter the frequency: concentration of mass near the nodes will increase the frequency and concentration of mass away from the nodes will reduce it. The designer may have little scope to alter load distribution and often this in any case may not be desirable. Alteration of load distribution may lead to a condition yielding high longitudinal bending stresses which would be unacceptable.

The second method of altering ship frequencies is to alter the value of I, since frequency is proportional to $\sqrt{I}$. In general this can only mean increasing I, since a reduction in this quantity would result in a reduction in the strength of the ship which would be unacceptable. A 5% increase in I would only raise the frequency by 2.5% approximately and a considerable amount of additional material would have to be incorporated in the structure to achieve this. The material would be most effective in the uppermost deck and in the bottom of the ship, but even so this could mean loss of deadweight and may be uneconomical.

Synchronism between exciting forces and ship frequencies can only be avoided in the case of wave induced vibration by altering the frequency of encounter of the waves with the ship. This can be done by either altering course or reducing speed. Once again this may not be a very economical procedure if continued for any lengthy period and should only be resorted to in the event of very severe vibration of this type.

Where synchronism cannot be avoided the only other course is to use some sort of vibration damper. The use of these devices has been dealt with fully by Inglis[31] and Todd.[13] The term 'vibration damper' is really a misnomer in the sense that they do not owe their effectiveness to an increase in damping of the system, but rather to the principle known as 'secondary resonance'. If a periodic force is applied to a system such as a bar in flexure and the frequency of the force is equal to the frequency of the bar then it is possible to reduce the amplitude of the vibrations of the bar by attaching to it another spring-mass system having the same frequency. What happens is that the energy is really transmitted to the second system,

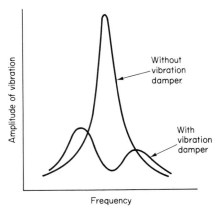

Figure 11.24

which then vibrates. The type of response curve which would be obtained is that shown in *Figure 11.24*. The high resonance peak is eliminated and two much reduced peaks are created.

Although Inglis's paper on this subject was published more than 40 years ago, little interest in these devices for ships has been shown in the intervening time until recently, when with the incidence of wave induced vibration there seems to have been a resurgence of interest. The Ship Division of the National Physical Laboratory has in the last two years developed a type of vibration damper[32] shown in *Figure 11.25*. It consists of a column of fluid which compresses a volume of air, thus providing the spring part of the system while the mass of the column provides the mass. The mass of the water involved is about one-thousandth of the mass of the ship. Experi-

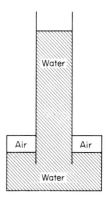

Figure 11.25

ments appear to show that this device is effective in reducing vibration, but according to Todd there are many difficulties to be overcome in its useful operation.

REFERENCES

1. Schlick, O., 'Vibration of steam vessels', *Trans. Instn. of Naval Architects*, 1884.
2. Taylor, J. L., 'The theory of longitudinal bending of ships', *Trans. North East Coast Instn. of Engrs. & Shpbdrs.*, 1924–1925
3. Taylor, J. L., 'Ship vibration periods', *ibid.*, 1927–1928
4. Johnson, A. J., 'Vibration tests on all welded and riveted 10,000 ton dry cargo ships', *ibid.*, 1950–1951
5. Crawford, L., 'Theory of long ships' superstructure', *Trans. Society of Naval Architects and Marine Engineers, U.S.A.*, 1950
6. Caldwell, J. B., 'The effect of superstructures on the longitudinal strength of ships', *Trans. Instn. of Naval Architects*, 1957
7. Muckle, W., 'The influence of proportions on the behaviour of partial superstructures constructed of aluminium alloy', *ibid.*, 1955
8. Muckle, W., 'Superstructures with large side openings: a comparison between theory and experiment', *Trans. Royal Instn. of Naval Architects*, 1966
9. Johnson, A. J., 'Stresses in deckhouses and superstructures', *Trans. Instn. of Naval Architects*, 1957
10. Chapman, J. C., 'The interaction between a ship's hull and a long superstructure', *ibid.*, 1957
11. Paulling, J. R., and Payer, H. G., 'Hull-deckhouse interaction by finite element calculations', *Trans. Society of Naval Architects and Marine Engineers, U.S.A.*, 1968
12. Burrill, L. C., 'Ship vibration: simple methods of estimating critical frequencies', *Trans. North East Coast Instn. of Engrs. & Shpbdrs.*, 1934–1935
13. Todd, F. H., *Ship Vibration*, Edward Arnold, 1961
14. Lewis, F. M., 'The inertia of the water surrounding a vibrating ship', *Trans. Society of Naval Architects and Marine Engineers, U.S.A.*, 1929
15. Landweber, L., and de Macagno, M. C., 'Added mass of two dimensional forms oscillating in a free surface', *Journal of Ship Research, Society of Naval Architects and Marine Engineers, U.S.A.*, 1957
16. Taylor, J. L., 'Vibration of ships', *Trans. Instn. of Naval Architects*, 1930
17. Townsin, R. L., 'Virtual mass reduction factors: J values for ship vibration calculations derived from tests with beams including ellipsoids and ship models', *Trans. Royal Instn. of Naval Architects*, 1969
18. Morrow, J., 'On the lateral vibration of bars of uniform and varying sectional area', *Phil. Mag.*, July 1905
19. Todd, F. H., 'Some measurements of ship vibration', *Trans. North East Coast Instn. of Engrs & Shpbdrs.*, 1931–1932
20. Todd, F. H., 'Ship vibration—a comparison of measured with calculated frequencies', *ibid.*, 1932–1933
21. Tobin, T. C., 'A method of determining natural periods of vibration of ships', *Trans. Instn. of Naval Architects*, 1922
22. Nicholls, H. W., 'Vibration of ships', *ibid.*, 1924
23. Den Hartog, J. P., *Mechanical Vibration*, McGraw Hill Book Co., 1956
24. Townsin, R. L., 'The calculation of higher mode frequencies in ship vibration', *Shipbuilder & Marine Engine-builder*, Sept. 1961
25. Richards, J. E., 'An analysis of ship vibration using basic functions', *Trans. North East Coast Instn. of Engrs. & Shpbdrs.*, 1951–1952

26. Vedeler, G., 'The torsion of ships', *Trans. Instn. of Naval Architects*, 1924
27. de Wilde, G., 'Structural problems in ships with large hatch openings', *International Shipbuilding Progress*, 1967
28. Johnson, A. J., and Ayling, P. W., 'Graphical presentation of hull frequency data and the influence of deck houses on frequency prediction', *Trans. North East Coast Instn. of Engrs. & Shpbdrs.*, 1956–1957
29. Lewis, F. M., *Principles of Naval Architecture* (Chapter on 'Hull Vibration of Ships'), Society of Naval Architects and Marine Engineers., U.S.A.
30. Goodman, R. A., 'Wave excited main hull vibration in large tankers and bulk carriers', *Trans. Royal Instn. of Naval Architects*, 1971
31. Inglis, C. E., 'A suggested method for minimising vibrations in ships', *Trans. Instn. of Naval Architects*, 1933
32. 'Hydrodynamic damper to reduce wave-excited vibration', *The Naval Architect*, Royal Instn. of Naval Architects, Apr. 1971

12

Ship design

The various aspects of naval architecture which have been discussed in previous chapters are all facets of the design problem. The determination of the scantlings of the structure so as to ensure that the ship is sufficiently strong to withstand the imposed loads is concerned with structural design. Similarly the development of a suitable form which will give as low a value of the resistance as possible is another aspect of design. What is generally referred to as ship design, however, is the synthesis of all these different considerations in order to produce a ship of given dimensions and suitable layout to fulfil specified requirements. The starting point of any design must therefore be these specified requirements, and generally they are not under the control of the ship designer.

Design requirements

The ship designer will usually be supplied with a specification from the prospective owner which sets out his requirements. How these requirements are arrived at is not within the scope of the present work, beyond some brief remarks on the subject.

The requirements of a ship or a fleet of ships could be looked at from an overall point of view as the transport of a given quantity of a particular commodity from one place to another. For example, suppose that Country A has a consumption of X tonnes of oil per annum and this quantity is to be obtained from Country B. The overall question is, how can this transportation problem be solved most economically? In other words, how many ships should be employed and what should be their dimensions and speeds so that transportation costs can be minimised? The problem could be further complicated by the fact that there may be more than one source of the product. Also, there may be constraints such as limitations on draughts at the terminal ports, in which case the design of the ships may be influenced by these factors and the total problem becomes one of transport, of which the design of the ship is only one part. This is well illustrated in the case of the container ship, where the design of suitable port facilities is necessary to enable the rapid turn round of the ship made possible by the use of containers.

The shipowner, in deciding what type, size and speed of ship he wants, will not usually be faced with this overall problem but will come to a decision on more limited considerations. One will be the particular market which he proposes to enter or in which he intends to expand his fleet or replace obsolete tonnage. For instance, the demand for oil has expanded enormously over the last 25 years so that it could be expected that this would be a market in which an owner could profitably participate. It appears then that a market survey is an important study for a prospective owner.

Having decided upon the market for which the ship is to be constructed (which really means the commodity which is to be carried) the owner has still to settle what size of ship, as far as deadweight or deadmass is concerned, and where the ship is to trade in the world if there is a choice of sources for the particular commodity concerned. There are many answers to this problem depending upon the amount of money which is available for investment. Generally there is an economy of size, as has been illustrated by the increasing size of oil tankers over the last 25 years.

The question of the speed of the ship must also be given careful consideration. High speed has been seen to require high power, with correspondingly high fuel consumption. This means that the weights of fuel and machinery will also be high and the consequence is that less of the available displacement will be available for cargo, with a reduction in earning capacity. The most economical speed, i.e. that which would give the greatest annual return on the capital investment, is not easily determined and would require a detailed economic analysis. Often, for non-perishable cargoes the speed could be quite low, but there is a trend nowadays towards higher speeds in cargo ships. For ships carrying perishable cargoes such as meat, fruit, etc., there is an obvious advantage in speed and the speeds of such ships have always tended to be higher than in other types of dry cargo ships.

To sum up, as far as the cargo ship is concerned the basic information which will be available to the designer is the type of ship, the required deadweight or deadmass, the speed and the radius of action. In passenger ships additional information which is required will include the number of passengers to be catered for and the facilities to be provided, which will dictate the size and number of public rooms.

Influence of nature of cargo on ship type

Various ship types have already been discussed in Chapter 1 and it will be seen that the nature of the cargo to be carried has decided to a large extent the type of ship. Thus, a distinct type has developed for the carriage of oil in bulk and while the ordinary dry cargo ship, the 'tramp', which was usually of the shelter deck type, carried bulk cargoes a type has developed in recent years which has been specifically designed for the carriage of these cargoes and is appropriately called the 'bulk carrier'. These ships like the oil tanker have machinery aft and from outward appearance there is little to distinguish one from another. Other cargoes which would determine the type of ship are refrigerated goods such as meat or fruit, which would require insulated holds with refrigerating machinery, and liquid gas carriers which require to be fitted with special tanks capable of maintaining the liquid gases at low temperatures.

Determination of principal dimensions

Having decided the type of ship to be built the next step as far as the designer is concerned is to determine the principal dimensions so that the ship is capable of carrying the specified cargo. The total mass of the ship is made up of lightmass and deadmass. The former is simply the mass of the ship which in turn can be split up into three major parts, the steel mass, the mass of the machinery and the mass of the outfit and fittings. The deadmass will include the cargo mass, the fuel mass and the masses of stores, fresh water, crew and effects, and passengers and baggage if any. This may be written in the form of an equation:

Total mass = lightmass + deadmass
= steel mass + machinery mass + mass of outfit and fittings + cargo mass + fuel mass + masses of stores, fresh water, crew and effects and passengers and baggage   (12.1)

It will be seen that most of the items, with the exception of the cargo, stores, fresh water, crew and effects, and passengers and baggage, are themselves dependent upon the principal dimensions of the ship. It follows that as a starting point in order to find the total mass of the ship it is necessary to have some idea as to the fractions which each of these items is of the total mass. The designer is assisted in this by recording data for ships which have already been built. If such data are systematically recorded and classified for different types of ships then it is possible to choose suitable co-efficients which can be applied to the type under consideration. By this means it is possible to obtain a preliminary estimate of the total mass of the ship.

Suppose that the required mass is M, then in terms of the principal dimensions and block coefficient this can be written

$$M = \rho L B d C_B \tag{12.2}$$

where

ρ = density of seawater
L and B = length and breadth of the ship
d = draught
C_B = block coefficient.

There is an infinite number of solutions to this simple equation and the problem is to choose the most suitable values for the variables.

Considering the block coefficient first, this will depend mainly upon the speed of the ship. If the mass of the cargo to be carried is the primary consideration it is clear that the fuller the block coefficient the greater will be the available displacement. However, as block coefficient increases it has been seen that the resistance of the ship and hence the power required to drive it will also be increased, so there is some limit beyond which it becomes un-economical to go, because the mass of the machinery and fuel increases at a rate which will absorb the increase in displacement arising from the increased block coefficient. This economical block coefficient is very difficult to define. Generally, however, very full block coefficients are often used in modern deadweight carriers, in some cases exceeding 0.80.

Having decided upon a suitable block coefficient the three dimensions, length, breadth and draught, have to be decided upon. Breadth in relation to draught will depend upon stability considerations mainly. A breadth/draught ratio below about 2.25 will not usually be possible from a stability point of view but the actual value will depend very much on the ship type, and particularly if there are loads to be carried high up in the ship resulting in a high position of the centre of gravity (or centre of mass).

Suppose, now, the volume can be expressed in terms of the length L and the ratios of B and d to L. Hence $B/L = k_1$ and $d/L = k_2$. It follows that $M = \rho L^3 \times (B/L)(d/L)C_B = L^3 \times k_1 k_2 C_B$. Now the ratio of breadth to draught is $B/d = k_1 L/k_2 L = k_1/k_2 = k_3$ say, or $k_2 = k_1/k_3$.

Then

$$M = \rho L^3 k_1 \times (k_1/k_3) \times C_B = \rho L^3 (k_1^2/k_3)C_B \qquad (12.3)$$

This shows that having decided on the breadth/draught ratio k_3 and if the ratio of length to breadth is fixed, then all the dimensions can be determined. The choice of a suitable length/breadth ratio will probably depend upon resistance considerations. A high value of L/B will lead to a slim ship, while a low value will result in a stumpy ship.

As an example, consider a ship whose total mass is to be 50 000 t, the block coefficient being 0.80. Suppose that the length/breadth ratio is 7.5 and the breadth/draught ratio is fixed at 2.75. Then equation 12.3 becomes

$$50\,000 \times 1000 = 1025 \times L^3 \times (1/7.5^2)(0.80/2.75)$$

Hence

$$L^3 = (50\,000/1.025) \times (7.5^2 \times 2.75/0.8) = 9.432 \times 10^6$$

$$L = 211\,\text{m}$$

$$B = 211/7.5 = 28.13\,\text{m}$$

$$d = 28.13/2.75 = 10.23\,\text{m}$$

Should there be restrictions on draught these preliminary dimensions may have to be modified.

Having determined draught it is necessary next to fix the depth of the ship. This will be governed largely by freeboard considerations. It has been seen that the statutory freeboard required by the Load Line Regulations depends upon the length of the ship and the length/depth ratio. The depth which will just give the required draught can only be obtained by a process of trial and error, making use of the freeboards given in the Load Line Regulations. Having arrived at the depth, the length/depth ratio should be checked since if it is too large it may be necessary to put an excessive amount of material in the structure to obtain the required section modulus of the midship section.

Depending upon the cargo to be carried, the cubic capacity which it occupies may be more important than its actual mass. Thus, while in ships such as ore carriers or ships carrying grain the mass may be the important quantity, in ships carrying refrigerated cargoes, for example, the cubic capacity available is the important factor. Ships of this type may often not be loaded down to the maximum draught available under the Load Line Regulations when all the holds are full. Such ships are often designed for a reduced draught, which means that lighter scantlings are possible thus leading to a reduction in the steel mass.

Calculation of steel and outfit masses

Having obtained preliminary dimensions, the next stage in the design process is to obtain more accurate estimates of the masses making up the lightmass of the ship. The major items which are the concern of the designer are the steel and outfit masses. It is often useful when calculating these quantities to have a rough general arrangement of the ship showing the approximate layout of the various spaces such as cargo holds and machinery space, master and crew accommodation and passenger accommodation if any.

The steel mass lends itself fairly easily to calculation and to facilitate such calculations data collected for previous ships are very useful. It is the practice to list for finished ships the invoiced weights from the steel mills of all the plates and sections making up the total mass of the ship. Various methods can be used for estimating the steel mass for the design. Depending upon the degree of accuracy required the total steel mass for a basis ship may be proportional for differences in dimensions from the design, with corrections for features in the one which do not exist in the other. Alternatively, if a higher degree of accuracy is required each individual item such as shell, decks, bulkheads, etc., may be proportioned on the basis of dimensions and scantlings. This will generally necessitate preparing an approximate structural midship section for the design and comparing the scantlings with those of the basis ship. However this calculation is carried out the mass or weight so obtained will be that of the material as received in the shipyard. From this has to be deducted a scrap percentage to allow for the material cut off to give the finished sizes as fitted in the ship. This percentage can vary considerably and can only really be obtained by estimating the total mass fitted in a ship and comparing this with the invoiced mass. In riveted ships a figure which was often used was 10%, but in modern ship construction a figure much lower could be expected. Methods for calculating steel weight or mass have been discussed by Telfer.[1]

The calculation of the mass of the outfit and fittings is generally much more difficult. Many of the items making up the total are bought into the shipyard from subcontractors and the designer is dependent on information they supply for these items. Lifeboats, davits, winches, windlass, etc., are items which come into this category. Others such as wood, decks, floor coverings, etc., can be estimated from the plans of the ship. Once again the collected data for ships which have been built are useful in making estimates of the mass of outfit and fittings.

Machinery mass

The calculation of the mass of the machinery will depend upon the machinery type, and the power. A decision must first of all be made as to the type and then an estimate made of the power required. Knowing the principal dimensions, the block coefficient and the speed it is possible to make reasonable estimates of the power which the machinery will be required to develop, as has been shown in Chapters 8, 9 and 10.

The machinery mass can then be calculated from accumulated data, usually on the basis of power per tonne. Associated with the power requirements is the amount of fuel to be allowed for, which can be estimated from the specific fuel consumption for the particular type of machinery, the power, and the range which the ship is required to cover without refuelling.

Check on preliminary dimensions

Having made estimates of all the masses making up the total mass of the ship it is now possible to check if the dimensions originally fixed are satisfactory. It is unlikely that an exact balance would be achieved at this stage, so that some changes might have to be made in the original dimensions. From these, further estimates of the various masses could be made and again the dimensions checked. It will be seen that the process is an iterative one, several iterations being necessary to achieve exact agreement between available buoyancy on given dimensions and fullness and the sum of the various masses.

The ship form

When the designer is satisfied that the dimensions and fullness decided upon will provide enough buoyancy to carry the various loads, the ship form can then be proceeded with. This means the design of a lines plan. The usual starting point is to obtain a curve of immersed sectional areas and a curve of half-ordinates of the load waterline. When these two curves have been fixed the shapes of the body sections are very largely determined.

The curve of immersed sectional areas in conjunction with the load waterline will also dictate the shapes of the waterline at the ends. It is not proposed to enter into details as to how these curves are arrived at, but they can be produced from parent forms which have been tank tested and have proved to be efficient from a resistance point of view.

The area under the curve of cross-sectional areas gives the total volume of displacement and the longitudinal position of the centroid of this area is the longitudinal position of the centre of buoyancy. The former quantity must of course give the volume required to carry the various loads. The latter quantity, i.e. the longitudinal position of the centre of buoyancy, has an influence on resistance and ideally it should be located in the position which will give minimum resistance, as was discussed in Chapter 8. However, the position of the centre of buoyancy has an important influence on the trim of the ship and it is often this consideration which fixes its position. For a ship to float on an even keel, which is the desirable condition, the centre of gravity (or centre of mass) and the centre of buoyancy should be in the same longitudinal position.

From the basic information discussed here it is possible to develop a preliminary lines plan for the ship, from which in turn it is possible to determine all the hydrostatic data discussed in Chapters 4 and 5. Depending upon the amount of detail that it is desired to go into at this stage, the form so developed may be tank tested to make a more accurate estimate of power and to investigate any possible improvements which may be obtained.

Stability and trim

Two of the most important matters which must be checked when the form of the ship has been derived are the stability of the ship and the trim in all conditions of loading. It is necessary to know the vertical and horizontal positions of the centre of gravity of the ship and also the volumes and centroids of all the spaces on board the ship where loads are to be carried.

The centre of gravity of the lightship can be obtained by direct calculation, but this is a long and tedious process, needing the determination of the mass and centroid of every item making up the total mass. It is often useful, especially in the preliminary stages of a design, to make use of basis ships of similar type. It was seen in Chapter 5 that the vertical and horizontal positions of the centre of gravity of a ship are obtained when it is near completion by carrying out an inclining experiment and observing the draughts at which the ship is floating. The position of the centre of gravity for a new design can be obtained by correcting these data for a basis ship for differences between the two and for differences in dimensions. It is thus possible to obtain fair estimates of the positions of the centre of gravity without recourse to detailed direct calculation. Sometimes, however, the direct method is necessary for ships of novel design where no basis ship data are available.

The volumes and centroids of all the spaces in the ship which carry cargo, fuel, etc., can readily be obtained when the ship form has been settled. It is thus possible to determine the position of the centre of gravity of the ship in any desired condition of loading. Making use of the hydrostatic data obtained for the ship form, the initial stability as represented by the metacentre height can also be calculated, together with the trim. Should stability be insufficient it will generally be necessary to increase the beam of the ship. If the original beam is B_1 and B_2 is the modified beam, then if the volume of displacement remains unchanged:

$$\text{Increase in } GM = \text{BM} \{(B_2^3/B_1^3) - 1\} = \delta GM$$

where BM is the height of the metacentre above the centre of buoyancy. Supposing that there is no change in draught, the block coefficient would be reduced in the ratio B_1/B_2. If on the other hand the volume is allowed to increase

$$\text{Increase in } GM = \text{BM} \{(B_2^2/B_1^2) - 1\}$$

for the same draught.

The increase in beam would involve some small increase in weight and if the buoyancy provided by the increased volume was not absorbed by this weight then either the draught would be reduced or the block coefficient would have to be made finer. Adjustments of this sort might involve redoing some of the earlier calculations.

Should the trim be unsuitable in some condition of loading then this could be adjusted by redistributing the cargo or by altering the position of the centre of buoyancy. Often it may not be possible to do the former, so that the longitudinal position of the centre of buoyancy has to be adjusted to suit the centre of gravity. Thus, the centre of buoyancy may be moved from the optimum position required for resistance and this is one instance where a compromise has to be reached between conflicting requirements. It should be noted that moving the centre of buoyancy will move slightly the centre of gravity of the lightship and will also move the centre of gravity of the cargo, because of the changes involved in the sectional areas of the ship.

Determination of scantlings of the structure

When the ship being designed is built to the rules of a classification society the scantlings of the various parts of the structure can be obtained from the rules of the particular society. When ships of novel design are being considered which do not come within the scope of the rules it may be necessary to carry out a full strength investigation along the lines described in Chapter 7. However, even when classification society rules are applicable it will be found that with most societies a still water bending moment calculation will have to be carried out. When the scantlings of the structure have been determined and approved by the classification society it is possible to make a final check on the mass of the structure and make any adjustments to the value used in determining the principal dimensions of the ship.

Watertight subdivision

It has been seen that in passenger ships the transverse watertight bulkheads are required to be arranged so that the ship could float if any one compartment was damaged. This is a minimum standard. It was also shown that this requirement could have an influence on the freeboard assigned to other types of ships. The correct disposition of the watertight bulkheads is therefore a matter of importance when designing a ship.

The total number of bulkheads to be fitted is specified by classification societies but how they are disposed is left to the designer. The requirements of subdivision will then decide where they are to be placed. The correct positioning of bulkheads may have a considerable effect on the general layout of the ship, so that this is a matter which should be investigated before the general arrangement of the ship is proceeded with. Depending upon the degree of watertight subdivision required it may be necessary to fit more bulkheads than required by the classification society concerned.

General arrangement

It was stated earlier that a preliminary general arrangement was useful when making estimates of steel mass and the masses of the other items making up the total mass of the ship. When dimensions have been finally fixed a detailed general arrangement is prepared for the proposed ship. This shows the layout of all spaces including cargo holds or tanks, ballast tanks, machinery space and space allotted to accommodation for master, crew and passengers.

The design of suitable accommodation is a study in itself and requires careful consideration so that the crew who have to work continuously in the ship are satisfactorily housed. There are certain minimum standards laid down for such accommodation by the Department of Trade in the UK, but it will be generally found that the accommodation provided in most modern ships is far in excess of this.

As far as passenger accommodation is concerned, it also requires careful consideration so that the necessary standard of comfort is provided and that easy access is possible to public rooms. The catering facilities are not the least important in this respect.

The space allotted to the machinery in a ship is usually decided by the ship designer and the marine engineer who is building the machinery. From the ship designer's point of view this space should be as small as possible. The marine engineer, however, will require adequate space to be provided for the efficient working of the machinery, so that the final size of the machinery space is often a compromise between the two.

The above are only a few of the items which have to be considered when preparing the general arrangement of a ship.

Cost

When the technical features of the design of a ship have been decided upon and the general arrangement drawn, the question of cost has to be considered. It is a very specialised task and one to which only the briefest reference will be made here. Some general points are, however, of interest.

When considering innovations in the design of a ship their effectiveness should be evaluated against the additional cost, if any, which will be involved. In this category are such things as the introduction of new and more costly materials such as aluminium alloys or glass reinforced plastics into the structure. What is to be gained should always be set against the additional expenditure. The same can be said of the fitting of special pieces of equipment, such as stabilisers in passenger ships. These are now considered to be essential in ships of this type and the additional cost is justified on the grounds of passenger comfort which makes the ship more attractive.

In recent years much attention has been given to 'design for production', which really means design for ease of production. This is particularly applicable in the structural field where modification in the details of some of the structural parts can enable them to be produced more cheaply.

The total cost of a ship can be divided into two parts: the cost of the items bought in by a builder from subcontractors, and the cost of working material, such as steel, and assembling the various items in the shipyard. The former part represents in most ships the major fraction of the total cost and is thus not to any great extent under the control of the shipbuilder. He does not, therefore, have much room to manoeuvre, but a great deal can be done to ensure that the preparation and assembly of parts is carried out in the most efficient manner possible. This can not only reduce the cost of production but also speed up production time. Much study has gone into these shipbuilding problems in the last few years and has led to substantial reductions in cost and building time.

Design optimisation

The general procedure for the development of a design to meet specified requirements has been discussed in this chapter and it has been shown that there is a variety of answers to the basic design problem. The question then arises as to whether the solution proposed is the best one, or in other words is the design the optimum one?

Optimisation may be considered from different points of view. First of all the overall problem can be considered, in which the question might be posed, is the design produced within the limits of the specified requirements the one which will give the shipowner the maximum return for the capital which has been invested? By varying some of the features of the design it may be possible to improve the financial return. To investigate this it is necessary to carry out a detailed economic analysis and determine the influence of varying features of the design such as length, breadth, draught, speed, etc. An attempt to do this was discussed by Tutin many years ago,[2] but to carry out the investigation proposed in this paper a great deal of calculation was involved and this could rarely be done as an ordinary design routine. However, in the time since then the advent of the computer has enabled a great range of calculations to be carried out very quickly. At the present time it is thus possible to investigate the influence of changes in the various parameters of the design and ascertain if the original design can be improved. Such an investigation would be of use not only to the shipowner but also to the shipbuilder in enabling him to produce a design at a competitive price. Other papers on the subject of economic design to which the reader is directed are by Goss,[3,4] Gilfillan,[5] and Snaith and Parker.[6]

Apart from the overall problem of optimum design, attention has been directed to optimisation in more limited fields. Structural optimisation is one problem in which there is a great deal of scope for investigation. The problem here is to obtain a given strength of structure for the least mass or weight of material, or alternatively for least cost. It is clear that least weight of structural material is related to least cost, although least weight does not necessarily mean least constructional cost. Investigations on structural optimisation have been carried out by Moe[7] and Smith and Woodhead.[8] In these investigations modern techniques such as linear and non-linear programming have been employed. Investigation of this sort can lead to substantial savings in material and cost.

REFERENCES

1. Telfer, E. V., 'The structural weight similarity of ships', *Trans. North East Coast Instn. of Engrs & Shpbdrs*, 1955–1956
2. Tutin, J., 'The economic efficiency of merchant ships', *Trans. Instn. of Naval Architects*, 1922
3. Goss, R. O., 'Economic criteria for economical ship design', *Trans. Royal Instn. of Naval Architects*, 1965
4. Goss, R. O., 'The economics of automation in British shipping', *ibid.*, 1967
5. Gilfillan, A. W., 'The economic design of bulk cargo carriers', *ibid.*, 1969
6. Snaith, G. R., and Parker, M. N., 'Ship design with computer aids', *Trans. North East Coast Instn. of Engrs. & Shpbdrs.*, 1972–1973
7. Moe, J., and Lund, S., 'Cost and weight minimization of structure with special emphasis on longitudinal strength members of tankers', *Trans. Royal Instn. of Naval Architects*, 1968
8. Smith, G. K., and Woodhead, R. G., 'A design scheme for ship structures', *ibid.*, 1973

Index

Index